Molecular mechanisms of drug action

Second Edition

Christopher J. Coulson

Glaxo Group Research, UK

UK Taylor & Francis Ltd, 4 John St., London WC1N 2ET

USA Taylor & Francis Inc., 1900 Frost Road, Suite 101, Bristol, PA 19007

Reprinted 1995

British Library Cataloguing in Publication Data
A catalogue record for this book is available from the British Library.

ISBN 0–7484—0068–0 (cloth)
ISBN 0–7484–0078–8 (paper)

Library of Congress Cataloging-in-Publication Data are available

Cover design by Russell Beach
Typeset by Photo·Graphics
Printed in Great Britain by Burgess Science Press, Basingstoke, on paper which has a specified pH value on final manufacture of not less than 7.5 and is therefore @acid free'.

Molecular mechanisms of drug action

Contents

Preface

As my experience is that of an industrial biochemist who has worked in the pharmaceutical industry for over 20 years, this book is written from a practical standpoint. It includes the mechanism of drug action, encompassing drugs for infectious disease, as well as for 'endogenous conditions' such as cancer, arthritis, heart disease, schizophrenia etc. It is written from the point of view of targets - whether enzymes in pathways, receptors or ion channels in membranes.

The alternative procedure, where disease is made the central focus, has been used in a number of other publications and so I have approached the subject in a different way which was better able to link the practical aspects of drug targeting with the biochemical and pharmacological background. My aim was to bring together our present state of knowledge with respect to drug mechanisms (information that is otherwise scattered throughout the literature) in a readable and accessible form, which could assist both those teaching the subject, and students who wish to study it. I felt that there was a gap between academia and industry which such a book might help to close.

The introductory chapter is concerned with the basic principles that cover enzyme inhibition and receptor binding by drugs. The rest of the book is divided into two sections; the next seven chapters deal with drugs that modulate biochemical pathways, both of synthesis and breakdown, while the last four are concerned with organizational structures of the cell. Chapters 2 to 5 are concerned with the biosynthesis of DNA, protein, carbohydrate and cell walls. Much of the chemotherapy of cancer, viruses and bacteria is to be found within these chapters. Next are two chapters on lipid biosynthesis; the first one covering the pathways of sterol and steroid interconversions and the second on the various pathways that lead from arachidonic acid. Although not a pathway, I have next included a discussion on inhibitors of zinc metalloenzymes as they form a coherent group.

In the second section the agonists and antagonists at neurotransmitter receptors are discussed first, followed by those agents that interfere directly with membranes. In these two chapters I have deliberately covered membrane enzymes alongside ion channels and receptors. Microtubule ligands form the subject of Chapter 11 and hormone modulators are discussed in Chapter 12. In the Appendix, I discuss the basis for the measurement of binding constants for ligand/macromolecule binding.

Some readers may feel that there are notable lacunae in the coverage of pharmaceuticals. This has occurred because I have tried to cover those drugs that show recognized principles in their mode of action; some have not been included because their mode of action is not known and the work done on them cannot be related in a coherent fashion in a textbook of this sort. No attempt has been made to cover all drugs. Others are included, even though their mode of action is not fully understood, if their development illustrates a useful point. The use of cromoglycate in asthma is such a case. Other drugs of interest have been discussed if they have been designed particularly for a given condition and may be approaching the market - although not yet fully launched.

I have not tried to cover drug delivery systems, metabolism, side-effects etc., except where these are germane to the mechanism. The quantity of material available in these areas is so great that the length of the book would have reached unmanageable proportions if it had been included. Furthermore, the amount of coverage does not relate to market value or quantity of drug sold, but rather to scientific interest, so that a widely used drug whose mode of action is clearly defined may not warrant a long coverage, while an interesting but less used drug may have a greater coverage.

The book is intended for third, and possibly second, year students studying subjects or modules in microbiology, pharmacology, biochemistry, pharmacy and medicine. It may also be useful for medicinal scientists in the drug industry who are changing fields and need a quick entrée into a fresh area. Comprehensive reviews have been listed, if available, to help the reader to follow up points of interest and to go more deeply into the subject. These are largely in accessible journals and the occasional book.

I am deeply indebted to Charles Ashford, John Foreman, Ian Kitchen, Hugh White, Alan Wiseman, Helen Wiseman and David Williams for reading all of part of the manuscript and who made many useful suggestions for improvement and corrections.

Preface to second edition

A few new drugs (and some old ones) have been added if their mode of action is sufficiently clear. Other drugs that have been withdrawn from the market have been omitted. As a result of responses to the first edition, questions have been added at the end of each chapter which it is hoped will help in studying the subject. Some questions can be answered by reading the chapter alone, while others will require some background reading from the review articles mentioned.

A much greater emphasis has been placed on chirality as it is becoming a major feature of drug development. If the information on drug isomers developed in earlier years is not available, this point is usually noted in the text. The techniques of molecular biology have permitted enormous advances in our understanding. I have made references to these wherever appropriate.

Some drugs are known by different generic or non-proprietary names in different parts of the world. A glossary of those that differ markedly between Britain and the USA is included to clarify the situation for the reader. As a general rule the USA adopted name includes the counterion of the salt if appropriate but the British approved name does not. The name used here is that commonly accepted.

Glossary

British Approved Name (BAN)	United States Adopted Name (USAN)
Adrenaline	Epinephrine
Azidothymidine	Zidovudine
Cromoglycate	Cromolyn
Frusemide	Furosemide
Glibenclamide	Glyburide
Isoprenaline	Isoproterenol
Lignocaine	Lidocaine
Mepyramine	Pyrilamine
Noradrenaline	Norepinephrine
Rifampicin	Rifampin
Salbutamol	Albuterol

Chapter 1

General principles

1.1 Background

It would have been impossible to write a book of any length detailing mechanisms of drug action more than 20 years ago, because very little was known at that time. Penicillin, for example, had been on the market for 20 years, but we were still far from detailing its precise mode of action. The philosophical attitude, prevalent in those days and still common now, was that drugs were just used because they worked: enquiry into their mode of action was deemed unnecessary. The fact that, at that time, our detailed understanding of biological processes was very limited, may not be unconnected. Many of us entering the drug industry in the 1960s hoped, however, to be able to design drugs more effectively on the basis of molecular structure, whether it be of the enzyme or the receptor.

There have been a number of major advances in the intervening years. It was, after all, only 23 years ago that the connection between aspirin and prostaglandin biosynthesis was realized by Vane (1971), which led in its turn to a much greater understanding of the biosynthesis of prostanoids. In fact, drugs have frequently provided researchers in the medical sciences with tools to unravel a 'knotty' scientific and/or medical problem.

1.2 Do drugs have a specific mode of action?

Perhaps one question that should be considered is whether drugs do have a singular mechanism of action that can be reduced to molecular terminology. The concept of a drug as a 'magic bullet' that can penetrate through a myriad of biochemical systems, acting at only one specific site to have the desired effect, is perhaps reductionist to the point of absurdity. Nevertheless, in practical terms the history of drug development suggests that, in the initial analysis, it is possible to view the process in this fashion. The basis for this assertion lies in the linking of enzyme inhibition, receptor binding or ion channel blocking with expected effects on substrate levels or events inside and outside the cell. These must lead to organ effects or death of parasitic microorganisms, and improvement of the patient's condition. These factors must all be consistent with the proposed mode of action.

Nevertheless, drugs frequently act at sites other than the intended one - molecular, cellular and/or organ. As a consequence, they show undesirable side-effects. Indeed, the process of developing a drug is intended to reduce undesirable activities to an acceptable level. Unless the aspect of the structure of the drug molecule that gives rise to the activity also gives rise to the side-effect, it is usually possible to reduce the undesirable activity by making minor changes to the structure.

In the last analysis the crucial factor is the therapeutic ratio. This is the ratio between the dose required to treat the condition and the dose which gives rise to unacceptable side-effects. There are occasions when the side-effect is apparently random (rare but often lethal), e.g. the development of aplastic anaemia associated with some drugs, notably chloramphenicol (see Chapter 3). Difficult decisions then have to be made, weighing up the possible risk to the patient against the benefit.

Another factor that affects such a decision is the condition for which the drug is prescribed. A drug for a condition that is frequently lethal (such as cancer) can be tolerated with a lower therapeutic ratio than one intended for reducing blood pressure. For viral diseases very few drugs are available; permission might therefore be given to use a drug with a low safety margin if there was no other drug available for a certain condition.

We must not forget that it is a patient who has to take the selected drug. The side-effects, therefore, need to be minimal and not of such a nature as not to be in any way distressing. At present, our need is for effective, non-toxic drugs to treat a variety of conditions in a preventive or palliative manner.

1.3 Basic processes

Drugs can be used to help the body reject an invading pathogenic organism (whether parasite, fungus, bacterium or virus) or to modify some aspect of the metabolism of the body that is functioning abnormally. In the former case, a drug is normally used which is toxic to the pathogenic organism but not to the host, and stimulation of the body's normal processes for combating invaders may play a part. In the latter case, the approach is usually more subtle, with a modulation of a process as the requirement.

In all cases, the drug can have biological activity only by interacting with the molecules of the target organ or organism. These molecules are usually proteins - enzymes that catalyse reactions essential for the functioning of the organism, or proteins called receptors which transmit signals by interacting with messenger molecules such as hormones and neurotransmitters. The drug has its effect by binding to either an active site or to a secondary site that influences the active site on the enzyme or receptor and thereby prevents access by the normal substrate or ligand (inhibitor or antagonist) or it provokes a signal where none was wanted (substrate or agonist).

The interaction of the drug with the protein molecule can be measured in terms of the strength of inhibition of an enzyme reaction or the strength of drug binding to a receptor. The former may be quantified as I_{50} or IC_{50}, the concentration of inhibitor required to reduce the rate of a reaction or the binding of a ligand by one-half. This, however, varies with the amount of substrate or ligand available to the enzyme or receptor, thereby making comparisons between data obtained under different conditions almost impossible.

With more data, a binding constant for ligand attaching to receptor or inhibitor constant for enzyme reaction may be determined. These quantities are usually expressed in terms of the dissociation constant of the ligand–receptor complex (K_d) or enzyme–inhibitor complex (K_i). For the ligand–receptor interaction, this constant is equal to the ligand concentration at which 50 per cent of the receptors are occupied by ligand and it is assumed, although this is not always correct, that the measured response has fallen to one-half of its maximum value. For typical drug–receptor interactions, the dissociation constants are of the order of 10^{-7} to 10^{-10} M.

1.4 Drug binding to enzymes

Inhibitors can bind to enzymes in several ways. They may bind either reversibly or irreversibly, and with either the active site or with another part of the enzyme. In the latter case, the binding can cause a conformational change in the enzyme, thus producing distortion of the active site. Different effects of inhibitor concentration upon the kinetics of the enzyme in these various cases allow evidence to be obtained on the type of binding involved in any particular situation.

Reversible inhibition can occur in various ways including competitive, non-competitive and uncompetitive binding. If the inhibitor binds to the enzyme alone, the inhibition is said to be competitive. In this case, the inhibitor constant (K_i) is defined as the dissociation constant of the enzyme/inhibitor complex (see Appendix for a more detailed discussion of K_i). In principle, increasing the substrate concentration can eventually restore the reaction rate to what it was in the absence of inhibitor. Frequently, the substrate competes with the inhibitor for the substrate binding site, but there are cases in which competitive inhibition is observed although substrate and inhibitor bind to distinct sites.

In non-competitive inhibition the inhibitor can bind to the enzyme–substrate complex as well as to the enzyme itself, at a site separate from the substrate binding site. In this case, increasing the substrate concentration may reduce but cannot eliminate the inhibition. Uncompetitive inhibition is found when the inhibitor will only bind to the enzyme–substrate complex.

The binding constants noted above are, in principle, independent of ligand or substrate concentration and so results obtained in different laboratories can be compared with some confidence. They do not, however, necessarily bear any relation to the I_{50} values mentioned above. (See Appendix 1 for a discussion.)

The links between drug and target are usually of a reversible nature and so the drug will eventually be released from the complex. The complex is usually formed instantaneously, although this can take many hours if the binding constant is of the order of 10^{-9} M or smaller. In some cases, however, the drug and target will react covalently, and this is often characterized by time-dependent inhibition and by lack of enzyme recovery after dialysis. If the ligand is bound reversibly, it dissociates during dialysis and enzyme activity is recovered. If irreversible inhibition has occurred, the activity is not recoverable; instead synthesis of fresh protein will be needed to restore the original level of the target. Some reversible inhibitors whose binding is extremely tight may cycle on and off the enzyme so slowly that for all practical purposes the enzyme is inactivated by their presence.

Tightly binding inhibitors therefore need a considerable time for preincubation for the inhibition to reach its maximum effect. They do not satisfy the normal kinetic criteria of enzyme reactions in that the equilibrium is not set up rapidly and the concentrations of inhibitor and substrate are not very much higher than that of the enzyme. In fact, the inhibitor and enzyme concentrations are often very similar and in some cases stoichiometric or one-to-one. Under these conditions the Lineweaver–Burk plots (double reciprocal plots of reaction velocity against substrate concentration) will have both curved and linear portions while the progress curves will be aptly named because, in effect, a portion of the enzyme is being put gradually out of action (Morrison, 1969, 1982). An example is methotrexate binding to dihydrofolate reductase with a K_d of 10^{-11} M (Chapter 2).

Irreversible inhibition is particularly useful if the step in a pathway chosen as the target for drug action is not the rate-limiting one. Reversible inhibitors are likely to be ineffective in these situations because substrate levels increase and, even with non-competitive inhibitors, the inhibition is reduced. Most drugs are administered at least on a daily basis. Irreversible inhibitors, however, probably do not need to be administered daily, since it may take several days for enough enzyme to be resynthesized to return to steady state levels. A dosage regimen in which the drug is only given on alternate days is likely to be more useful than one requiring daily dosage. Otherwise the effect of the drug is likely to become magnified to the point of producing side-effects. A case in point is the use of monoamine oxidase inhibitors (Chapter 9), which were given on a daily basis and eventually lowered the level of the enzyme in the stomach wall to such a level that tyramine and dopamine ingested from food were not deaminated and caused hypertensive crises that were sometimes lethal. This was known as the 'cheese effect' because cheese contains a considerable amount of tyramine.

A special type of irreversible inhibition occurs when the drug is chemically unreactive but the enzyme converts it to a highly reactive species that inactivates the enzyme. This has been given the name of mechanism-based irreversible inhibition and the drugs are called suicide substrates or K_{cat} inhibitors because they require the catalytic activity of the enzyme. Examples of this include allopurinol inhibiting xanthine oxidase and 5-fluorodeoxyuridylate inhibiting thymidylate synthetase (both in Chapter 2), clavulanate inhibiting β-lactamase (Chapter 5) and the monoamine oxidase inhibitors (Chapter 9). For a fuller discussion see Walsh (1984).

Another class of enzyme inhibitors is described as transition state inhibitors. Every reaction proceeds through an intermediate state in which the structure attached to the enzyme is neither substrate nor product but an intermediate form. The transition state is the structure of highest energy on the pathway from substrate to product. As enzymes catalyse a specific reaction, compounds that interfere with the transition state are likely to be highly selective in their action. One example of a transition state inhibitor is 2-deoxycoformycin which inhibits adenosine deaminase (Chapter 2) (Lienhard, 1973).

An inhibitor does not always bind to the active site of an enzyme. It may modulate the enzyme activity by binding to another part of the protein, and is called an allosteric effector - allosteric is derived from Greek words meaning 'other shape'. Allosteric effectors are thus distinguished from those competitive inhibitors which resemble the substrate in shape, and are termed isosteric. Inhibition of this type is also known as negative cooperativity, whereas activators that bind allosterically give rise to positive cooperativity. An enzyme that behaves in this way is normally composed of more than one subunit. Ribonucleotide reductase is a case in point where the effectors are the purine and pyrimidine trinucleotides (Chapter 2).

1.5 Drug binding to receptors

It is important to be clear about what constitutes a receptor. For the binding of a ligand to a receptor to be a genuine physiological phenomenon and not just non-specific binding, the ligand binding should:

1. be saturable, in which case a plot of amount of drug bound against drug concentration will level off and reach a plateau;
2. be high affinity (binding constants less than 10^{-6} M) and low capacity for the binding to be regarded as specific;
3. obey the Law of Mass Action (Appendix); and
4. be linked to a pharmacological response characteristic of the particular ligand.

If these criteria are not fulfilled, the data may indicate, fortuitously, the existence of a binding protein for a particular ligand but not a receptor with physiological significance. It is also advisable, if possible, to have a series of agonists and antagonists, preferably of different structural types, that are specific for a given receptor and can thus define its role.

An agonist is an agent that interacts with the receptor to produce a clearly defined change in the target cell. An antagonist also binds to the receptor, but lacks the structural features essential to initiate any further changes and so can block the action of an agonist. In a cell-free system, therefore, both agonist and antagonist will bind, but it will be impossible to distinguish between them without data on activity on a cellular or tissue basis. A partial agonist shows activity at a receptor but not to the same extent as the natural agonist.

The measurement of the reversible binding of a radioactive ligand to a receptor preparation (radioligand binding) has greatly increased our understanding of receptors and, provided it is understood that the binding occasionally may have to be discounted because it is an artefact of the preparation, the knowledge gained has been of great value in identifying and, in some cases, isolating the receptor. One of the greatest challenges in modern pharmacology is (a) to link the binding data with pharmacological effect and (b) to understand how in molecular terms the signal is transduced in a given cell to produce a given response (Chapter 9).

1.6 Further considerations

The attachment of a drug to its target is only part of the process that leads to an effective drug. Anti-bacterial drugs have to be able to reach the bacterium in sufficient quantity to be able to kill it (bactericidal action). In some cases the drug merely prevents the bacterium from growing and does not kill it (bacteristatic action), but the host may still recover from the infection by the anti-bacterial action of the immune system. Similarly, some anti-fungal drugs merely stop fungal growth, but here the situation is more serious because fungal infections frequently occur in individuals with impaired immune systems, for example in cancer patients or in patients who are receiving immune suppressant drugs after an organ transplant. It is therefore most important to have fungicidal drugs.

The activity of anti-bacterial and anti-fungal drugs is usually measured by the minimum concentration at which the drug completely inhibits the growth of the microorganism. This concentration is known as the minimum inhibitory concentration (MIC).

Furthermore, absorption, serum binding and metabolism must all be taken into account in developing an effective drug. The drug, if given by mouth, must be able to cross the stomach wall, reach the bloodstream, and be transported in sufficient quantity to the site of action. Serum proteins may play a part in this process - notably albumin, which is particularly effective at carrying drugs that exist in anionic form at pH 7.4, such as the non-steroidal anti-inflammatory drugs (Chapter 7). Albumin binding can, however, if too strong, prevent a compound that shows good activity *in vitro* from demonstrating efficacy *in vivo* (e.g. clorobiocin activity against Gram-negative bacteria is a case in point, Chapter 2). Other compounds may fail to show activity *in vivo* because they are metabolized to inactive metabolites.

1.7 Partition coefficient

A partition coefficient is a measure of how a solute distributes itself between two immiscible solvents. Often denoted *P* and expressed as the logarithm (log *P*) it is particularly useful in describing the potential of a drug for reaching the brain. The brain (and the spinal cord) is surrounded by a fatty sheath and a considerable degree of lipid partitioning or lipophilicity (i.e. preferential transfer into lipid, not just solubility) is required for the drug to cross into the brain. Molecular weight and charge are also of importance. The concept of blood–brain transport is of crucial importance if a drug is expected to act within the brain, as with an anti-psychotic agent, or to handle infections in the membrane that surrounds the brain and spinal cord (meninges), such as meningitis. Conversely, it is essential to avoid side-effects of a mental nature, such as drowsiness or hallucinations, when the drug is intended to act elsewhere in the body.

The partition coefficient is measured by allowing the compound to come to equilibrium, assisted by shaking, between an aqueous and an organic medium, usually *n*-octanol or hexane. The quantity of material in each phase at equilibrium is determined by a physical method, such as ultraviolet absorption, and the ratio between the concentration in the lipid phase and that in the aqueous phase is known as the partition coefficient.

The partition coefficient, as measured above, views the compound as an uncharged species. The distribution coefficient (*D*) is a measure of how a drug distributes itself between aqueous and lipid phases at a given pH, usually close to neutrality, which may be more relevant conceptually to biologists since the blood plasma and intracellular pH are normally about pH 7.4. This is particularly relevant to molecules which are capable of ionization since the charged species is less likely to be able to enter a lipophilic membrane (e.g. the local anaesthetics discussed in Chapter 10). The aqueous medium chosen for the measurement is often phosphate buffer at pH 7·4.

For ionizable drugs, the tendency to ionize is measured in terms of a dissociation coefficient *K*, which, like hydrogen ion concentration, is often expressed as its negative logarithm: $-\log K$ or p*K*. If the p*K* of the drug is within one log unit of the pH at which the measurement is made, more than 5 per cent of the compound will be in the charged form. The p*K* is identical to the pH at which equal numbers of the drug molecules are charged and uncharged. pK_a refers to the molecule ionizing as an acid (releasing a proton) while pK_b is the similar position for a base (accepting a proton). Usually only the non-ionized form of the drug will move into the lipid phase.

It should be noted that this concerns passive transport across membranes. Active transport is a completely different phenomenon that requires the expenditure of energy, and can be saturated and inhibited in a similar fashion to enzyme reactions. There is an active transport system into the brain for the aromatic amino acids, for example. Note that there is often confusion in the literature about which is intended.

An interesting use of the partition coefficient alone to rationalize binding of ligands to macromolecules has been made in the case of various families of drugs and other compounds binding to serum albumin. This transport protein carries tryptophan, fatty acids and other naturally occurring materials around the body, and also a number of foreign molecules (xenobiotics) including drugs. The binding of members of a closely related group of compounds, such as penicillins, is directly proportional to log P; no other parameter appears to be necessary to explain the binding. This is because the same factors of hydrophobic versus hydrophilic interaction govern the albumin binding as determine the partition coefficient. Sulphonamide binding, on the other hand, may be defined by a term representing the pK_a of the sulphonamide side-chain as well as lipophilicity (see Jusko and Gretch, 1976 for a review).

Other attempts to rationalize the activity of families of related structures on biological systems (structure–activity analyses) have been made using various physical constants including log P, together with molecular size, electronic effects in aromatic rings and shape of substituents. Ligands may bind to macromolecules by ionic, hydrogen or covalent bonds or by hydrophobic or van der Waals interactions. The intention of this analysis is to make predictions for new, more active, structures and also to reduce toxicities. A detailed discussion on this subject is beyond the scope of this book but is available in the review by Hansch (1985).

1.8 Drug nomenclature

Drug nomenclature is often rather confusing so it may be helpful to explain the different names that may attach to a drug and how they were derived.

1. The detailed **chemical name** (according to IUPAC or other systematic nomenclature).
2. The name by which the drug is often known (the **generic or non-proprietary name**), which is usually a shortening of the chemical name and is usually the name approved by drug regulatory authorities. These can often vary from country to country. This name is useful for scientists and physicians to describe the drug.
3. The manufacturer's patented **trade name** by which the drug is sold, which is usually short and 'punchy' so that it is easily remembered by hard-pressed general practitioners. There may be several of these if different drug companies are marketing the same product, possibly in differing formulations.
4. There is also a number given by a drug company to any compound, synthesized by the organic chemists or extracted from a fermentation culture medium, that is usually of the form of an initial letter or letters followed by a number which may extend to six digits. Usually when the decision is taken to progress the compound towards the market, it is given a generic name. Earlier publications may have the number only, and it may be necessary to correlate name and number from later publications.

An example of these names is provided by the following structure of a drug which is used for the treatment of infections caused by anaerobic microorganisms (Chapter 4).

1. The detailed chemical name is 1-(2′-hydroxyethyl)-2-methyl-5-nitroimidazole.
2. The generic name is metronidazole. We can see how this name was arrived at from the detailed chemical name with the relevant letters in capitals:

   1-(2′-hydroxyethyl)-2-METhyl-5-NitROimIDAZOLE

   (there has been a slight change in the order to make a more harmonious sound).
3. The trade name of metronidazole is Flagyl.
4. The original company number was 8823 RP (short for Rhône-Poulenc).

Metronidazole

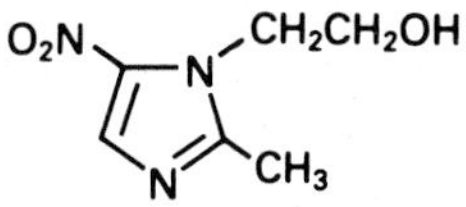

Generic names are not always entirely based on the chemical name but in some cases the use of the drug is taken into account. The anti-herpesvirus drug, originally numbered BW 248U, was subsequently named acyclovir, although it might have been given the generic name acycloguanosine since guanosine is the purine base in the structure (Chapter 2). Zovirax is the trade name.

Acyclovir

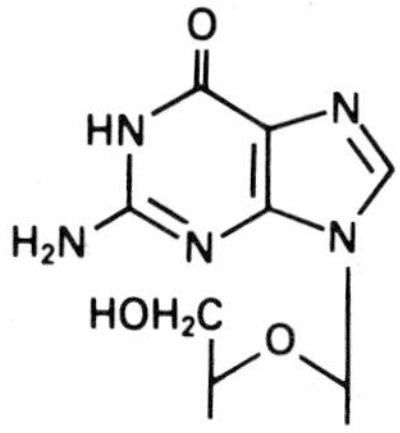

Throughout this book the generic name of the drug is used, except in one or two instances when the best known trade name and generic name are both quoted.

1.9 Stereochemistry

The interaction of biological molecules is frequently a question of asymmetry, whereby for example L- and D-amino acids and sugars are respectively active or inactive in the natural biochemical pathways, and L-noradrenaline is the ligand for adrenergic receptors. The L- and D-forms of the amino acids are known as stereoisomers (IUPAC, 1970).

If a carbon atom has four different substituents attached, it is impossible to superimpose one form of the molecule on the other. One form rotates the plane of polarized light in one direction and the other in the opposite direction to an equal degree - hence the use of the terms + and − for isomers, known as optical isomers. An equal mixture of the two stereoisomers does not rotate the plane of polarized light and is known as a racemic mixture or racemate. Use of the terms D and L, noted above, is known as the Fischer convention and indicates the configuration in relation to glyceraldehyde. This convention is used mainly for amino acids, sugars and other related molecules of importance in biochemistry and pharmacology.

There is another system of terminology named after Cahn, Prelog and Ingold which classifies the stereoisomers into *R* (rectus or right-handed) and *S* (sinister or left-handed), derived from the order of precedence in atomic weight of the substituent groups on the asymmetric carbon atom. The substituent atom with lowest atomic number is imagined as lying behind the asymmetric centre. The other substituents project towards the viewer. If the sequence of decreasing priority is clockwise then *R* is assigned, if it is anti-clockwise, *S*. This system is in more general use in organic chemistry and is applied to drugs.

There is no particular relationship between the different systems. Spectroscopic techniques, such as circular dichroism (CD) or optical rotatory dispersion (ORD) which depend on the spatial relationship of the substituent groups, must be used to determine the absolute configuration of a molecule. Receptors and enzymes will interact with the form for which they are selective and are likely to ignore the other.

Furthermore, the presence of a double bond in the molecule may also give rise to stereoisomerism (but not optical isomerism). The two forms which arise from the substituents being on the same or opposite sides of the double bond are known as *cis* and *trans*. An example is the carboxylic acids, fumaric and maleic - only fumaric acid, the *trans* isomer, is recognized as a substrate by the Krebs cycle enzyme, fumarase. *Cis/trans* isomerism can also occur when two substituents project out of a saturated ring (e.g. pilocarpine, Chapter 9).

Many drugs are prepared as racemates or mixtures of isomers, frequently a pair, sometimes four or more if there are several optically active centres. One or more of the individual compounds may be active, may contribute to the toxicity or may just be inactive ballast (Ariens *et al.*, 1988). Chemical synthesis normally produces an equal quantity of each individual isomer, and in the past it has not seemed sufficiently cost-effective to attempt to separate them. Recently, however, there has been more of an effort to prepare the isomers, especially where one isomer is responsible for the toxicity and another for the activity as in the case of lamivudine, the anti-HIV drug (Chapter 2).

The importance the medicinal scientific community attaches to stereoisomerism is shown by the launching of a journal (*Chirality*) specifically devoted to the biological effects of chirality. Furthermore, the attitude of the regulatory authorities such as the Federal Drug Agency in the USA is leaning towards the regulation of optically pure drugs; data for each stereoisomer must be provided. Wherever possible, I have included data on the stereoisomers, although in many cases it is still not available.

1.10 The future

We may be standing on the threshold of an explosion of biological knowledge which will allow us to treat the major diseases of today. Such a leap forward is obviously to be desired. One of the most important messages that this book is intended to convey is that most of the major diseases are still, despite the advances of recent years, neither fully understood, nor curable.

Thus, most of the drugs for cancer are nearly as toxic to the host as to the tumour. The inflammation of arthritis may be suppressed but the underlying disease process still cannot be halted and is poorly understood. In the case of mental illness, we treat schizophrenics with drugs that largely sedate the patient and do not cure, although they may partially arrest the course of the disease. The human immunodeficiency virus (HIV) now presents a major challenge. In spite of our hard-won success in dealing with bacterial disease, we still have to be eternally vigilant (a) to cope with mutants that are resistant to many antibiotics and (b) to devise new drugs to kill fungi colonizing the areas rendered vacant by the destruction of bacteria. Nature does, indeed, abhor a vacuum!

References

Ariens, E.J., 1986, *Trends in Pharmacol. Sci.*, **7**, 200–5.
Ariens, E.J., Wuis, E.W. and Veringa, E.J., 1988, *Biochem. Pharmacol.*, **37**, 9–18.
Hansch, C., 1985, *Drug Metab. Rev.*, **15**, 1279–94.
IUPAC, 1970, *J. Org. Chem.*, **35**, 2849–63.
Jusko, W.J. and Gretch, M., 1976, *Drug Metab. Rev.*, **5**, 43–140.
Lienhard, G.E., 1973, *Science*, **180**, 149–54.
Morrison, J.F., 1969, *Biochim. Biophys. Acta.*, **185**, 269–86.
Morrison, J.F., 1982, *Trends in Biochem. Sci.*, **7**, 102–5.
Vane, J.R., 1971, *Nature New Biology*, **231**, 232–6.
Walsh, C., 1984, *Ann. Rev. Biochem.*, **53**, 493–536.
Wolfenden, R., 1976, *Adv. Biophy. Bioeng.*, 271–306.

Chapter 2

Nucleic acid biosynthesis and catabolism

2.1 Introduction

As nucleic acids are a fundamental part of any organism and have to be synthesized early on in the growth cycle, it is not surprising that many of the drugs in clinical use for the chemotherapy of neoplastic disease and microbial infection have the enzymes of nucleic acid synthesis as their primary target. A number of enzymes involved in viral and bacterial replication are either unique to that organism, or are sufficiently different from the host enzyme which catalyses a similar reaction for that difference to be exploited.

Cancer cells are characterized by rapid, uncontrolled growth; there do not seem to be any major differences between tumour enzymes and those from normal cells. It is, therefore, the rate of synthesis of DNA that forms the target for chemotherapeutic agents. Consequently, toxicity problems do arise by virtue of inhibition of these enzymes in rapidly growing non-tumour tissue - hence the practice of delivering anti-tumour drugs in 'cocktails' of three or four drugs at a time, working on different pathways, so that the toxicity of any one drug does not become overwhelming. An outline scheme of the drugs and their targets discussed in this chapter is presented in Table 2.1.

Another difference between microbial infection and cancer lies in the efficacy of the immune system. Usually with bacterial and viral infection we rely on the immune system to assist in, and eventually take over, the healing process. In cancer, and also in some fungal infections, the immune system is weakened and, therefore, a complete kill of the tumour or fungal cells is required; drugs that merely arrest the growth of these cells are frequently not sufficient to cure the disease.

The type of interference with enzymes that we find in this area is not only the straightforward competitive inhibition of the rate-determining step in a pathway - although that may occur. Other approaches include the concept of false substrates whereby one enzyme accepts a foreign substrate and yields a product that can react with another enzyme; in the cancer field this is known as the anti-metabolite approach. Furthermore, the synthesis of a macro-molecular product allows the possibility of incorporation of false substrate into the

Table 2.1 Drugs discussed in Chapter 2

Target	*Drug*	*Disease*
2.2 Nucleotide biosynthesis		
Dihydropteroate synthase	Sulphonamides	Bacterial infection
Dihydrofolate reductase	Trimethoprim	Bacterial infection
	Pyrimethamine	Malaria
	Methotrexate	Cancer
Ribonucleotide reductase	Hydroxyurea	Cancer
Orotate phosphoribosyltransferase (thymidylate synthetase)	5-Fluorouracil	Cancer
Cytosine deaminase (thymidylate synthetase)	5-Fluorocytosine	Candidosis
Inosine monophosphate dehydrogenase	Ribavirin	Influenza
Dihydroorotate dehydrogenase	Atovaquone	Malaria
2.3 Nucleic acid biosynthesis		
Thymidine kinase (viral DNA polymerase)	Acyclovir	Herpesvirus infection
	Vidarabine	Herpesvirus infection
Deoxycytidine kinase (mammalian DNA polymerase)	Cytarabine	Cancer
DNA gyrase	Ciprofloxacin	Bacterial infection
	Norfloxacin	Bacterial infection
	Novobiocin	Bacterial infection
DNA topoisomerase II*	Doxorubicin	Cancer
	Etoposide	Cancer
DNA polymerase (RNA primed), reverse transcriptase	Azidothymidine	Infection by human immunodeficiency virus (AIDS)
RNA polymerase	Rifampicin	Tuberculosis
2.4 Nucleotide catabolism		
Adenosine deaminase	2-Deoxycoformycin	Cancer
Xanthine oxidase	Allopurinol	Gout
Hypoxanthine phosphoribosyl transferase	Allopurinol	Leishmaniasis
Guanylate cyclase	Isosorbide nitrate	Angina pectoris

Where two enzymes are recorded as targets, the first enzyme uses the drug as a false substrate while the enzyme in brackets is the ultimate target.

*This assignation remains controversial.

polymer chain, resulting in either chain termination or the production of a faulty macromolecule.

We also find the occasional occurrence of a suicide substrate (allopurinol, section 2.4.2) and transition state analogues (2-deoxycoformycin, section 2.4.1) which have aroused considerable interest (see Chapter 1 for an outline of these concepts). The recent concentration on the widespread screening of secondary microbial metabolites (i.e. those compounds formed that are not on the normal pathways required for cell growth) and of the products of organic chemical synthesis, has greatly increased the scope of chemical structures that are under investigation.

2.1.1 Overall scheme

Nucleic acids in mammalian organisms are constructed from nucleotides which in turn are made up from a purine or pyrimidine base, a sugar (ribose in ribonucleic acid or deoxyribose in deoxyribonucleic acid) and a phosphate group attached to the 5′-position of the ribose ring (Fig. 2.1). The synthesis of these moieties may be *de novo* (Fig. 2.2) - i.e. constructed from small molecular building blocks - or via salvage pathways which effectively recycle a pre-formed base. Purines are synthesized by attaching an amino group to a ribose ring and then using this as the anchor from which to build up an imidazole ring to yield 5-aminoimidazole-4-carboxamide ribotide (AICAR).

One more step leads to the formation of inosine monophosphate (the first purine nucleotide) from which the pathways diverge to yield guanosine and adenosine monophosphates. The monophosphates have a second and third phosphate group added to give the triphosphates that are required for RNA biosynthesis. The dinucleotides may be reduced to deoxynucleotides, which after conversion to the trinucleotides, are used for DNA biosynthesis.

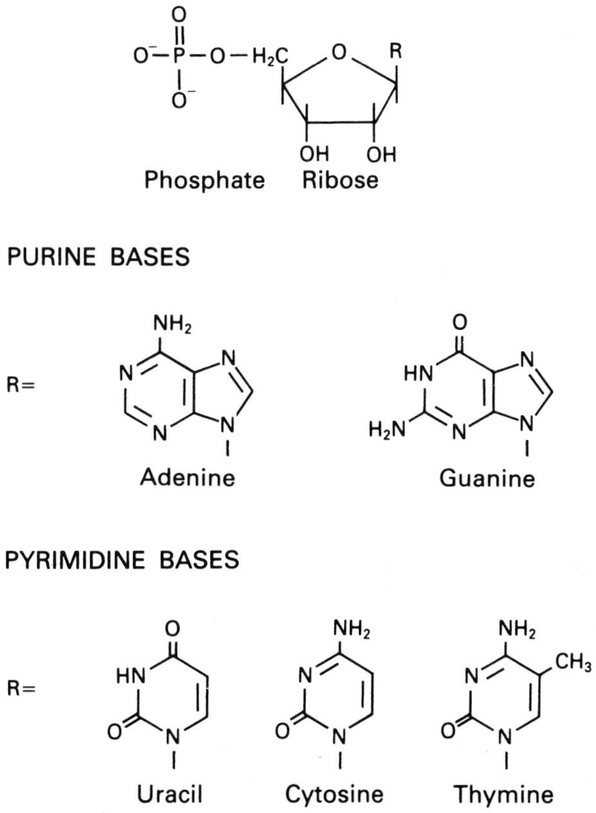

Phosphate Ribose

PURINE BASES

R=

Adenine Guanine

PYRIMIDINE BASES

R=

Uracil Cytosine Thymine

Figure 2.1 Nucleotide structure.

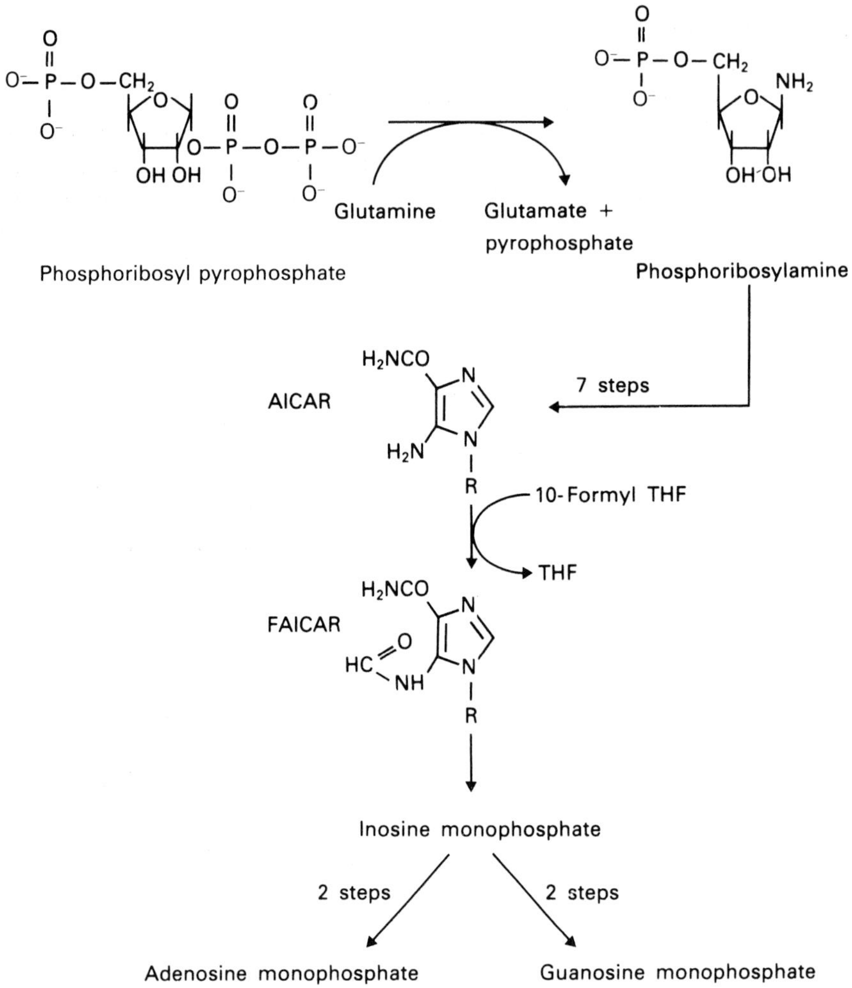

Phosphoribosyl pyrophosphate

Glutamine

Glutamate + pyrophosphate

Phosphoribosylamine

7 steps

AICAR

10-Formyl THF

THF

FAICAR

Inosine monophosphate

2 steps

2 steps

Adenosine monophosphate

Guanosine monophosphate

R, ribose 5′-phosphate; AICAR, 5-aminoimidazole-4-carboxamide ribotide; FAICAR, 5-formamidoimidazole-4-carboxamide ribotide.

Figure 2.2 Outline of purine nucleotide biosynthesis.

Folic acid analogues are essential cofactors for the transfer of some one-carbon units. The particular conversions of interest in the nucleotide biosynthetic pathways are the addition of a formyl (-CHO) group to glycinamide ribotide and subsequently to AICAR by 10-formyltetrahydrofolate in the purine pathway, and the methylation of deoxyuridine monophosphate (UMP) to deoxythymidine monophosphate (dTMP) catalysed by thymidylate synthetase in pyrimidine biosynthesis. The first step in the biosynthesis of folate is the formation of dihydropteroate from *p*-aminobenzoate and 2-amino-4-hydroxy-6-methylpteridine (Fig. 2.4). This step is the target for the sulphonamide antibacterials (hence the build-up of AICAR in cells treated with sulphonamides - see below).

The heterocyclic ring of pyrimidines, on the other hand, is constructed before the ribose is attached (Fig. 2.3). Indeed, in mammals the first three steps to yield dihydroorotate are all catalysed by the same protein molecule, although the activities are on separate proteins in bacteria. Orotate is converted to the first pyrimidine nucleotide, orotidine 5-phosphate, which loses a carboxyl group to form uridine 5-monophosphate. The addition of two more

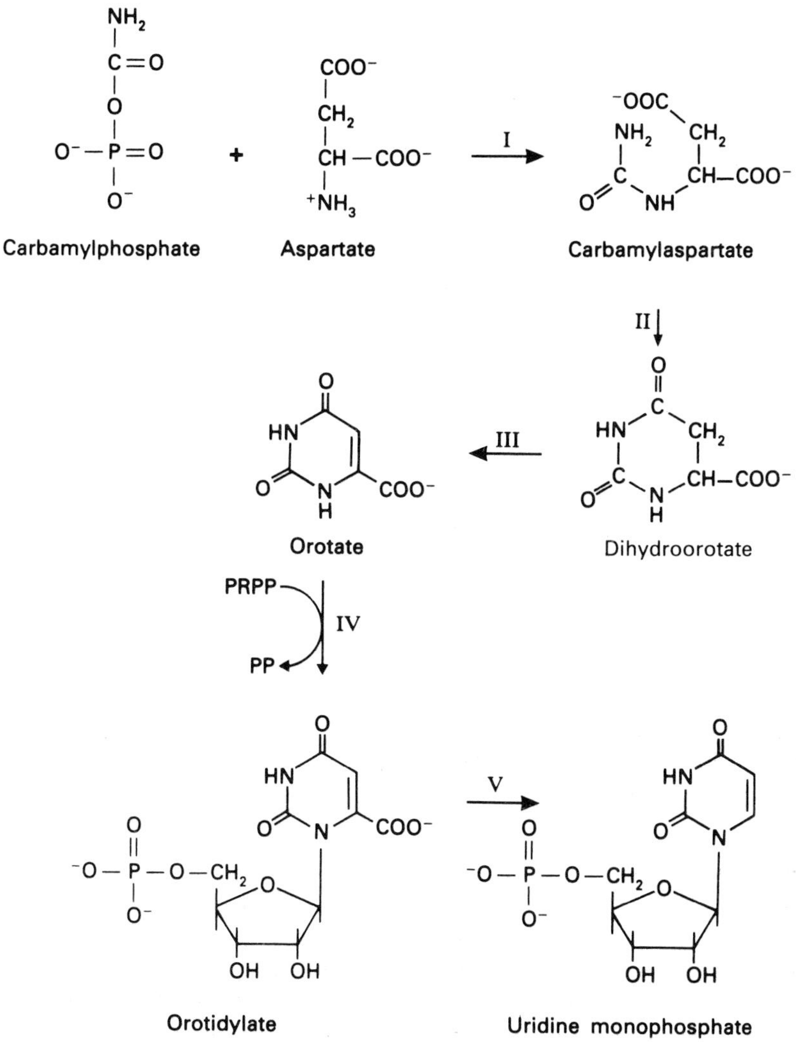

Carbamylphosphate + Aspartate —I→ Carbamylaspartate

—II→ Dihydroorotate

—III→ Orotate

PRPP → PP, IV

Orotidylate —V→ Uridine monophosphate

I, Aspartate transcarbamylase; II, dihydroorotase; III, orotate dehydrogenase; IV, orotidylate pyrophosphorylase; PRPP, phosphoribosyl pyrophosphate; PP, pyrophosphate

Figure 2.3 Pyrimidine nucleotide biosynthesis.

phosphate groups forms uridine triphosphate from which cytidine triphosphate can be made. Thymidine monophosphate is formed by the methylation of uridine monophosphate catalysed by thymidylate synthetase. Again, as with the purine nucleotides, it is the ribo*di*nucleotides which are reduced to the deoxy form.

The action of DNA polymerases, using the four deoxyribotrinucleotides, produces DNA. RNA is synthesized from a DNA template in a process (transcription) requiring the four ribotrinucleotides, and messenger and transfer RNA are employed to generate protein on the ribosome (translation, Chapter 3). Exceptions to this sequence are found with viruses that contain RNA as the genetic code. Two examples are the influenza virus, where the RNA acts as its own template, and the virus that produces the disease known as acquired immune deficiency syndrome (AIDS), where the RNA acts to form DNA, catalysed by a viral enzyme known as reverse transcriptase (section 2.3.5).

Purine nucleotides are first broken down to nucleosides by removal of the phosphate group and ribose groups to produce inosine. Subsequent changes include oxidative steps to hypoxanthine, xanthine and uric acid (Fig. 2.10). Pyrimidine nucleotides are also converted to the free bases, uracil and 5-methyluracil), that eventually yield malonate (from cytosine) and 2-methyl malonate (from thymine).

2.2 Nucleotide biosyntheses - enzyme targets of drugs

2.2.1 Dihydroorotate dehydrogenase - atovaquone as antimalarial

Atovaquone is a naphthoquinone under development for the treatment of protozoal infections in man, notably malaria caused by *Plasmodium falciparum* and toxoplasmosis caused by *Toxoplasma gondii*. Quinones are unusual among marketed drugs, probably because of their innate reactivity. The drug may also be effective against *Pneumocystis carinii*, a microorganism originally classed as a protozoon but now believed to be phylogenetically closer to a fungus. Infections by this agent have become more serious with the increasing number of patients whose immune systems are suppressed, e.g. in AIDS.

The malarial parasite depends entirely on *de novo* biosynthesis of pyrimidine nucleotides for nucleic acid biosynthesis, unlike man where salvage pathways from preformed nucleosides play a significant part. Atovaquone acts by inhibiting the membrane-bound complex dihydroorotate dehydrogenase - the third enzyme in the pyrimidine biosynthetic pathway - see Fig. 2.3. Although the enzyme is not the rate-limiting step, which in most pathways is the first enzyme (in this case aspartate transcarbamoylase), the inhibition of the target enzyme is irreversible; a build-up in the precursors such as carbamoylaspartate and a reduction in the end-products such as UTP results. Furthermore, the inhibition could not be reversed by adding uridine (Hammond *et al.*, 1985).

Dihydroorotate dehydrogenase is found on the inner side of the mitochondrial membrane and passes electrons to ubiquinone in the electron transport chain. The specific molecular target is a polypeptide between cytochromes *b* and *c*1 in mitochondria isolated from *P. falciparum* (Fry and Pudney, 1992). Atovaquone is active as the *trans* isomer.

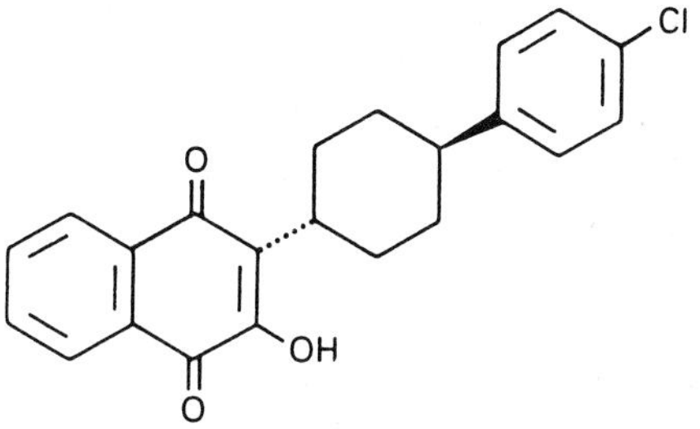

Atovaquone
Relative stereochemistry.

The IC_{50} for various strains of *P. falciparum* is between 0·7 and $4{\cdot}3 \times 10^{-9}$ M, which is more potent than the other antimalarial drugs tested (amodiaquin, mefloquin, pyrimethamine, chloroquin and quinine; Hudson *et al.*, 1991).

2.2.2 Dihydropteroate synthetase - sulphonamides as anti-bacterials

One family of drugs which was recognized early to have anti-bacterial properties was the sulphonamides. These have been in medical use since the 1940s – indeed Winston Churchill was given one of the earliest compounds, M & B 693, in North Africa in February 1943 when he had an attack of pneumonia, and it is likely that it played a major role in his recovery (Churchill, 1951). These drugs interfere with the purine biosynthetic pathway; 5-aminoimidazole-4-carboxamide ribotide (AICAR) accumulates in bacterial cells treated with sulphonamides. The mechanism of interference is indirect in that AICAR accumulates not because the next enzyme in the pathway is inhibited, but because there is insufficient folate cofactor to service the enzyme. Mammals, however, require preformed folate in the diet as an essential vitamin, and so inhibition of this step is not likely to lead to mammalian toxicity.

The biosynthesis of folate requires *p*-aminobenzoic acid as a starting material and sulphonamides are sufficiently close in structure to act not only as inhibitors of dihydropteroate synthetase but also as false substrates (reviewed in Woods, 1962) (Fig. 2.4). Any agent that can penetrate a bacterium which needs a functioning enzyme is very likely to have activity against that bacterium.

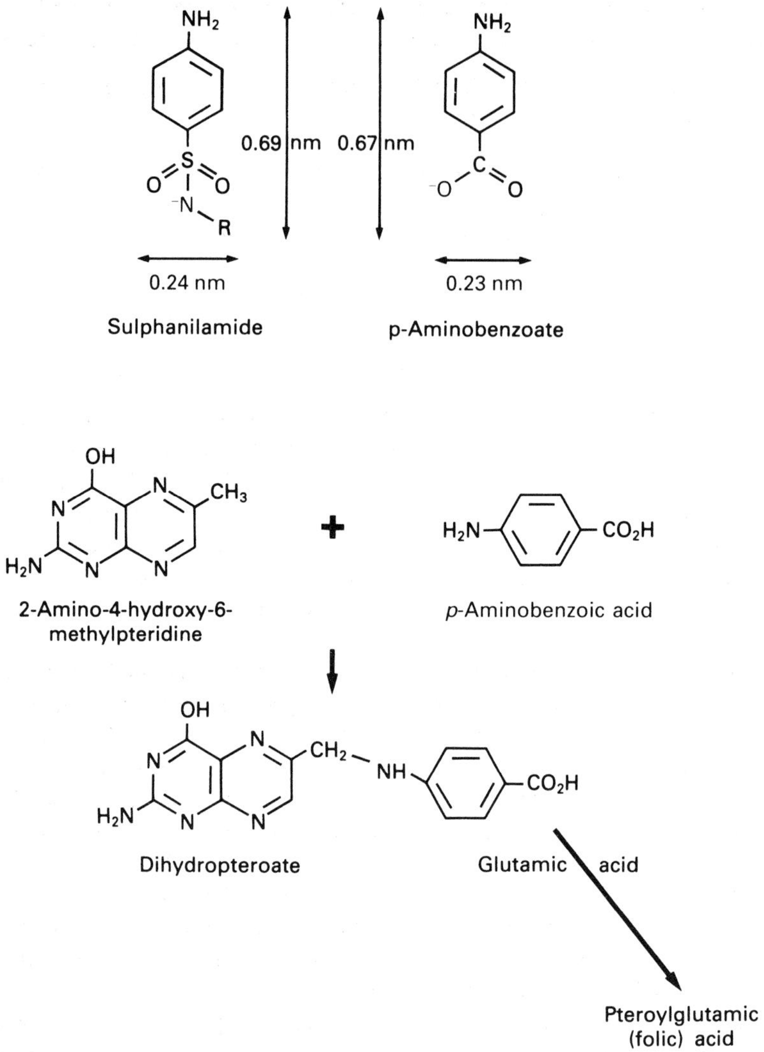

0.69 nm 0.67 nm

0.24 nm 0.23 nm

Sulphanilamide p-Aminobenzoate

2-Amino-4-hydroxy-6-methylpteridine + *p*-Aminobenzoic acid

Dihydropteroate

Glutamic acid

Pteroylglutamic (folic) acid

The conversion of 2-amino-4-hydroxy-6-methylpteridine to dihydropteroate and to folic acid is shown.
A typical sulphonamide is compared with *p*-aminobenzoate and is shown to be structurally similar.

Figure 2.4 Dihydropteroate synthetase reaction.

Sulphonamides are not now usually used alone to treat infections in man as there are many other more effective families of anti-bacterials currently available, although sulphonamides may be used in conjunction with inhibitors of dihydrofolate reductase (see below). An exception to this is dapsone, which is sometimes used alone for the treatment of leprosy (Shepard *et al.*, 1976).

2.2.3 Dihydrofolate reductase (DHFR) - trimethoprim and pyrimethamine as anti-bacterials, methotrexate as anti-cancer

The reaction carried out by this enzyme is crucial to the recycling of the folate cofactors involved in methylation of the DNA bases:

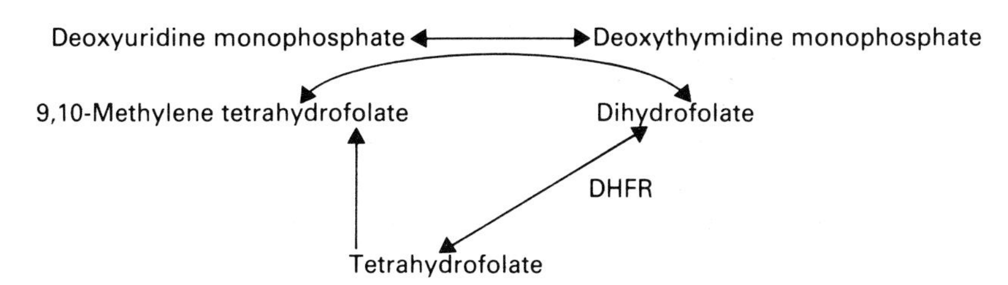

DHFR has turned out to be an effective target for both anti-microbial and anti-cancer drugs, possibly because one folate molecule has to be reduced for every deoxythymidine monophosphate molecule synthesized, and in situations of rapid growth this requirement could be considerable. The mechanism of thymidylate synthetase requires that the one-carbon fragment is transferred as a methylene group to the 5-position of the uracil ring of dUMP. This is then reduced to a methyl group by the pteridine ring of the coenzyme to yield dihydrofolate. DHFR is required to reform tetrahydrofolate.

Unlike the case of dihydropteroate synthase above, mammals carry out the same interconversion and so selectivity becomes of prime importance. In this instance trimethoprim, used as an anti-bacterial agent, and pyrimethamine, an effective agent for killing the protozoal parasites (plasmodia) that cause malaria, are far less inhibitory towards a mammalian liver DHFR than for the enzyme from their respective microbial targets. This is shown in Table 2.2 (the effect of the drugs on the parasite, *Plasmodium berghei*, that infects mice is shown for comparison) and see Matthews *et al.* (1985).

DHFR has been confirmed as the target for trimethoprim by the identification of an altered, and largely uninhibited, DHFR in resistant organisms (reviewed in Hitchings and Smith, 1980).

These agents are able to penetrate their target organisms by non-ionic diffusion. Methotrexate, a diaminopyrimidine which is used for choriocarcinoma and trophoblastic tumours (Hammond *et al.*, 1981), strongly inhibits DHFR in tumour cells with a K_i of 1×10^{-9} M. The drug forms a ternary complex with enzyme and NADPH, as shown by X-ray crystallography (Matthews *et al.*,

Table 2.2 Concentration for 50 per cent inhibition of DHFR ($\times 10^{-9}$ M)

	Rat liver	*E. coli*	*P. berghei*
Pyrimethamine	700	2500	0·5
Trimethoprim	260 000	5	70

(from Ferone *et al.*, 1969)

1978). The action of the drug *in vivo* relies on the greater need of the tumour cell for purine nucleotides compared with normal cells, and so gives rise to considerable toxicity in rapidly dividing cells such as in the bone marrow. Methotrexate is transported into mammalian cells by the active transport system for folate, to which it bears a marked resemblance. Microorganisms do not have such a transport system and, although methotrexate is as effective against the microbial enzymes as against the mammalian, it has little antimicrobial activity (Ferone *et al.*, 1969).

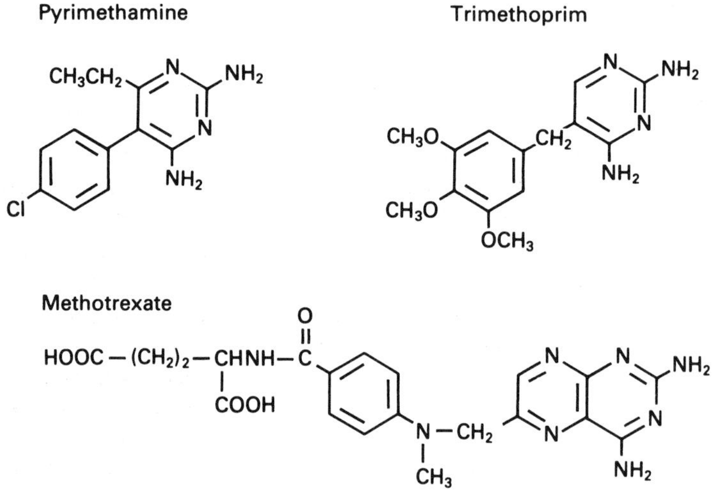

Pyrimethamine

Trimethoprim

Methotrexate

As the targets for sulphonamides and diaminopyrimidines lie on different pathways that intersect, it seemed appropriate to test combinations of these drugs to establish whether they could be used in conjunction more effectively, i.e. whether they showed synergy. Synergy was found to occur both *in vitro* (Moody and Young, 1975) and *in vivo* (Ramachandran *et al.*, 1978) when trimethoprim was combined with sulphamethoxazole against bacterial infections. With pyrimethamine similar effects have been observed against human malaria, but in this case the value of such combinations is neither to reduce toxicity nor to obtain high potency, but rather to lower the dose of pyrimethamine such that there is less likelihood of resistance developing. Resistance is a serious problem with the malarial parasite at present.

Combinations of trimethoprim with other sulphonamides such as sulphadiazine (Hurly, 1959) or dapsone (Lucas *et al.*, 1969) were shown to be very effective in man. Sulphadoxine currently appears to be the sulphonamide of choice for synergy with trimethoprim as it has a particularly long half-life for a sulphonamide - seven to nine days in plasma. This is clearly of great value for reducing the number of doses required for treatment.

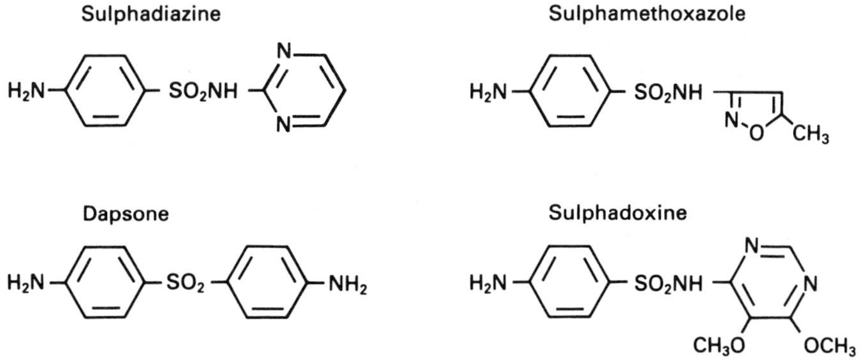

Sulphadiazine

Sulphamethoxazole

Dapsone

Sulphadoxine

2.2.4 Ribonucleotide reductase - hydroxyurea as anti-cancer

The reduction of the nucleoside diphosphates to their respective deoxy counterparts is the first committed step in deoxyribonucleotide biosynthesis, and is catalysed by ribonucleotide reductase. The activity of this enzyme has been found to follow closely the regulation of DNA biosynthesis, although this connection may not be any more than a correlation (Thelander and Reichard, 1979). The same enzyme reduces all four ribonucleotide diphosphates by directly replacing the 2-hydroxyl group on the ribose ring by a hydrogen atom. With the mammalian enzyme, magnesium and ATP are also required for activity and a number of allosteric regulators have been identified (see Chapter 1 for a discussion on allosterism). In order to carry out the reduction, the enzyme requires the iron–sulphur protein thioredoxin, together with thioredoxin reductase which uses NADPH to regenerate thioredoxin. Thioredoxin contains two free sulphydryl groups positioned in such a way that a disulphide bond can be formed between them:

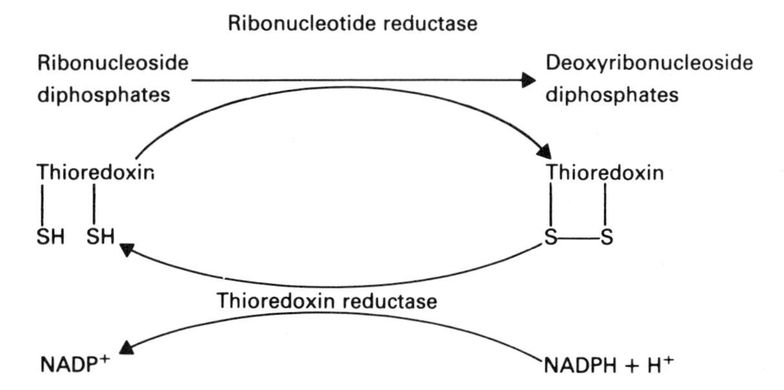

The *E. coli* enzyme on which most of the work has been done is a heterodimer composed of two subunits B1 and B2. The larger subunit contains two equivalent polypeptides of 86 kDa carrying the substrate binding site and allosteric effector sites. The smaller subunit composed of two equivalent polypeptides of 45 kDa carries a unique prosthetic group consisting of two catalytic, but non-equivalent, high spin ferric iron atoms stabilizing a tyrosyl radical. The oxygen of the phenolic hydroxyl formally carries the radical in a one-electron oxidized state. The tyrosine (Tyr-122) is buried in the protein about 1 nm from the surface and is surrounded by hydrophobic side-chains. Each iron atom is associated with one polypeptide chain and the pair are anti-ferromagnetically coupled to each other (i.e. their spins are opposed or anti-parallel) and are bridged by an oxygen atom and two carboxylate residues. These groups are located in the active site at the interface between the two subunits, together with two cysteine thiols which are reduced by the NADPH–thioredoxin system (reviewed in Stubbe, 1989; Elgren *et al.*, 1991; Fig. 2.5).

The mechanism of this unusual reaction involves removal of a hydrogen atom from position 3 of the sugar by the tyrosyl radical, followed by acid-catalysed cleavage of the 2′ C–OH bond to yield a radical cation. This intermediate is then reduced by addition of two hydrogens from the two thiols. Finally the same hydrogen atom that was originally abstracted from the 3′ position is returned to the product and the tyrosyl radical is regenerated (Stubbe, 1989; Fig. 2.6).

Since the activity of ribonucleotide reductase correlates well with the level of DNA biosynthesis it might be expected to be a logical target for chemotherapy (reviewed in Elford *et al.*, 1981), but it is perhaps surprising that so far only one drug on the market, hydroxyurea, owes its effectiveness to inhibition of this enzyme. This drug is used for the treatment of chronic granulocytic leukaemia and malignant melanoma (Kennedy and Yarbro, 1966; Ariel, 1970). The use of hydroxyurea suffers from the drawback that frequent large doses are required to maintain effective concentrations *in vivo*. This is understandable in view of the high drug concentration, 5×10^{-4} M, required to produce 50 per cent inhibition of the enzyme (Elford *et al.*, 1979). Hydroxyurea originally presented a challenge with regard to its mechanism of action. Recent studies, however, show that the drug quenches the tyrosyl radical by

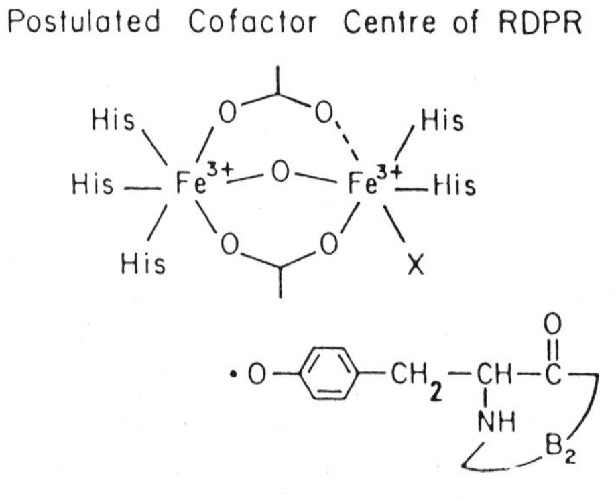

Postulated Cofactor Centre of RDPR

Figure 2.5 Proposed structure of the binuclear iron centre of ribonucleotide reductase.

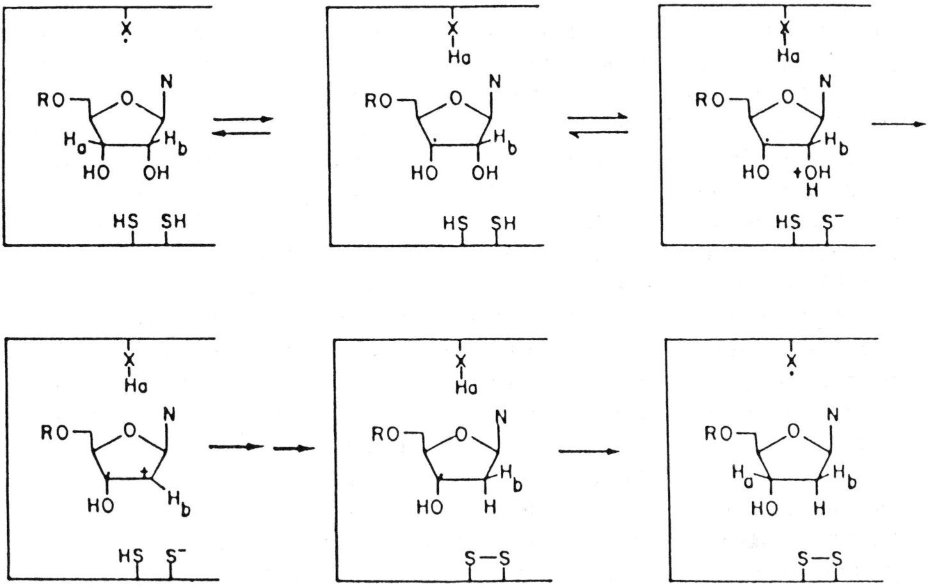

X = protein

Figure 2.6 Postulated mechanism of reduction of nucleotides to deoxynucleotides catalyzed by ribonucleotide reductase.

the transfer of an electron, thus forming transient nitroxide-like free radicals which have a very short half-life. The enzyme is thereby inactivated. Hydroxyurea is poor as a general free radical scavenger, however, but is much more specific for ribonucleotide reductase than are other free radicals, possibly because it is small enough to pass along the 1 nm cleft in the protein (Lassmann *et al.*, 1992).

2.2.5 Thymidylate synthetase - 5-fluorouracil as anti-cancer, 5-fluorocytosine as anti-fungal

The methylation of uridine monophosphate to yield thymidine monophosphate is a crucial step in the provision of pyrimidine bases for the synthesis of DNA, and so is another possible target for cancer chemotherapy. 5-Fluorouracil is believed to exert its anti-tumour effect, after conversion to ribo- and deoxyribonucleotide, partly on thymidylate synthetase and partly on RNA biosynthesis (reviewed in Heidelberger *et al.*, 1983).

5-Fluorouracil (5-FU) is converted to 5-fluorouridine monophosphate (5-FUMP), probably in one step catalysed by orotate phosphoribosyltransferase using phosphoribosyl-1-pyrophosphate as co-substrate. 5-FUMP is phosphorylated to the diphosphate (5-FUDP), reduced in the presence of ribonucleotide reductase to 5-FdUDP and then hydrolysed to 5-FdUMP. This latter compound forms a covalently bound ternary complex with the folate cofactor $N^{5,10}$-methylene-tetrahydrofolate and thymidylate synthetase, which closely resembles the transition state of the normal reacton. 5-FdUMP acts as a false substrate since the folate methylene group is transferred to the 5-position of the uracil ring, but further conversions cannot take place because of the presence of the fluorine atom at position 5 of the uracil ring. As a consequence, a covalent link remains between the methylene group and either positions 5 or 10 of the pteridine ring of the cofactor (see Fig. 2.7). The inhibitor, therefore, behaves as a suicide substrate and the synthesis of thymidylate is shut off.

Other studies, however, have tended to suggest that thymidylate synthetase may not be the only target. The incorporation of 5-FU into RNA to yield a faulty ribosomal RNA may also play a part in the drug's efficacy and both mechanisms may be important to varying extents in different situations (Heidelberger *et al.*, 1983). 5-FU is used with good effect in certain types of carcinoma of the breast and of the gastrointestinal tract (Bonadonna and Valogussa, 1981; Kisner *et al.*, 1981).

A closely related compound, 5-fluorocytosine (5-FC), has found favour as a treatment for fungal infections, namely candidosis and cryptococcosis, particularly when used in conjunction with amphotericin B (Cohen, 1982). 5-FC is not recommended as a single medication in view of the rapid development to resistance. Up to 30 per cent of the patients in one trial treated with 5-FC alone developed resistance (Block *et al.*, 1973) - hence the concomitant use of amphotericin. 5-FC is converted to 5-FU by a fungal enzyme, cytosine deaminase; similar metabolic conversions to those for 5-FU in mammalian cells follow. 5-FU is initially converted to 5-FUMP by UMP pyrophosphorylase (an enzyme that normally acts to salvage uracil in *Candida albicans*). This appears to be the point at which resistance is induced, since resistant *C. albicans* has little or no detectable enzyme (Whelan and Kerridge, 1984). Cytosine deaminase is believed to be largely absent from the host, although Diasia *et al.* (1978) found some evidence of 5-FU production in man.

2.2.6 Inosine monophosphate dehydrogenase - ribavirin as antiviral

Ribavirin (virazole) has shown variable clinical effectiveness against a variety of viruses, including those with RNA as the genetic material, such as influenza, and those containing DNA, for example, herpes, measles and some adenoviruses (reviewed in Chang and Heel, 1981). There is some uncertainty about the effectiveness of the drug, but, at least in the case of influenza, the delivery of the drug by aerosol is more effective than by the oral route. The latter may deliver insufficient drug to the lungs for activity to be shown (McClung *et al.*, 1983). Ribavirin sufficiently resembles adenosine to be phosphorylated intracellularly by host adenosine kinase, and is subsequently converted to the di- and trinucleotides by other host enzymes. The drug structure can also be drawn to resemble guanosine (Gilbert and Knight, 1986).

REACTION INTERMEDIATE
(PROPOSED):

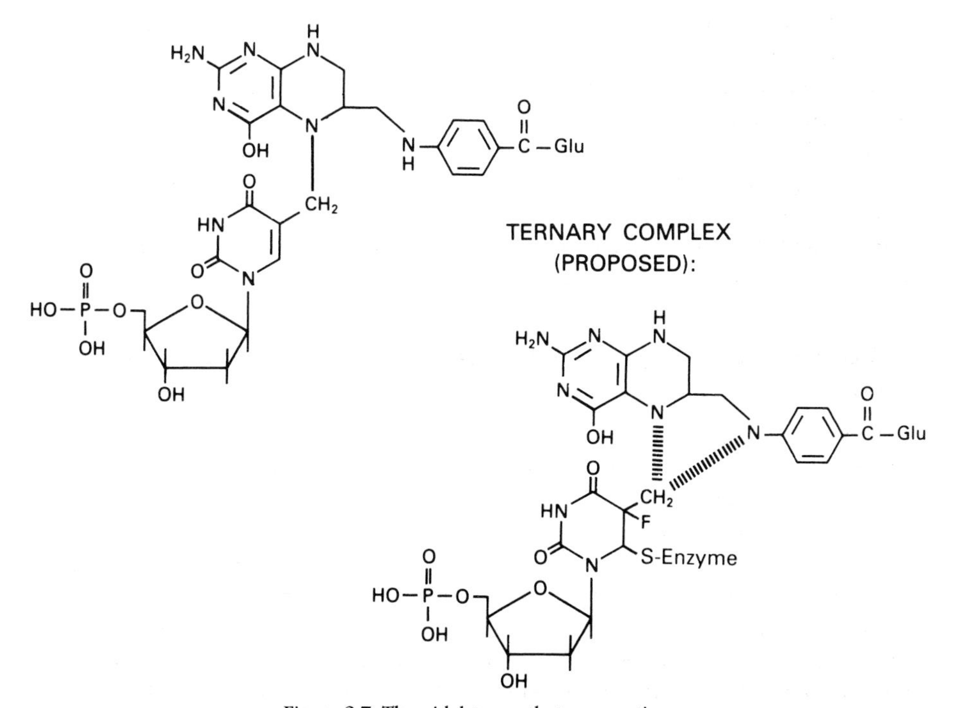

Figure 2.7 Thymidylate synthetase reaction.

5-FU 5-FC

Intracellular pools of GTP are reduced by up to 45 per cent in influenza-infected cells treated with ribavirin (Wray *et al.*, 1985). This results from inhibition of the first committed enzyme in *de novo* synthesis of GMP; the NAD-linked inosine monophosphate dehydrogenase, by the monophosphate of ribavirin (Streeter *et al.*, 1973). This enzyme converts to IMP to xanthosine monophosphate which is then transformed into GMP by GMP synthetase (see Figure 2.8). The inhibition is competitive with a K_i of $2{\cdot}5 \times 10^{-7}$ M against IMP dehydrogenase from Ehrlich ascites tumor cells (K_m for IMP is $1{\cdot}8 \times 10^{-5}$ M).

This inhibition accounts for part of the antiviral effect, but the fact that guanosine can only partly reverse the inhibition of proliferation of influenza virus in canine kidney cells suggests that other mechanisms may be involved. This is supported by the observation that the lowering of the GTP levels reaches a minimum at $2{\cdot}5 \times 10^{-5}$ M ribavirin but the anti-viral effect continues to increase up to $1{\cdot}4 \times 10^{-4}$ M drug (Wray *et al.*, 1985). Other possible targets include ribavirin triphosphate inhibition of (a) the capping reaction of viral messenger RNA (GTP is added to the 5′ end of freshly synthesized viral mRNA to render it effective), and (b) the viral RNA polymerase complex (see review by Gilbert and Knight, 1986).

The nature of the virus under consideration may well play a part in the mechanism of action. Patterson and Fernandez-Larson (1990) believe that inhibition of the viral RNA polymerase complex is the mechanism by which ribavirin inhibits the growth of two other RNA viruses: La Crosse and vesicular stomatitis. Multiplicity of mechanisms of action may also explain why it is very difficult if not impossible to obtain ribavirin-resistant virus strains.

2.3 DNA biosynthesis

So far we have been considering drugs that interfere with the process of nucleotide biosynthesis. We now turn to the next step in the formation of nucleic acid; the magnesium-dependent process in which the four nucleoside triphosphates are utilized in the polymerization of DNA. In mammalian cells more than one enzyme is required to do this conversion; two enzymes, α and β, are responsible for the formation of chromosomal DNA, while γ catalyses the synthesis of mitochondrial DNA.

The viral genome, on the other hand, codes for a number of enzymes which it needs to synthesize a fresh virus particle. Foremost among these is the RNA

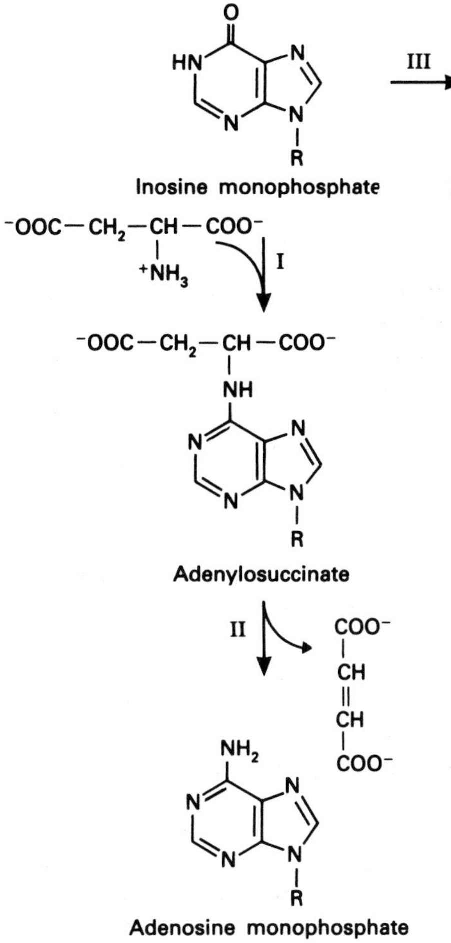

I: Adenylosuccinate synthetase
II: Adenylosuccinate lyase
III: IMP dehydrogenase
IV: GMP synthetase
R: Ribose

Figure 2.8 The biosynthesis of GMP.

Ribavirin

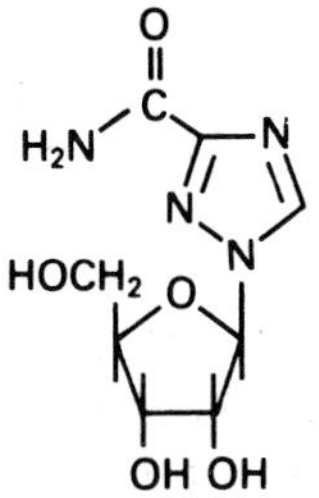

or DNA polymerase with which, after subversion of the host cell's protein synthesizing machinery, more genome is synthesized. Although all DNA polymerases require magnesium, the enzyme coded by herpesviruses 1 and 2 differ from their mammalian counterparts in a number of ways, including a fivefold activation by 6×10^{-2} M ammonium sulphate while the host cell enzymes are completely inhibited at this concentration (Mar and Huang, 1979).

A few drugs that interfere with DNA biosynthesis have been developed for the treatment of cancer and for viral disease. Indeed, the main drugs used to treat viral infection may be found in this section, which is interesting considering how many other activities of the virus could be targets, such as the binding of the virus to the outside of the target cell; the removal of the protein outer coating of the virus and the penetration of the nucleic acid into the cell; the induction of protein as well as nucleic acid biosynthesis by the cellular apparatus; assembly of protein and nucleic acid into infectious virus particles followed by their exit from the cell.

This process destroys the cell, whether bacterial or mammalian, and can lead to a fresh round of virus replication if other cells are nearby. A zone of lysis, known as a plaque, is then formed wherever replicating viral particles are located. If host cells are grown on a suitable nutrient in a plate, plaque formation can be used to quantify the amount of active virus present.

2.3.1 Herpesvirus DNA polymerase - acyclovir and vidarabine as anti-virals

9-[(2-Hydroxyethoxy)methylguanine (generic name, acyclovir), an analogue of a purine nucleoside, is the first of a new family of drugs for the treatment of infections caused by the family of DNA viruses, known as herpesvirus, which range in severity from the ubiquitous mouth ulcer, through serious cases of blindness, to life-threatening disease (reviewed in Whitley and Gnann, 1992). Herpesvirus infections in humans can be divided into four types: herpesvirus type 1 which is usually responsible for mouth ulcers and eye damage; type 2 which mainly gives rise to serious genital infections; varicella zoster virus which produces chicken pox and shingles; cytomegalovirus (large-cell-producing virus) which gives rise to a type of pneumonia, particularly in immunocompromised conditions such as AIDS.

These viruses have the ability to become latent, i.e. after an infection the virus migrates to the nervous system where it remains dormant until reactivated by stimuli such as stress. When latent, herpesvirus 1 resides in the trigeminal ganglion (a relay station for the fifth cranial nerve) near the inner ear. There is a need not only to treat viral infections but also to block either latency or reactivation of these viruses.

Acyclovir is converted to a trinucleotide which takes part in the reaction catalysed by viral DNA polymerase by incorporating the false purine base into the template primer. The addition of the next nucleotide (dCTP) produces a dead-end ternary complex that brings DNA synthesis to a premature end, since acyclovir does not have the 3′-hydroxyl group necessary for chain elongation (Reardon and Spector, 1989).

Another virally coded enzyme, thymidine kinase, is responsible for converting the drug to a nucleotide analogue. In practice, the enzyme is unfortunately named because it is not specific for pyrimidines, and also shows thymidylate kinase activity. Nevertheless, acyclovir is a poor substrate for the enzyme from herpesvirus 1 with a K_m of 4×10^{-4} M and a V_{max} of $6{\cdot}1 \times 10^{-12}$ mol/min/μl of enzyme preparation. The respective figures for thymidine, in contrast, are $8{\cdot}5 \times 10^{-6}$ M and 218 (Ashton *et al.*, 1982). Acyclovir monophosphate is converted to the diphosphate by the action of viral thymidine kinase or by cellular guanosine monophosphate kinase and a number of kinases can add a third phosphate (Miller and Miller, 1982). Acyclovir is not toxic to uninfected cells because they do not convert it to a triphosphate analogue.

Acyclovir

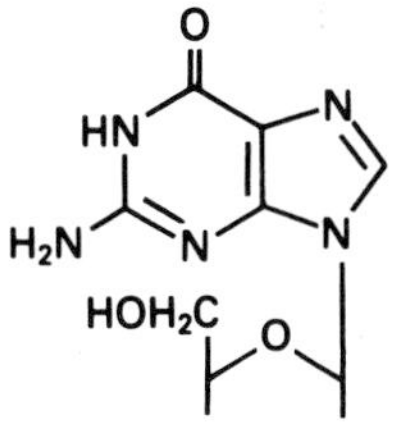

Acyclovir is active to a similar extent against types 1 and 2 virus with 50 per cent inhibition of viral plaque formation at concentrations of 0·046 to $1{\cdot}8 \times 10^{-6}$ M for type 1 and 0·65 to $1{\cdot}8 \times 10^{-6}$ M for type 2 (Parris and Harrington, 1982). Since there are two viral enzymes required to convert acyclovir into an antiviral agent, resistance can and has occurred in all three possible ways: an altered thymidine kinase that phosphorylates thymidine but not acyclovir through mutations in the thymidine binding region, a virus lacking thymidine kinase (which is not an essential enzyme) and single base mutations in the ultimate target DNA polymerase (Chatis and Crumpacker, 1992).

Acyclovir is also used to treat varicella zoster infections, albeit at a higher dose (Ljungmann *et al.*, 1986). Cytomegalovirus is resistant as it lacks a thymidine kinase. Ganciclovir, the drug of choice for cytomegalovirus infections, is phosphorylated by host kinases to the triphosphate which inhibits the viral DNA polymerase. Infections by these viruses are not normally life-threatening – except in immunocompromised conditions such as AIDS (Chatis and Crumpacker, 1992).

Vidarabine (arabinosyladenine or Ara-A) has both anti-tumour and anti-viral activity. The drug is active against keratinization of the cornea and encephalitis caused by herpesvirus 1 in adults and neonates, and against herpes zoster infections in immunocompromised patients (reviewed in Whitley *et al.*, 1980). The nucleoside is phosphorylated to the monophosphate by both host and virus-induced thymidine kinase; further enzymes convert it to the triphosphate. Ara-ATP is more inhibitory to herpesvirus than to host DNA polymerase; a K_i

of $1{\cdot}4 \times 10^{-7}$ M has been reported for herpes hominis DNA polymerase from infected cells (K_m for dATP was $1{\cdot}37 \times 10^{-5}$ M) while DNA polymerase α from rabbit kidney cells was inhibited with a K_i of $7{\cdot}4 \times 10^{-6}$ M (K_m for dATP was $6{\cdot}4 \times 10^{-6}$ M) (Muller *et al.*, 1977). Furthermore, Ara-A is incorporated into viral DNA and slows the rate of DNA elongation, but can be removed by the exonuclease activity associated with viral replication (see Chatis and Crumpacker, 1992).

All these inhibitions were competitive and were measured with activated DNA as the template. Clearly, low doses of the drug will inhibit viral replication, while higher doses will inhibit the mammalian enzyme and may lead to toxicity as the drug is not absolutely selective for viral replication (Muller *et al.*, 1977). Furthermore, Ara-A is incorporated into viral DNA, and it is possible that this effect contributes to its anti-viral action.

The development of Ara-A as an anti-tumour agent has been greatly limited by its rapid deamination to the inactive metabolite arabinosylhypoxanthine by adenosine deaminase, an enzyme that is widely distributed in body tissues and fluids. The drug is, therefore, insufficiently active in the clinic for practical use. Attempts have been made to circumvent this deamination by the use of 2-deoxycoformycin, a powerful adenosine deaminase inhibitor, which enhances and prolongs the activity of Ara-A in man (Major *et al.*, 1983), but this practice has not yet found general acceptance.

2.3.2 Mammalian DNA polymerase - cytarabine as anti-leukaemic

Arabinosylcytosine (Ara-C) is a drug that is commonly used for the treatment of acute leukaemia, both myelogenous and lymphocytic, and non-Hodgkin's lymphoma (Rubin, 1981). It is specific for the S-phase of the cell cycle. In order to exert its anti-tumour effect, Ara-C must first be converted to the monophosphate. This is performed by deoxycytidine kinase; subsequently deoxycytidylate kinase and nucleoside diphosphate kinase catalyse the conversion of the nucleotide to Ara-CTP. This metabolite interferes with DNA synthesis by inhibiting DNA polymerase competitively with dCTP (K_i for Ara-CTP is $8{\cdot}7 \times 10^{-6}$ M and K_m for dCTP is $9{\cdot}0 \times 10^{-6}$ M (Graham and Whitmore, 1970).

Resistance to the drug can occur either by virtue of reduced deoxycytidine kinase levels or by deamination to inactive uracil metabolites by cytidine deaminase. In addition, it has been shown that Ara-C is incorporated into DNA not so much at terminal, but rather at internucleotide linkages. This suggests that chain elongation is not stopped by the presence of the false sugar (Graham and Whitmore, 1970). More recent work with higher concentrations of drug has shown that a greater proportion of Ara-C may be found at the 3′ terminus, which implies that chain elongation is greatly slowed by the drug (Major *et al.*, 1982). This is understandable since the 2′-hydroxyl of arabinose is *trans* to the 3′-hydroxyl, in contrast to ribose, and will cause steric hindrance to the rotation of the pyrimidine base about the nucleoside bond. The bases of polyarabinonucleotides cannot, therefore, stack in the same way as do deoxyribonucleotides, and slowing of DNA elongation could well be the consequence.

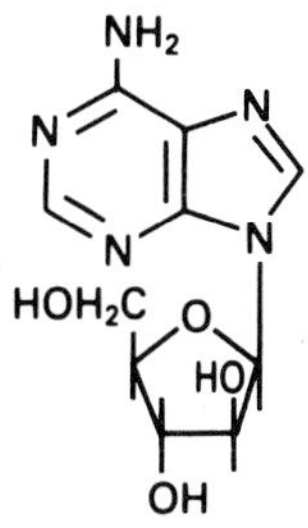

Ara-A 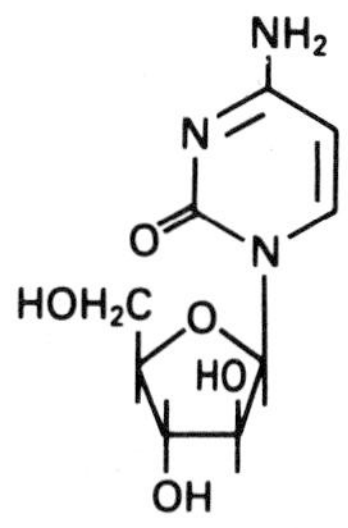 Ara-C

2.3.3 DNA topoisomerases

Another recently discovered family of enzymes is closely involved in the replication of DNA. DNA as normally found in the cell, whether prokaryotic or eukaryotic, is supercoiled, i.e. not only do the nucleoside bases form a double helical coil but the helix itself is twisted on itself to form supercoils. The topoisomers that result from a difference of two superhelical turns have different mobility on SDS–polyacrylamide gels and can be visualized, after ethidium bromide staining, in the form of a 'staircase'.

In the bacterial cell, chromosomal and plasmid DNA is in the form of a closed circle; supercoiling greatly reduces the overall volume of DNA allowing it to be packaged more tightly in the cell. DNA as normally extracted from bacterial cells has fewer supercoils than expected and is called negatively supercoiled. The enzyme-catalysed unwinding of supercoils is referred to as relaxation.

In the mammalian cell, the DNA is supercoiled on the histone proteins in a long helical coil which restrains the supercoils, i.e. prevents them from relaxing. In both prokaryotic and eukaryotic cells, supercoiled DNA is required for the major processes in which DNA is involved; replication, transcription, conjugation, etc.

A class of enzymes called topoisomerases has been recently discovered that can catalyse the interconversion of these topoisomers; they do not catalyse the net formation or cleavage of covalent bonds, but rather modify the three-dimensional structure or topography of DNA. In the mammalian cell, the arrangement of the DNA into a long helical coil requires an enzyme to remove the twists that are produced when DNA is replicated. Otherwise there will be a build-up of positive supercoils in front of the replication fork and negative supercoils behind (see Wang, 1991).

In the presence of magnesium and ATP, topoisomerases break and reform the strands of the DNA; topoisomerase I breaks one strand by forming a bond with the 4-hydroxy group of a tyrosine to the 3′ terminus of the DNA, while topoisomerase II breaks both strands but with the covalent link to the 5′ position (Rowe *et al.*, 1986). In both cases the other strand or strands is passed through the break and the DNA chain is resealed, introducing two supercoils.

2.3.3.1 Bacterial topoisomerase II - quinolone carboxylic acids as antibacterials

Often in the history of medicine a drug has been used for many years without any precise knowledge of its mode of action, e.g. nalidixic acid and its analogues. They first went on the market in the early 1960s for the treatment of urinary tract infections caused by Gram-negative organisms such as *Escherichia coli* and *Proteus* species. The drug has the drawback that resistance rapidly appears in previously sensitive organisms.

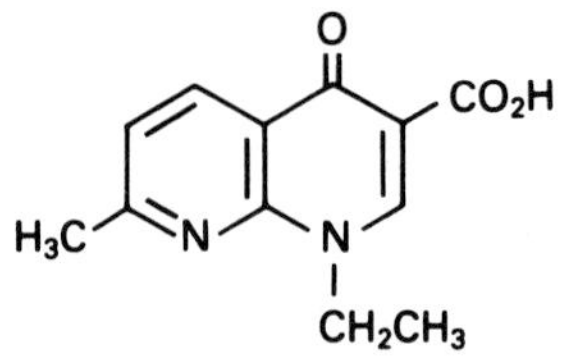

Nalidixic acid

Recently, the advent of analogues, the fluoroquinolones, with a greatly broadened spectrum of action has occurred, notably ofloxacin, and ciprofloxacin; the clinical use of these agents has been reviewed by Hooper and Wolfson (1991).

It became clear early on that nalidixic acid was interfering with DNA biosynthesis, but it required the discovery of DNA gyrase before the precise enzyme target was identified (Gellert, 1980). Many of the activities that bacterial DNA undergoes, including recombination, conjugation, replication and repair, require DNA to be supercoiled. The enzymes that largely control supercoiling are topoisomerases I and II (Drlica, 1992) - the latter is known as DNA gyrase. They play a crucial role in the bacterial cell.

Like the eukaryotic topoisomerase II, DNA gyrase catalyses the interconversion of topoisomers differing by units of two. Gyrase is unique, however, in that it can introduce negative supercoils into DNA in the presence of ATP and magnesium. Without ATP, it can relax both negatively and positively supercoiled DNA. DNA gyrase is composed of two subunits in the form of a tetramer, A_2B_2. Like the mammalian enzyme, the linkage with DNA is made through the 4-hydroxy group of a tyrosine (position 122; Horowitz and Wang, 1987); furthermore, nalidixic acid and the newer fluoroquinolones bind to the A subunit in the presence of DNA, the cleavable complex, to form a ternary complex. The genetic evidence strongly confirms this view since almost all drug-resistant bacteria contain a point mutation in the A subunit. These drugs may also bind direct to DNA (see Reece and Maxwell, 1991).

There is a paradox, however, in that the level of cellular supercoiling and the inhibition of *in vitro* growth is inhibited by drug concentrations that do not inhibit the enzyme. Normally one would expect a drug to inhibit a target enzyme at a lower concentration than an organism because the drug has to cross the cell wall and membrane to reach the target enzyme. To explain this paradox, a 'poison' hypothesis has been proposed, whereby the enzyme required to relax the DNA in front of the replication fork is poisoned by the enzyme–DNA–drug ternary complex while the rest of the cellular gyrase is not affected. Replication would thus be blocked but there would be no effect on cellular supercoiling (see Reece and Maxwell, 1991).

The fluoroquinolones are broad spectrum anti-bacterials effective against a wide range of Gram-negative and some Gram-positive bacteria. These drugs are amongst the preferred agents for the treatment of infections of the gastro-intestinal and urinary tracts, but lack activity against blood-borne infections of *Streptococcus* and *Enterococcus* (Hooper and Wolfson, 1991).

Ofloxacin is a chiral agent, the *S*-isomer being 10 to 100-times more active than the *R*-isomer, indicating the asymmetry of the binding site. The *S*-isomer happens to be laevorotatory and is under development as laevofloxacin both for the treatment of bacterial infections and HIV (Wentland *et al.*, 1988). The mode of action is unclear in the latter case.

4-Hydroxycoumarin antibiotics, of which novobiocin is the only one to have reached the market, have been isolated from *Streptomyces* species and are mainly active against Gram-positive species. Coumarins interfere with ATP-induced supercoiling but not ATP-independent relaxation by binding to the B subunit (Mizuuchi *et al.*, 1978). Mutations in the B subunit block their effect. These antibiotics also inhibit cellular levels of supercoiling at levels consistent with inhibition of gyrase. Steady-state kinetic experiments suggest that coumarins are competitive inhibitors of ATP hydrolysis and DNA supercoiling, but this does not imply that they are necessarily binding at the same site as ATP (Sugino and Cozzarelli, 1980). Structurally they are not particularly similar to ATP, although they do have a negative charge at neutral pH from the 4-hydroxy group and a sugar ring.

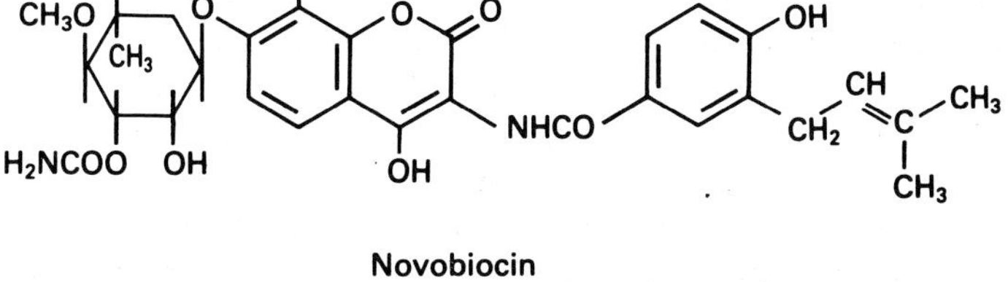

Novobiocin

2.3.3.2 Mammalian DNA topoisomerase II

No enzyme equivalent to DNA gyrase has been found in mammalian cells, probably because mammalian DNA is negatively supercoiled by being wrapped around histones. There is no need, therefore, to 'wind up' the nucleic acid, merely to unwind it with the help of ATP. Topoisomerase II works down an energy gradient using the Mg–ATP complex by removing positive or negative supercoils (relaxation). Unlike DNA gyrase it does not introduce supercoils. Topoisomerase II recognizes topological structure rather than base sequences, unlike other DNA-interacting enzymes, and so acts at asymmetric sites (Osheroff *et al.*, 1991).

In addition, in mammalian cells topoisomerase II is an integral part of the chromosome scaffold. This structure is seen most clearly in mitosis when the highly compacted DNA forms loops anchored at intervals to the chromatid axis. If the histones are removed, the overall structural form of the chromosome still remains and one of the major proteins at the base of the chromatid axis is topoisomerase II. The function of the enzyme is presumably, amongst other things, to untangle loops in the sister chromosomes at anaphase. Mutations to produce a non-functional enzyme result in a lethal event, as the cells cannot separate their chromosomes at anaphase (Earnshaw, 1988).

Any interference with this structure is likely to be crucial to the functioning of the cell and at least two classes of anti-tumour drugs owe their action to a binding to the cleavable complex of enzyme and DNA; the epipodophyllotoxins and the anthracyclines.

Doxorubicin (adriamycin), produced by *Streptomyces peucetius* var. *caesius*, is an example of the anthracyclines which were originally believed to act by slotting in (intercalating) between nucleic acid bases like actinomycin and thereby interfering with DNA and RNA biosynthesis. Doxorubicin, however, also forms a complex with topoisomerase II and DNA in the same fashion as the antibacterial drugs and DNA gyrase, and this is an essential part of its mechanism of action. Studies with closely related structures have shown that cleavage of DNA is related to cytotoxicity. Although the precise molecular detail of the interaction has not been worked out, a covalently linked protein is attached to the 5′ end of a DNA strand. The four (planar) rings of the drug intercalate with DNA while the hydroxyl group on the side-chain interacts with topoisomerase II (Bodley *et al.*, 1989).

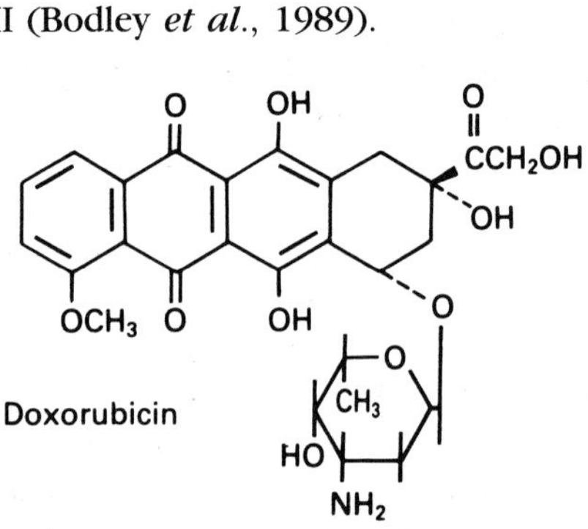

Doxorubicin

Doxorubicin is effective in the treatment of acute leukaemias and malignant lymphomas and also in a variety of solid tumours. The drug is part of a cocktail for the treatment of carcinoma of the ovary, breast and small cell carcinoma of the lung, and a wide range of sarcomas (Calabresi and Chabner, 1990).

Etoposide is the most important member of the epipodophyllotoxins and its derivation from the plant product podophyllotoxin makes an interesting story (Stahelin and Von Wartburg, 1991). It is an example of how the process of drug discovery actually takes place, rather than the sanitized version that is often presented. The process took at least 20 years and involved the classic situation of a very active impurity present in low quantity.

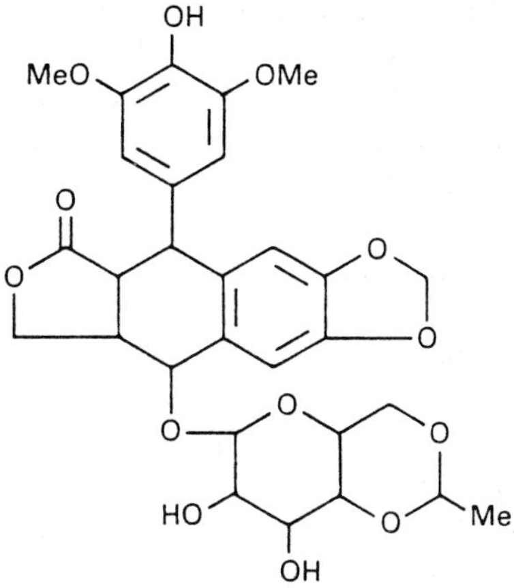

OH MeO OMe O O O O O O O O HO O Me OH

Etoposide

Podophyllotoxin inhibits the growth of cancer cells by acting as a spindle poison and arresting the cells in metaphase with clumped chromosomes. In an attempt to improve on the cytotoxic activity of podophyllotoxin crude *Podophyllum* glucoside fraction was treated with benzaldehyde to give the benzylidene derivatives. The mixture had activity against the 11210 leukaemia cell that could not be explained by the known major constituents. After considerable effort at purification, a material was discovered that turned out to have a different mechanism of action. It did not interact with microtubules, but prevented cells from entering mitosis by arresting them in the G_2 phase. Using this information a number of analogues were synthesized, one of which was etoposide (Stahelin and Von Wartburg, 1991).

Etoposide does not intercalate with DNA and owes its anti-tumour action entirely to formation of ternary cleavage complexes with DNA and topoisomerase II. Both single and double-stranded cleavage reactions appear to be stimulated while the religation reaction is inhibited (Osheroff, 1989). The 4′-hydroxyl is essential for both activity against the enzyme and cytotoxicity to tumour cells (Sinha *et al.*, 1990). Etoposide has been used in combination with other agents for small cell carcinoma of the lung and is being tried in other solid tumours, lymphomas and leukaemias (Henwood and Brogden, 1990). Sarcoma is the term given to tumours arising in bone, connective tissue or muscle, while carcinomas arise in covering or lining membranes. The latter are more commonly encountered than sarcomas.

2.3.4 Reverse transcriptase - azidothymidine for AIDS

A number of RNA viruses that can infect man and animals are known as retroviruses because they use their genomic RNA to prime an enzyme to synthesize complementary DNA. This enzyme is called reverse transcriptase and it has both RNA and DNA-dependent DNA polymerase activity. The RNA of the viral genome is replicated into a RNA/DNA duplex, the RNA is stripped off by RNAase activity also found in the enzyme and the DNA acts as a template to form double-stranded DNA (see Hostomsky *et al.*, 1992; Darby, 1992). Reverse transcriptase is a heterodimer composed of two subunits (66 and 51 kDa); a polypeptide cleaved from the carboxy terminus of the larger produces the smaller. Both subunits show nucleic acid polymerizing activity, derived from the N-terminal domain, but only the larger has RNAase activity, based in the C-terminus.

The viral genome encodes an enzyme known as an integrase to allow the DNA to be inserted into the host DNA and thus become latent, subsequently being transcribed by the host cell to produce virus particles.

If ever the medicinal scientist fraternity needed to be reminded that micro-organisms are capable of changing in such a way as to present new threats to the health of man, the development of acquired immune deficiency syndrome (AIDS) as a result of virus infection has provided such a reminder. The disease is caused by the virus known as human immunodeficiency virus (HIV) (earlier names were human T-cell lymphotropic virus III, HTLV-III, and lymphadenopathy-associated virus, LAV, which appears to be a recent import into humans possibly from the African green monkey.

The virus infects the T cells of the immune system and thus the host defences are seriously weakened. Consequently, the AIDS patient frequently dies of other diseases that are normally controlled by the T cells, for example pneumonia, caused by *Pneumocystis carinii* or cytomegalovirus, or an otherwise rare form of cancer, Kaposi's sarcoma (Shaw *et al.*, 1985). Furthermore, there may be degenerative changes in nervous tissue as a consequence of infection, but it is probably too early to be sure how serious these changes are.

A family of nucleosides, based on pyrimidine and purine bases, have been shown to prolong the life of AIDS sufferers, but not to induce a cure (Sandstrom and Oberg, 1993). Azidothymidine (thymidine) was the first to be used, while didanosine (hypoxanthine), zalcitabine and lamivudine (cytidine) followed. These agents are all converted intracellularly by host kinases to the triphosphate which then interferes with viral nucleic acid synthesis in two ways: the triphosphate at concentrations below 10^{-7} M (Mitsuya *et al.*, 1985) itself binds to the nucleotide binding site preventing the natural nucleotide from binding. The enzyme can also treat the nucleotide as a false substrate by introducing it to the growing nucleotide chain, thereby causing the chain to terminate prematurely - there is no 3′-hydroxyl group to bind the next phosphate group to form a 5′-3′ diester (Yarchoan *et al.*, 1989; Hart *et al.*, 1992).

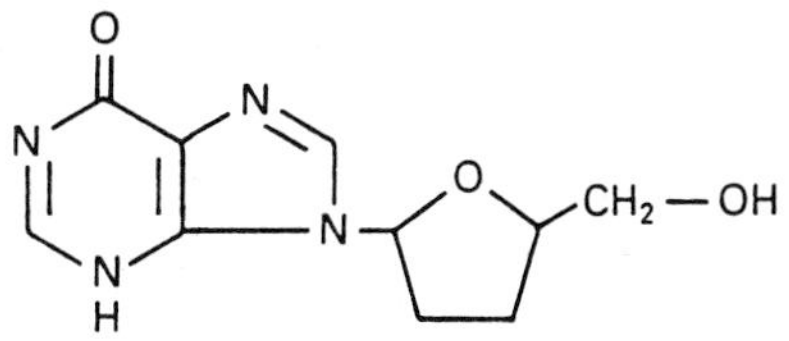

Didanosine

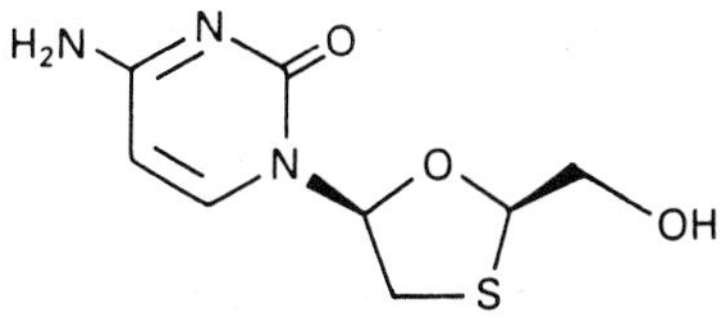

Lamivudine

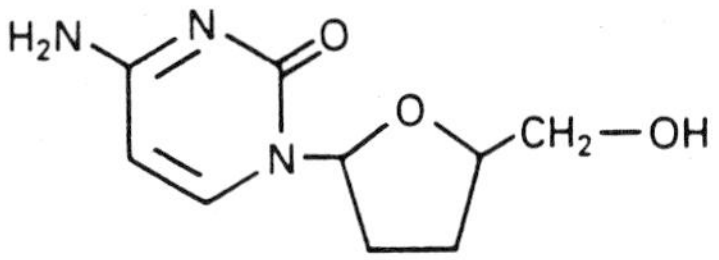

Zalcitabine

Azidothymidine

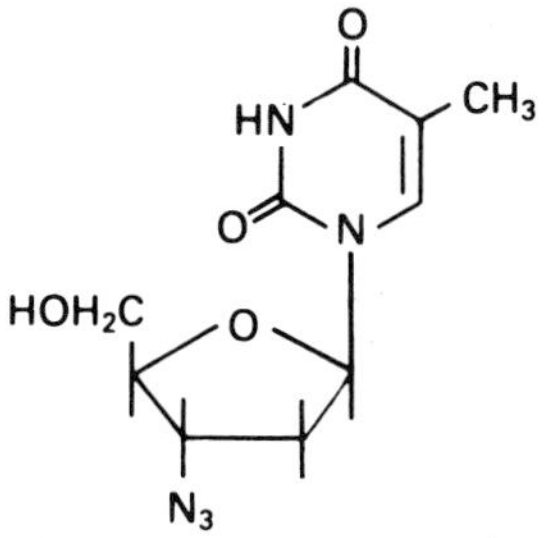

In the case of lamivudine and its 5-fluoro analogue, the more active agent is the 'unnatural' – or L-isomer while the D-isomer shows less activity but greater toxicity. The suggestion is that cellular enzymes, such as cytidine deaminase, recognize the 'natural' isomer thus inducing toxicity, whereas the 'unnatural' isomer is not so recognized (Hoong *et al.*, 1992).

Mammalian DNA polymerases differ in their sensitivity to these drugs, with type α much less (K_i 1–2 $\times 10^{-4}$ M), but types β and γ K_i 2·6–70 $\times 10^{-6}$ M and less than 4×10^{-7} M, respectively; (Yarchoan *et al.*, 1989). These inhibitory effects may be responsible for drug toxicity. The mammalian enzymes do not use these nucleotides as false substrates.

2.3.5 Bacterial RNA polymerase - rifampicin as antimycobacterial

Rifampicin, which is derived by chemical synthesis from the rifamycins, natural products of *Nocardia mediterranae*, inhibits bacterial, but not eukaryotic, DNA-primed RNA polymerase. The drug is more effective (i.e. has a lower MIC) against Gram-positive bacteria than Gram-negative, but this is probably because of permeability barriers in the latter. The major use of the drug is to treat infections produced by the mycobacteria (rod-shaped Gram-positive bacteria), so-called because approximately 40 per cent of their cell walls consists of mycolic acid - complexed to the peptidoglycan (see section 5.4). Mycolic acid is a hydroxy fatty acid with long hydrocarbon chain branches, the presence of which causes the bacterium to stain in a characteristic way known as acid-alcohol fastness. Important pathogenic mycobacteria are *M. leprae*, the cause of leprosy, and *M. tuberculosis* causing tuberculosis.

The target of rifampicin is the β subunit of RNA polymerase. The β subunit must be complexed in the trimer $\alpha_2\beta$ for specific binding to occur. With the complete pentameric enzyme ($\alpha_2\beta'\beta\sigma$) rifampicin binds with a binding constant of 10^{-9} M at 37°C (see Wehrli, 1983). If the β subunit alone is treated with a rifampicin quinone derivative it is non-specifically modified, whereas in the $\alpha_2\beta$ complex the binding site only is modified. This suggests that the binding site develops as a result of formation of the enzyme complex, presumably by allosteric interactions (Lowder and Johnson, 1987). Mutations in the β subunit alone, however, will generate resistance to rifampicin.

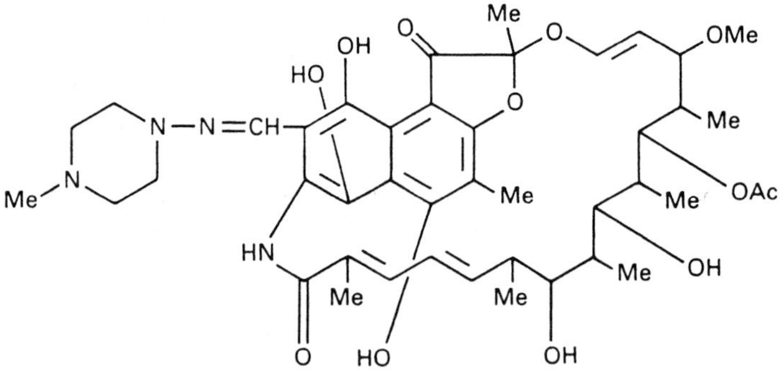

Rifampicin

Binding of the drug produces abortive chain initiation through the formation of the dinucleotide pppApU and thereby prevents further elongation of the RNA chain. This is probably because rifampicin interferes with the binding site for the growing RNA chain, thereby destabilizing the binding of the intermediate oligonucleotides to the DNA-enzyme complex (Wehrli, 1983).

Rifampicin has to be given as part of a cocktail of drugs because a small proportion of the naturally occurring enzyme is resistant. The need for drugs active against *M. tuberculosis* has become more urgent in recent years as the occurrence of tuberculosis is increasing.

2.4 Nucleotide catabolism

2.4.1 Adenosine deaminase - 2-deoxycoformycin as antileukaemic

Adenosine deaminase catalyses the deamination of adenosine and deoxyadenosine to inosine and deoxyinosine respectively. This enzyme is important in lymphocytic function and its absence through genetic defect leads to severe immune deficiency (Thompson and Seegmillar, 1980). 2-Deoxycoformycin is the most potent known inhibitor of the enzyme and acute lymphoblastic leukaemia of T-cell origin responds to such inhibition, probably because T cells are particularly rich in adenosine deaminase and the enzyme levels are raised even further in various forms of acute lymphocytic leukaemia (Dearden *et al.*, 1991). The enzyme plays an important role in the salvage pathway for purines since inosine may be phosphorylated to IMP which can, as we have seen, be converted into GMP. Inhibition of this salvage pathway in times of high purine requirement is likely to be critical to the cell.

The action of 2-deoxycoformycin as a powerful adenosine deaminase inhibitor may also be indirect. A build-up of dATP, derived by phosphorylation from adenosine and adenylate kinases, can inhibit ribonucleotide reductase which is particularly sensitive to the triphosphate balance. Secondly, accumulated adenosine can drive *S*-adenosylhomocysteine hydrolase into reverse to form *S*-adenosylhomocysteine (Hershfield *et al.*, 1983). This agent is a powerful inhibitor of the methylation of RNA and DNA, possibly by antagonizing the methyl transfer activity of *S*-adenosylmethionine.

2-Deoxycoformycin is a fermentation product of *Streptomyces antibioticus*. The mode of action is reviewed in Agarwhal (1982). The drug is slightly more tightly bound than its ribose analogue, coformycin (K_i of $2{\cdot}5 \times 10^{-12}$ M compared with 1×10^{-11} M). Since a similar relationship occurs with the substrates with adenosine (K_m $2{\cdot}5 \times 10^{-5}$ M) and deoxyadenosine (7×10^{-6} M), it is likely that the enzyme has a greater affinity for deoxyribose than ribose. The drug is a close structural analogue of the transition state of the enzyme reaction (see Chapter 1 for discussion on transition states). The six-membered ring of the purine nucleus has been expanded to a seven-membered ring by the addition of a carbon atom. There are two double bonds in that ring so the aromaticity is lost and the ring becomes puckered; the hydroxyl group is a secondary alcohol, and so the keto-enol tautomerism is also lost. This leads to a structure similar to the transition state intermediate in which a hydroxyl group has been added at position 6 to give a tetrahedral carbon atom (Wolfenden *et al.*, 1977). The reaction mechanism is shown in Fig. 2.9.

This type of inhibition leads to some rather unusual kinetic results. The binding of the inhibitor is to the form of the enzyme present in the transition not to the ground state, and so the inhibition takes time to develop. The initial velocity plots show only relatively moderate inhibition if the inhibitor is not pre-incubated with the enzyme. A double reciprocal plot demonstrates characteristics of competitive inhibition with a K_i of 1×10^{-8} M. With pre-incubation, however, the plot obtained indicated a non-competitive inhibitor with a K_i of $2{\cdot}5 \times 10^{-12}$ M. The situation is similar to what is obtained with irreversible time-dependent binding to an enzyme, namely that a proportion of the enzyme is put out of action by the inhibitor because the enzyme–inhibitor complex takes so long to dissociate (the half-life for dissociation is 25 to 30 hours). The activity observed is that of the remaining uncomplexed enzyme, and so the

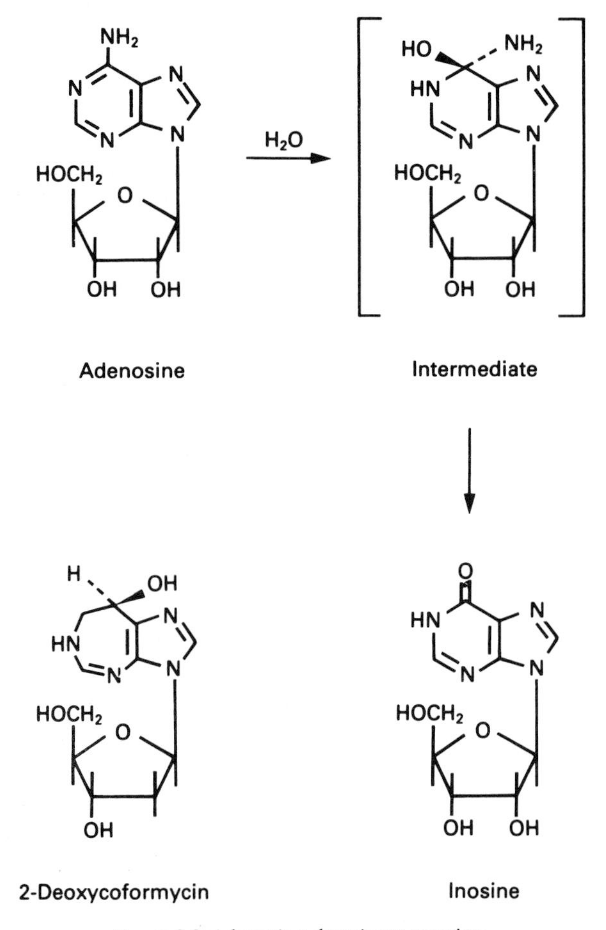

Figure 2.9 Adenosine deaminase reaction.

K_m remains the same while V_{max} is reduced - a classical example of non-competitive behaviour. The rates of association and dissociation of this complex are slow, possibly as a result of a slow conformational change in the enzyme as it is transformed from the ground state to the activated transition state (Agarwhal, 1982).

2.4.2 Xanthine oxidase - allopurinol for gout

Xanthine oxidase is the last enzyme on the breakdown pathway of purine bases in primates (Fig. 2.4) and it catalyses the conversion of hypoxanthine to xanthine and of xanthine to uric acid. The latter is normally excreted, although quantities of the other purines may also find their way into the urine. In some diseases, notably gout, the production of purines can be increased as a primary cause of the disease (Scott, 1980). Enzyme deficiencies with a genetic origin may play a part. One such case is deficiency of the salvage enzyme hypoxanthine phosphoribosyltransferase (HPRT) which leads to an elevated level of hypoxanthine phosphoribosylpyrophosphate. The latter stimulates *de novo* purine biosynthesis at the initial rate-limiting step of the formation of phosphoribosylamine (Fig. 2.2).

The consequence of increased purine synthesis is an increased throughput down the catabolic pathway to uric acid. When levels of the latter rise above saturation, crystals of monosodium urate form in the synovial fluid. The characteristic symptoms of gout derive from an inflammatory response to these crystals and thus closely resemble the painful joint swellings in rheumatoid arthritis. This may occur in one joint only or in several (Kelley and Wyngarden, 1974). In advanced gout, deposits (tophi) of sodium urate form on or near joints or tendon sheaths, which are soft initially but eventually harden (Scott, 1980).

For therapy the major need is to lower serum uric acid levels, although anti-inflammatory drugs will relieve the symptoms on a short-term basis. One of the most useful drugs in effecting a long-term cure is allopurinol. Xanthine oxidase is the target of the drug, and so serum and urine hypoxanthine and xanthine levels are raised while, more importantly, uric acid levels are lowered. In addition, the drug is useful when given in combination with anti-leukaemic drugs since serum urate levels can rise sharply as the leukaemia cells die. This is an example of secondary gout, secondary in that uric acid formation is increased as a consequence of other changes. In this case, the danger is not only that acute episodes of gout may occur, but also that sodium urate crystals may form in the distal tubule of the kidney (Scott, 1980).

Clearly, if a drug is metabolized by xanthine oxidase, its action is likely to be potentiated by allopurinol. For example, 6-mercaptopurine (a drug used for the treatment of leukaemia) is metabolized by xanthine oxidase to 6-thiouric acid, an inactive metabolite (Elion, 1967). The dose of mercaptopurine required when given in conjunction with allopurinol must therefore be reduced to avoid widespread toxicity which would otherwise occur if higher mercaptopurine levels were sustained for longer periods of time.

Xanthine oxidase is a complex enzyme containing, in effect, a transport system involving molybdenum, flavin nucleotide and two iron–sulphur centres which convey electrons to oxygen to yield superoxide anion (O_2^-) (Hille *et al.*, 1993). Allopurinol inhibits the enzyme in a complex fashion, and may be regarded as one of the earliest examples of a suicide substrate (Spector and Johns, 1970; Massey *et al.*, 1970 - see Chapter 1 for a discussion on suicide substrates). If the inhibition is studied without pre-incubation of enzyme and inhibitor, allopurinol behaves as though it were a competitive inhibitor with a K_i of 7×10^{-7} M. With pre-incubation in the presence of air, the inhibition increases and is no longer competitive with substrate. Allopurinol is also a substrate for xanthine oxidase and the product of the reaction, oxipurinol (alloxanthine), is also an inhibitor. In the presence of xanthine as substrate and oxygen, or anaerobically without substrate, the enzyme is inactivated by oxipurinol. If the oxidation of xanthine, which requires the enzyme to cycle between reduced and oxidized forms, and for the enzyme to be in an anaerobic environment, both result in enzyme inactivation by oxipurinol, it is likely that the reduced form of the enzyme is sensitive to oxipurinol (Massey *et al.*, 1970). The dissociation constant of the oxipurinol–enzyme complex is $5{\cdot}4 \times 10^{-10}$ M (Spector and Johns, 1970). Inhibition can be reversed by prolonged dialysis or by allowing the complex to be reoxidized in the presence of air, thus confirming that it is the partly reduced form of the enzyme that is receptive to oxipurinol inhibition.

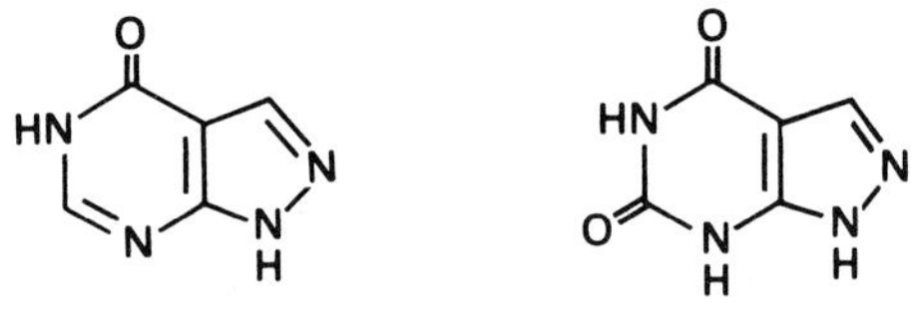

Allopurinol **Oxipurinol**

The inactivation of reduced xanthine oxidase by oxipurinol follows first-order kinetics by appearing to be dependent on the concentration of reduced enzyme. This may be the result of an internal rearrangement of the enzyme–inhibitor complex in a time-dependent fashion (Spector and Johns, 1970). The similarity between the tight or stoichiometric binding of oxipurinol to xanthine oxidase, and of coformycin to adenosine deaminase was noted by Cha *et al.* (1975).

Allopurinol has been found to be effective in the treatment of kala-azar (leishmaniasis) (Berman, 1988). In this instance the drug is acting as a false substrate for the parasite's hypoxanthine phosphoribosyltransferase (Fig. 2.10) - much more efficiently than for the human erythrocyte enzyme (Tuttle and Krenitsky, 1980). Subsequent enzymes convert the ribonucleotide into

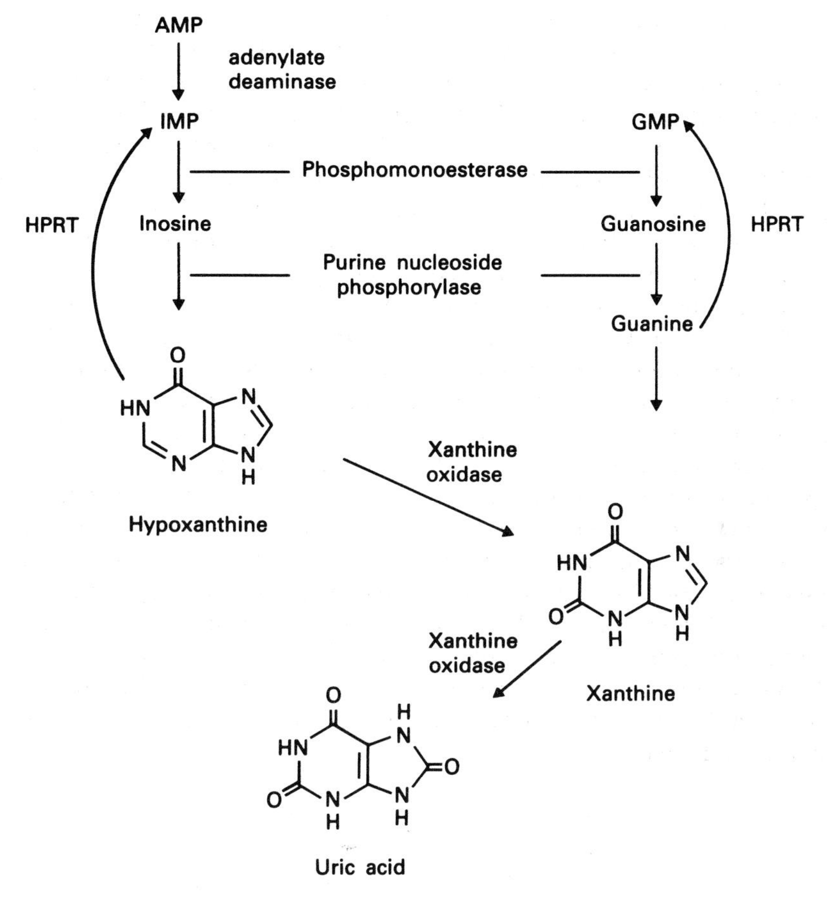

Figure 2.10 Purine nucleotide catabolism and salvage.

an analogue of ATP which is then incorporated into a faulty RNA (Nelson *et al.*, 1979).

2.4.3 Guanylate cyclase

Angina pectoris is characterized by sudden sharp pains in the chest which may spread out into the left arm, and is brought on by exercise, stress, emotion or eating. A reduction in the volume of the coronary vessels (atherosclerosis) is the cause of the usual angina, while spasm of the coronary vessels underlies variant (Prinzmetal's) angina. Both of these increase the demand for oxygen; lack of oxygen (ischaemia) brings on the pain which can, therefore, result from insufficient blood flow or increased requirement for oxygen or both.

A family of organic nitrates has been used for over a century in the treatment of angina pectoris and congestive heart failure amongst other cardiac conditions (Abrams, 1987). Glyceryl trinitrate was the first to be used in this way, while isosorbide dinitrate has also been favoured. These drugs dilate the peripheral blood vessels, reducing the strain and oxygen demand on the heart, and also dilate the coronary arteries in an area short of oxygen, thus increasing the oxygen supply (Ignarro, 1989; Berlin, 1987).

Despite this long use it is only recently that the molecular mechanism of their action has been identified. Nitric oxide is the key, and it may be formed in two ways: to make compounds more water-soluble, hepatic glutathione organic nitrate reductase catalyses the formation of nitrite ion which breaks down to nitric oxide and denitrated compound; alternatively, nitrates react with cysteine to produce *S*-nitrosocysteine which breaks down to nitric oxide. Nitric oxide activates soluble guanylate cyclase by binding to the central iron of the prosthetic heme (Ignarro, 1989).

Cyclic GMP levels activate a cyclic-GMP-dependent protein kinase which relaxes smooth muscle by reducing the levels of free calcium inside the cell, probably by binding to the endoplasmic reticulum (Feelisch and Noack, 1987; Kukovetz *et al.*, 1991). A direct correlation exists between the rate of formation of nitric oxide and the inhibition of guanylate cyclase and the elevation of cyclic GMP (Greenberg *et al.*, 1991).

Another more recent introduction, nicorandil which is a nitrate ester of nicotinic acid, also works through cyclic GMP but, in addition, seems to have another mode of action which is independent of cyclic GMP because a very high concentration of drug is required to activate guanylate cyclase. The activity of nicorandil varied greatly according to the tissue, unlike the other organic nitrates - a result which is not likely to be due to a potassium channel blockade because potassium chloride inhibits smooth muscle relaxation by organic nitrates and the differences between nicorandil and the other esters remained in KCl-contracted blood vessels (Greenberg *et al.*, 1991).

Questions

1. What four bases are present in DNA? How does RNA differ from DNA?
2. What does the term synergy mean? What classes of antibacterial drugs show synergy and why?
3. What part does the tyrosyl radical play in ribonucleotide reductase?
4. Why is the nucleic acid polymerase from retroviruses called reverse transcriptase?
5. Why does rifampicin not kill *E. coli*?
6. What is meant by supercoiling? What advantages to a bacterial cell does supercoiling confer?
7. How does topoisomerase I differ in the mechanism of action from topoisomerase II?

8. How do the fluoroquinolones interfere with bacterial topoisomerase II?
9. What mechanisms of action can be put forward to explain the action of ribavirin?
10. What is meant by a transition state complex? Name an enzyme that can be inhibited by a drug resembling such a complex.

References

Abrams, J., 1987, *Drugs,* **34**, 391-403.
Agarwhal, R.P., 1982, *Pharmacol. Ther.,* **17**, 399-429.
Ariel, I., 1970, *Cancer,* **25**, 705-14.
Ashton, W.T., Karkas, J.D., Field, A.K. and Tolman, R.L., 1982, *Biochim. Biophys. Res. Commun.,* **108**, 1716-21.
Berlin, R., 1987, *Drugs,* **33** (Suppl 4), 1-4.
Berman, J.D., 1988, *Revs. Infect. Dis.,* **10**, 560-79.
Block, E.R., Jennings, A.E. and Bennett, J.E., 1973, *Antimicrob. Ag. Chemother.,* **3**, 649-56.
Blum, R.H. and Carter, S.K., 1974, *Ann. Intern. Med.,* **80**, 249-59.
Bodley, A., Liu, L.F., Israel, M., Seshadri, R., Koseki, Y., Guiliani, F.C., Kirschenbaum, S., Silber, R. and Potmesil, M., 1989, *Cancer Res.,* **49**, 5989-93.
Bonadonna, G. and Valagussa, P., 1981, *New Engl. J. Med.,* **304**, 10-15.
Calabresi, P. and Chabner, B.A., 1990, p. 1243 in the *Pharmacological Basis of Therapeutics*, 8th edn, (eds A.G. Gilman, T.W. Rall, A.S. Nies and P. Taylor), Pergamon, New York.
Cha, S., Agarwhal, R.P. and Parks, R.E., 1975, *Biochem. Pharmacol.,* **24**, 2187-97.
Chang, T.-W. and Heel, R.C., 1981, *Drugs,* **22**, 111-28.
Chatis, P.A. and Crumpacker, C.S., 1982, *Antimicrob. Ag. and Chemother.,* **36**, 1589-95.
Chen, G.L., Yang, L., Rowe, T.C., Halligan, B.D., Tewey, K.M. and Liu, L.F., 1984, *J. Biol. Chem.,* **259**, 13560-6.
Churchill, W.S., 1951, In: *The Hinge of Fate*, Vol 4 in the Second World War (London: Cassell), p. 582.
Cohen, J., 1982, *Lancet,* **ii**, 532-7.
Cozzarelli, N.R., 1980, *Science,* **207**, 532-7.
Darby, G., 1992, *Biochem. Soc. Trans.,* **20**, 505-8.
Dearden, C., Matutes, C. and Gatovsky, D., 1991, *Brit. J. Cancer,* **64**, 903-6.
De Clerq, E. and Walker, R.T., 1984, *Pharmacol. Ther.,* **26**, 1-44.
Diasio, R.B., Bennett, J.E. and Myers, C.E., 1978, *Biochem. Pharmacol.,* **27**, 703-7.
Di Marco, A., 1975, *Cancer Chemother. Rep.,* **6**, 91-106.
Drlica, K., 1992, *Molec. Microbiol.,* **6**, 425-33.
Earnshaw, W.C., 1988, *Bioessays,* **9**, 147-51.
Elford, H.L., Wamplen, G.L. and Van't Riet, B., 1979, *Cancer Res.,* **39**, 844-51.
Elford, H.L., Van't, Riet B., Wamplen, G.L., Lin, A.L. and Elford, R.M., 1981, *Adv. Enzyme Regul.,* **19**, 151-68.
Elgren, T.E., Lynch, J.B., Juarez-Garcia, C., Munck, E., Sjoberg, B.-M. and Que, L., 1991, *J. Biol. Chem.,* **266**, 19265-8.
Elion, G.B., 1967, *Fed. Proc.,* **26**, 898-904.
Feelisch, M. and Noack, E.A., 1987, *Eur. J. Pharmacol.,* **139**, 19-30.
Ferone, R., Burchall, J.S. and Hitchings, G.H., 1969, *Molec. Pharmacol.,* **5**, 49-59.
Fry, M. and Pudney, M., 1992, *Biochem. Pharmacol.,* **43**, 1545-53.
Gellert, M., 1980, *Annu. Rev. Biochem.,* **50**, 879-910.

Gilbert, B.E. and Knight, V., 1986, *Antimicrob. Ag. Chemother.*, **30**, 201-5.
Graham, F.L. and Whitmore, G.F., 1970, *Cancer Res.*, **30**, 2636-44.
Greenberg, S.S., Cantor, E., Ho, E. and Walega, M., 1991, *J. Pharmacol. Exptl. Ther.*, **258**, 1061-71.
Hammond, C.B., Weed, J.C., Barnard, D.E. and Tyray, L., 1981, *Cancer*, **31**, 322-32.
Hammond, D.J., Burchell, J.R. and Pudney, M., 1985, *Molec. Biochem. Parasitol.*, **14**, 97.
Hart, G., Orr, D.C., Penn, C.R., Figueiredo, H., Gray, N.M., Boehme, R.E. and Cameron, J.M., 1992, *Antimicrob. Ag. Chemother.*, **36**, 1688-94.
Heidelberger, C., Danenberg, P.V. and Moran, R.G., 1983, *Adv. Enzymol.*, **54**, 57-119.
Henwood, J.M. and Brogden, R.N., 1990, *Drugs*, **39**, 438-90.
Hershfield, M.S., Kredich, N.M., Koller, C.A., Mitchell, B.S., Kurtzberg, J., Kinney, T.R. and Falletta, J.M., 1983, *Cancer Res.*, **43**, 3451-6.
Hille, R., Kim, J.H. and Hemann, C., 1993, *Biochemistry*, **32**, 3973-80.
Hitchings, G.H. and Smith, S.L., 1980, *Adv. Enzyme Regul.*, **18**, 349-71.
Hooper, D.C. and Wolfson, J.S., 1991, *N. Engl. J. Med.*, **324**, 384-94.
Hoong, L.K., Strange, L.E., Liotta, D.C., Koszalka, G.W., Burns, C.L. and Schinazi, R.F., 1992, *J. Org. Chem.*, **57**, 5563-5.
Horowitz, D.S. and Wang, J.C., 1987, *J. Biol. Chem.*, **262**, 5669.
Hostomsky, Z., Hostomska, Z., Fu, T.-B. and Taylor, J., 1992, *J. Virol.*, **66**, 3179-82.
Hsiang, Y.-H. and Liu, I.F., 1988, *Cancer Res.*, **48**, 1722-6.
Hudson, A.T., Dickins, M., Ginger, C.D., Gutteridge, W.E., Holdich, T., Hutchinson, D.B.A., Pudney, M., Randall, A.W. and Latter, V.S., 1991, *Drugs Exptl. Clin. Res.*, **17**, 427-35.
Hurly, M.G.D., 1959, *Trans. R. Soc. Trop. Med. Hyg.*, **53**, 412-13.
Ignarro, L.J., 1989, *Pharm. Res.*, **6**, 651-9.
Kager, P.A., Rees, P.N., Wellde, B.T., Hockmeyer, W.T. and Lyerley, W.T., 1981, *Trans. R. Soc. Trop. Med. Hyg.*, **75**, 556-9.
Kelley, W.N. and Wyngaarden, J.B., 1974, *Adv. Enzymol.*, **41**, 1-33.
Kennedy, B.J. and Yarbro, J.W., 1966, *J. Am. Med. Assoc.*, **195**, 1038-43.
Kisner, D.L., Schein, P.S. and Macdonald, J.S., 1981, *Rec. Results Cancer Res.*, **79**, 28-40.
Kukovetz, W.R.D., Holzmann, S. and Schmidt, K., 1991, *Eur. Heart J.*, **12** (Suppl E), 16-24.
Lassmann, G., Thelander, L. and Graslund, A., 1992, *Biochem. Biophys. Res. Commun.*, **188**, 879-87.
Ljungmann, P., Lonngvist, B., Gahrton, G., Ringden, O., Sundgvist, V.A. and Wahren, B., 1986, *J. Infect. Dis.* (Suppl B), 840-7.
Lowder, J.F. and Johnson, R.S., 1987, *Biochem. Biophys. Res. Commun.*, **147**, 1129-36.
Lucas, A.O., Hendrickse, R.G., Okubadse, O.A., Richards, W.H.G., Neal, R.A. and Kofie, B.A.K., 1969, *Trans. R. Soc. Trop. Med. Hyg.*, **63**, 216-29.
Major, P.P., Agarwhal, R.P. and Kufe, D.W., 1983, *Cancer Chemother. Pharmacol.*, **10**, 128-38.
Major, P.P., Egan, E.M., Herrick, D.J. and Kufe, D.W., 1982, *Biochem. Pharmacol.*, **31**, 2937-40.
Mar, E.-C. and Huang, E.-S., 1979, *Intervirol.*, **12**, 73-83.
Massey, V., Komai, H., Palmer, G. and Elion, G.B., 1970, *J. Biol. Chem.*, **245**, 2837-44.
Matthews, D.A., Alden, R.A., Bolim, J.T., Filman, D.J., Freer, S.T., Hamlin, R., Wim, G.J.H., Kislink, R.L., Pastore, E.J., Plante, L.T., Xuong, N. and Krant, J., 1978, *J. Biol. Chem.*, **253**, 6946-54.
Matthews, D.A., Bolin, J.T., Burridge, J., Filman, D.J., Volz, K.W. and Kraut, J., 1983, *J. Biol. Chem.*, **260**, 392-9.
McClung, H.W., Knight, V., Gilbert, B.E., Wilson, S.Z., Quarles, J.M. and Divine, G.W., 1983, *J. Am. Med. Assoc.*, **249**, 2671-4.

Miller, W.H. and Miller, R.L., 1982, *Biochem. Pharmacol.,* **31**, 3879-84.
Mitsuya, H., Weinhold, K.J., Furman, P.J., St. Clair, M.H., Nusinoff-Lehrman, S., Gallo, R.C., Bolognese, D., Barry, D.W. and Broder, S., 1985, *Proc. Nat. Acad. Sci. (U.S.A.),* **82**, 7096-100.
Mizuuchi, K., O'Dea, M.H. and Gellert, M., 1978, *Proc. Nat. Acad. Sci. (U.S.A.),* **75**, 5998.
Moody, M.R. and Young, V.M., 1975, *Antimicrob. Ag. Chemother.,* **7**, 836-9.
Muller, W.E.G., Zahn, R.K., Brittlingmaier, K. and Felke, D., 1977, *Ann. N.Y. Acad. Sci.,* **284**, 34-48.
Nelson, D.J., LaFon, S.W., Tuttle, J.V., Miller, W.H., Miller, R.L., Krenitsky, T.A., Elion, G.B., Berens, R.L. and Marr, J.J., 1979, *J. Biol. Chem.,* **254**, 11544-9.
Osheroff, N., 1989, *Biochemistry,* **28**, 6157-60.
Osheroff, N., Zechiedrich, E.L. and Gale, K.C., 1991, *Bioessays,* **13**, 269-75.
Parris, D.S. and Harrington, J.E., 1982, *Antimicrob. Ag. Chemother.,* **22**, 71-7.
Patterson, J.L. and Fernandez-Larsson, R., 1990, *Rev. Infect. Dis.,* **12**, 1139-46.
Ramachandran, S., Godfrey, J.J. and Lionel, N.D.W., 1978, *J. Trop. Med. Hyg.,* **81**, 36-9.
Reardon, J.E. and Spector, T., 1989, *J. Biol. Chem.,* **264**, 7405-11.
Reece, R.J. and Maxwell, A., 1991, *Crit. Rev. Biochem. Mol. Biol.,* **26**, 335.
Rowe, T.C., Chen, G.L., Hsiang, Y.-H. and Liu, L.F., 1986, *Cancer Res.,* **46**, 2021-6.
Rubin, R.N., 1981, *Clin. Ther.,* **4**, 74-94.
Sandstrom, E. and Oberg, B., 1993, *Drugs,* **45**, 488-508.
Scott, J.T., 1980, *Brit. Med. J.,* **281**, 1164-6.
Shaw, G.M., Harper, M.E., Hahn, B.H., Epstein, L.G., Gadjusek, D.C., Price, R.W., Navia, B.A., Petito, C.K., O'Hara, C.J., Groopman, J.E., Cho, E.-S., Oleske, J.M., Wong-staal, F. and Gallo, R.C., 1985, *Science,* **227**, 177-82.
Shepard, C.C., Ellard, G.A., Levy, L., Upromolla, V., Pattyn, S.R., Peters, J.H., Rees, R.J.W. and Waters, M.F.R., 1976, *Bull. WHO,* **53**, 425-33.
Sinha, B.K., Politi, P.M., Eliot, H.M., Kerrigan, O. and Pommler, Y., 1990, *Eur. J. Cancer,* **28**, 590-3.
Spector, T. and Johns, D.G., 1970, *J. Biol. Chem.,* **245**, 5079-86.
Stahelin, H.F. and Von Wartburg, A., 1991, *Cancer Res.,* **51**, 5-15.
Streeter, D.G., Witkowski, J.T., Khare, G.D., Sidwell, R.W., Bauer, R.J., Robins, R.K. and Simon, L.N., 1973, *Proc. Nat. Acad. Sci. (U.S.A.),* **70**, 1174-8.
Stubbe, J.A., 1989, *Annu. Rev. Biochem.,* **58**, 257-85.
Sugino, A. and Cozzarelli, N.R., 1980, *J. Biol. Chem.,* **255**, 6599-604.
Thelander, L. and Reichard, P., 1979, *Annu. Rev. Biochem.,* **48**, 133-58.
Thompson, L.F. and Seegmiller, J.E., 1980, *Adv. Enzymol.,* **51**, 167-210.
Tuttle, J.V. and Krenitsky, T.A., 1980, *J. Biol. Chem.,* **255**, 909-16.
Wang, J.C., 1991, *J. Biol. Chem.,* **266**, 6659-62.
Wehrli, W., 1983, *Rev. Infect. Dis.,* **5**, Suppl. 3, S407-11.
Wentland, M.P., Pemi, R.B., Dorff, P.H. and Rake, J.B., 1988, *J. Med. Chem.,* **31**, 1694-7.
Whelan, W.L. and Kerridge, D., 1984, *Antimicrob. Ag. Chemother.,* **26**, 570-4.
Whitley, R.J. and Gnann, J.W., 1992, *New Engl. J. Med.,* **327**, 782-9.
Whitley, R.J., Alford, C., Hess, F. and Buchanan, B., 1980, *Drugs,* **20**, 267-82.
Wolfenden, R.L., Wentworth, D.F. and Mitchell, G.N., 1977, *Biochemistry,* **16**, 5071-7.
Woods, D.D., 1962, *J. Gen. Microbiol.,* **29**, 687-702.
Wray, S.K., Gilbert, B.E., Noall, M.W. and Knight, V., 1985, *Antiviral Res.,* **5**, 29-37.
Yarchoan, R., Mitsuya, H., Myers, C. and Broder, S., 1989, *New Engl. J. Med.,* **321**, 726-38.

Chapter 3

Protein biosynthesis

3.1 Introduction

The synthesis of protein from individual amino acids is an essential feature of living organisms. Major differences exist between the process in eukaryotes and prokaryotes, and these have been exploited serendipitously to produce anti-bacterial agents. Subtle differences are found between higher and lower eukaryotes (for example mammalian and yeast cells), but these have not so far indicated a compound that could be used to kill pathogenic fungi. This chapter is, therefore, concerned solely with anti-bacterial agents.

Protein synthesis in eukaryotes and prokaryotes takes place in ribosomes, both in the cytoplasm and in mitochondria, but as no drugs are believed to owe their mechanism of action to interference with the mitochondrial process (although chloramphenicol does interfere with this process in a few individuals giving rise to side-effects), only the cytoplasmic system will be considered in this chapter. Protein synthesis can be divided into three stages: chain initiation, elongation and termination. Although almost all the medically useful inhibitors of protein synthesis act on the elongation stage, initiation will also be discussed in order that the whole process may be understood (reviewed in Kozak, 1983; Maitra *et al.*, 1982). Eukaryotic protein synthesis is reviewed by Merrick (1992).

An amino acid first has to be activated by combination with a transfer RNA (tRNA) that is specific for that amino acid. This is done in two stages, both catalysed by the same specific aminoacyl–tRNA ligase or synthetase: the amino acid first reacts with ATP and Mg^{2+} to form an enzyme-bound amino acid adenylate and then the amino acid is transferred to the appropriate tRNA, with the concomitant release of AMP. The triplet anti-codon on the tRNA recognizes the codon for that amino acid on the messenger RNA, this complex formation takes place on the ribosome in the presence of GTP (Fig. 3.1).

Protein synthesis occurs on the ribosome, a complex particle containing various proteins and RNA. In bacteria, the overall sedimentation coefficient of the ribosome is 70S, and it readily dissociates into unequal subunits of 30S (containing 21 proteins and 16S RNA) and 50S (34 proteins, and both 23S and 5S RNA), respectively. Ribosomes are normally linked together by a messenger

Table 3.1 Antibacterial drugs discussed in Chapter 3

Section	*Drug*	*Target*
3.2	Aminoglycosides	30S ribosomal subunit
3.3	Chloramphenicol	50S ribosomal subunit
3.4	Tetracycline	30S ribosomal subunit
3.5	Erythromycin; Azithromycin	50S ribosomal subunit
3.6	Clindamycin	50S ribosomal subunit

RNA (mRNA) molecule like beads on a wire (polysomes), with the growing peptide chain at increasing lengths according to the ribosomal position on the mRNA. The ribosome directs the addition of amino acyl groups to a growing peptidyl tRNA by a condensation reaction that forms the peptide bond, and then moves with its growing chain along the messenger RNA to the next codon to be read.

In prokaryotes, protein synthesis is always initiated by formylmethionyl-tRNA (fMet-tRNA), while in eukaryotes this role is carried out by unformylated methionine. Formylation of methionine takes place after the methionine is linked to a form of tRNA specific for chain initiation. Also required are three proteins or initiation factors called IF 1 to 3 and also GTP. IF 1 and 2 are needed to position the mRNA and fMet - tRNA on the ribosome, while the role of IF 3 is to recognize the mRNA. GTP is hydrolysed to GDP in the process.

In order to assemble the active complex, the three initiation factors bind to the 30S subunit of the ribosome. FMet-tRNA and mRNA bind to the complex with a specific start codon, AUG, as part of the binding site on the mRNA. GTP also binds at this stage. A 50S subunit joins the complex to form the functional 70S ribosome, and GTP is hydrolysed to GDP and phosphate by ribosome-bound IF 2. IF 1 and 2 are then released from the complex so that fMet-tRNA is unblocked ready for peptide bound formation.

Chain elongation requires two sites on the ribosome (Fig. 3.1): one called the A-site (acceptor) where the fMet-tRNA is initially located before it transfers to the P-site (peptide nearer the 5′ end of the nucleic acid). The A-site is located on the 3′ side of the P-site and is largely on the 30S subunit. The 50S subunit carries the P-site, although there are clearly interactions between the two subunits as the substrates bind. The incoming aminoacyl-tRNA binds to the A-site, and then the fMet is condensed with the second amino-acid residue to make a dipeptidyl-tRNA; a reaction which requires a protein known as elongation factor 1 together with GTP which is hydrolysed in the process. The reaction is catalysed by a specific protein on the ribosome itself, and the spent tRNA is released from the ribosome.

The deacylated tRNA vacates the P-site and the dipeptidyl-tRNA is transferred back to the P-site still hydrogen-bonded to its mRNA codon - a process which requires the intervention of another molecule of GTP and elongation factor 2.

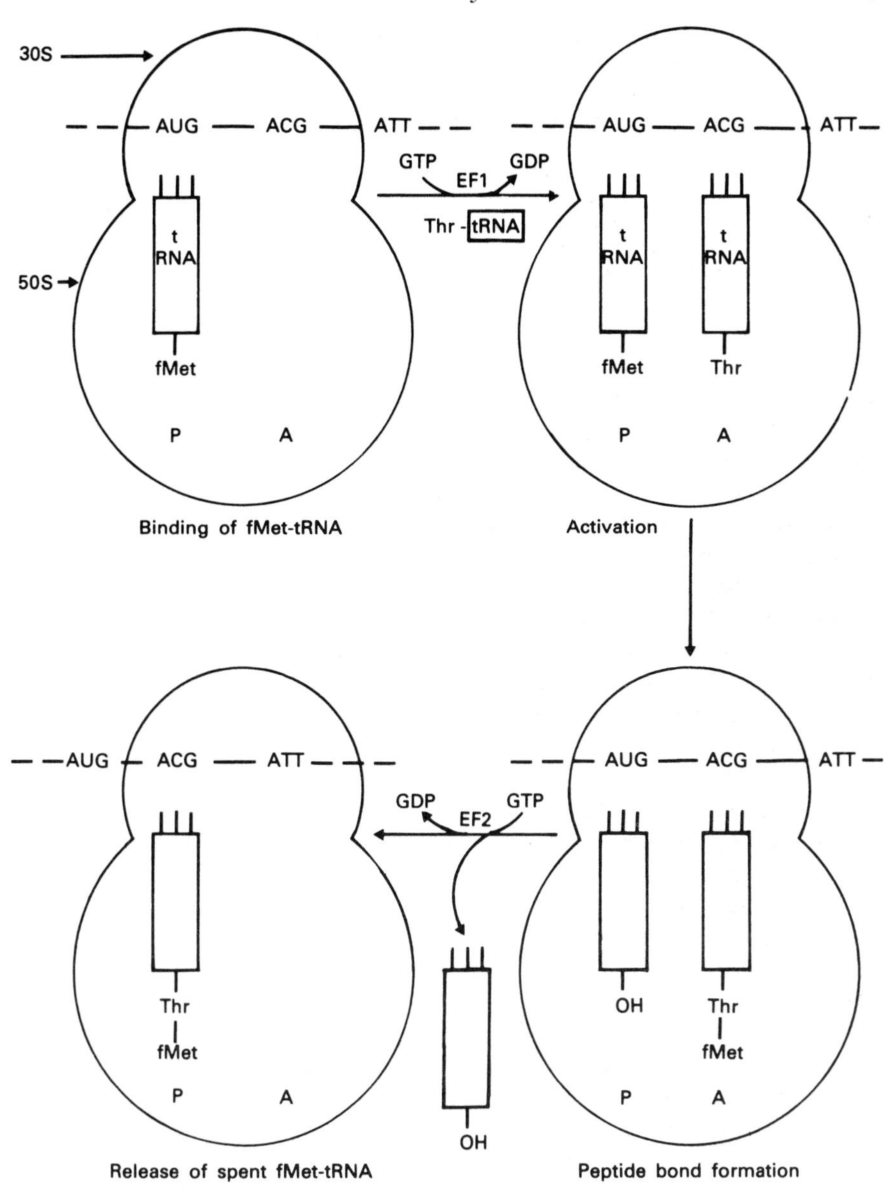

Figure 3.1 Chain initiation and formation of first peptide link.

This allows a third aminoacyl-tRNA defined by the mRNA codon to take up position in the A-site and the condensation process is repeated etc. (Fig. 3.1).

Protein release factors (RF1 and RF2) are needed in order to terminate the process. They recognize 'stop' codons: UAA, UAG and UGA. Termination is stimulated by GTP and probably requires the hydrolysis of GTP by one of the release factors.

The drugs that are discussed in this chapter, namely the aminoglycosides, tetracyclines, chloramphenicol, erythromycin and clindamycin, have their major effect on the chain elongation procedure. They are specific for protein synthesis in bacteria; apart from the tetracycline family which also inhibit eukaryotic protein synthesis. Nevertheless, despite this specificity, without exception they have undesirable side-effects and, in addition, their use over several decades has induced a considerable level of resistance among the bacterial population. For both these reasons, their use has been superseded, except in particular instances, by the much safer β-lactams (penicillins and cephalosporins) discussed in Chapter 5.

Bacteria may be divided into two main classes as to whether they take up the Gram stain; the cells are treated with crystal violet and then with iodine to fix the stain, followed by decolourization with ethanol or acetone. Finally the cells are exposed to safranine, a counterstain. Gram-positive microorganisms stain violet, whilst Gram-negative stain red. This difference in response to a histological stain is derived from major differences in the cell wall of the two types of bacteria; Gram-positive cells contain teichoic acids, polymers of glycerol or ribitol phosphate, unlike Gram-negative cells, which may explain the affinity of the Gram-positive cell wall for a basic dye such as crystal violet.

It is, however, not only a matter of chemistry but also of integrity as to whether a cell will take up the Gram stain because old, ruptured or dead Gram-positive cells may stain as though they were Gram-negative. Furthermore, Gram-positive cells contain very little lipid whereas Gram-negative are rich in lipid. In sharp contrast, one portion of the wall that both types have in common, but contributes more to the Gram-positive cell, is peptidoglycan (also known as murein or glycopeptide; Chapter 5).

3.2 Aminoglycosides

Aminoglycosides are a family of antibiotics produced mainly by streptomycetes (names ending in -mycin) or micromonospora (ending in -micin). Structurally, they consist of aminosugars connected to a central hexose/aminocyclitol – streptidine in streptomycin or L-deoxystreptamine in the other drugs mentioned in this chapter. Streptomycin was the first to be discovered but now is much less widely used than previously, while tobramycin is frequently regarded as the drug of choice when an aminoglycoside is indicated. Amikacin, a semisynthetic agent, is useful for the treatment of infections caused by aminoglycoside-resistant bacteria (Bailey, 1981).

Aminoglycosides are most useful therapeutically for infections caused principally by Gram-negative organisms such as the Enterobacteriaceae and pseudomonads, and are largely ineffective against Gram-positive agents such as streptococci (DeTorres, 1981). In aerobic environments, these drugs are bactericidal, i.e. they kill the bacterium (in certain instances a drug might only prevent the bacterium from growing and multiplying and would be termed bacteriostatic). Aminoglycosides, unlike other antibacterials, are taken up by the bacterium in three phases. The first phase is the binding of the drug to the outer surface of the cell wall and is rapid and passive. In the second phase, the drug passes across the cytoplasmic membrane by a fairly slow process, dependent on the energy derived from electron transport which is rate-limiting. The membrane potential is negative inside and drives the uptake (the aminoglycosides are cationic and thus attracted inwards). The third phase is fast and is probably the uptake of drug into the ribosome (see Amyes and Gemmell, 1992).

Aminoglycosides

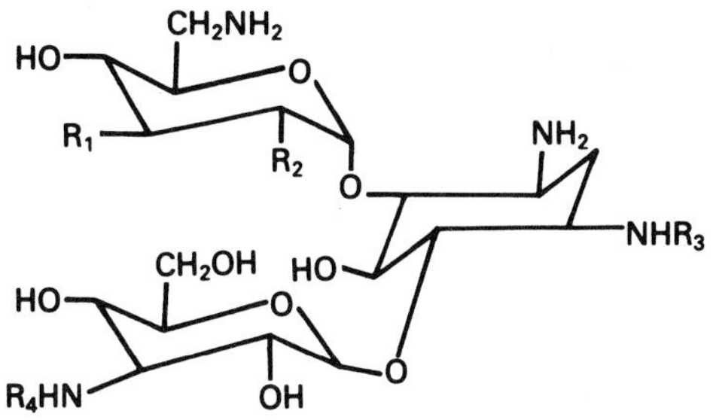

	R_1	R_2	R_3	R_4
Amikacin	OH	OH	$COCH(OH)(CH_2)_2NH_2$	H
Tobramycin	H	NH_2	H	NH_2

The primary intracellular site of action of the aminoglycosides is the 30S ribosomal subunit which consists of 21 proteins and a single 16S molecule of RNA. A single substitution of asparagine for lysine in protein S12 (i.e. protein 12 of the smaller ribosomal subunit as numbered by position of the protein on two-dimensional gel electrophoresis (Kaltschmidt and Wittman, 1970) of *Escherichia coli* prevents streptomycin binding to the ribosome (Birge and Kurland, 1969) but it is doubtful whether this is the only protein involved (Lando *et al.*, 1976; Davies and Courvalin, 1977).

Dihydrostreptomycin binds to the 30S subunit with a binding constant of $2{\cdot}7 \times 10^{-7}$ M (Lando *et al.*, 1976). Competition is observed between streptomycin and dihydrostreptomycin, not surprisingly, but not with gentamicin; indeed gentamicin appears to bind to the 50S subunit since protein L16 (protein 16 of the large ribosomal subunit) is altered in a gentamicin-resistant mutant (Buckel *et al.*, 1977). Tobramycin appears to bind to both subunits of the ribosome and at several different sites, since no saturation curve kinetics can be obtained when increasing concentrations of [^{3}H]tobramycin are used. Nevertheless tobramycin antagonism of acetyltobramycin binding suggests a binding constant of $1{\cdot}42 \times 10^{-6}$ M for the most tightly bound site (LeGoffic *et al.*, 1979).

One major effect of streptomycin binding is to cause misreading of the message so that the wrong amino acids are inserted into the growing polypeptide chain (Tai *et al.*, 1978). The aminoglycosides vary in their ability to cause misreading of the genetic code, presumably as a result of their varying affinity for different ribosomal proteins. This effect is found both in cell-free systems and in whole bacteria. Streptomycin can also interfere with chain elongation by inhibiting, in a partial and reversible manner, protein synthesis by the full complex of ribosomes attached to messenger RNA (polysomes).

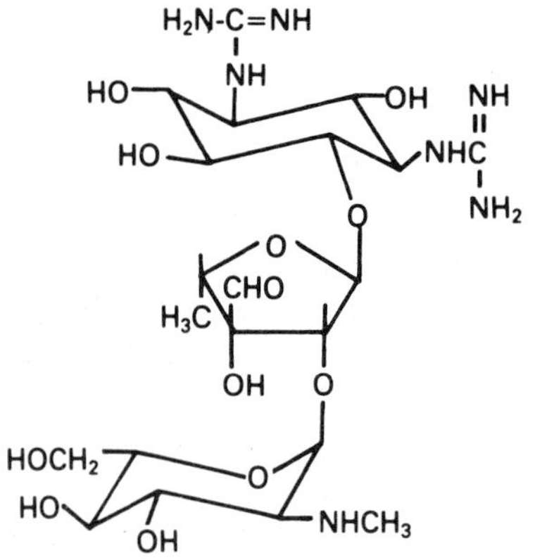

Streptomycin

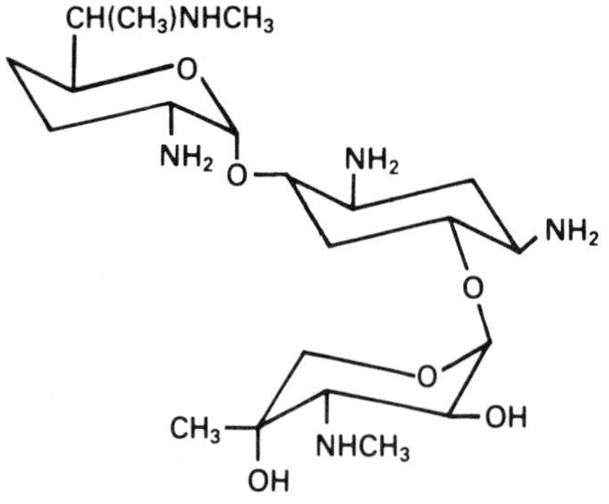

Gentamicin C_1

The drug does not, however, affect the process of initiation. It merely ensures that the initiation complex formed in its presence is less stable and less likely to engage in appreciable protein synthesis. The presence of the full complex would appear to mask the major streptomycin binding site, however, as streptomycin will additionally bind irreversibly to uncomplexed ribosomes, and thus cause the ribosomes to fall off the messenger RNA prematurely. The drug-bound ribosomes are impaired in IF-3-dependent dissociation and become engaged in abortive initiation and elongation (Wallace *et al.*, 1973). These two types of effect may cause the observed phenomena of partial inhibition of synthesis at lower concentration and total inhibition at higher concentration leading to cell death, in that at low concentrations streptomycin will not be able to block all the initiation sites in a cell but merely interfere transiently with polysomal complexes (Wallace *et al.*, 1973).

Drug resistance can occur, as one might expect, by alteration of the ribosome. Furthermore, in an anaerobic environment where the energy of oxidative phosphorylation is not available to facilitate the active transport of the drug across the outer membrane into the periplasmic space, otherwise sensitive bacteria become resistant. Clinically, however, the most important mechanism of resistance is by the development of metabolizing enzymes, secreted into the periplasmic space, that inactivate the antibiotic. Acetylation of amino groups and phosphorylation or adenylation of hydroxyl groups can all occur. Plasmids carry the genes for no fewer than 20 such enzymes (Amyes and Gemmell, 1992; Bryan, 1988).

The use of aminoglycosides is restricted by a narrow therapeutic margin since they are in varying degrees toxic to the ear and kidney (Phillips, 1982). Streptomycin, in particular, damages the auditory and vestibular branches of the eighth cranial nerve. This drug is still in use for the treatment of tuberculosis, although isoniazid and rifampicin are sometimes preferred.

3.3 Chloramphenicol

Chloramphenicol is a natural product isolated from the culture filtrate of *Streptomyces venezuelae*. It is a molecule of low molecular weight, characterized by the presence of a nitrobenzene group, unusual both for a natural product and also for a drug. The drug has a broad spectrum of activity against both Gram-positive and -negative organisms, and is able to penetrate freely into host tissues. In general, chloramphenicol is believed to be bacteriostatic, although it can be bactericidal to some species - notably *Haemophilus influenzae* which causes meningitis (Turk, 1977).

The drug binds primarily to the 50S subunit of the bacterial ribosome and, *in vitro*, protein L16 appears to be the target (Nierhaus and Nierhaus, 1973). When monoiodoamphenicol is incubated with whole cells, however, proteins S6 and L24 are labelled in addition to L16, suggesting that all three proteins are involved with the receptor site (Pongs and Messer, 1976). The binding constant for chloramphenicol binding to free ribosomes is 6×10^{-7} M (Fernandez-Munos *et al.*, 1971). L-Chloramphenicol is the active form.

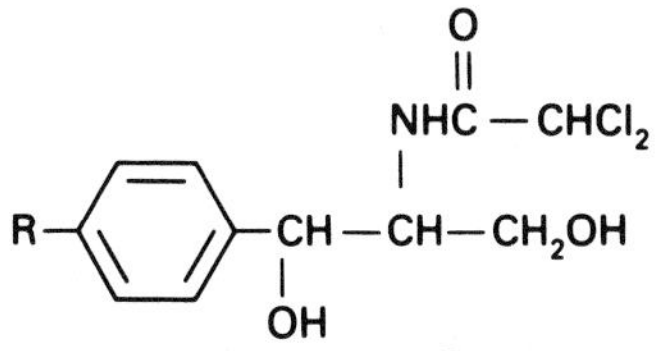

O
||
NHC—$CHCl_2$
|
R–C₆H₄–CH—CH—CH_2OH
|
OH

	R
Chloramphenicol	NO_2
Thiamphenicol	SO_2CH_3

Chloramphenicol also interferes with a rather interesting reaction called the puromycin fragmentation reaction. Puromycin is a natural product that resembles the terminal adenosine of an aminoacyl- (or peptidyl-) tRNA (Fig. 3.2). Puromycin can block chain elongation both by substituting for an incoming aminoacyl-tRNA at the A-site, and by accepting the peptidyl-tRNA bound at the P-site. The peptidyl-puromycin thus formed is incapable of further reaction as it contains an amide link attached to the ribose ring rather than

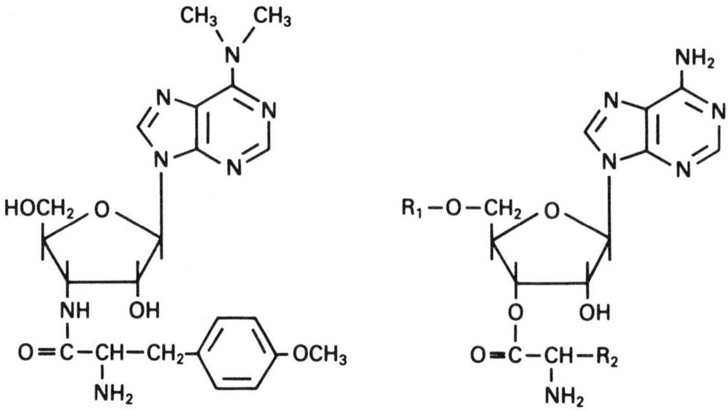

Puromycin

Terminal adenosine of an aminoacyl (or peptidyl)-tRNA

R_1 represents transfer RNA
R_2 is the side-chain of an amino acid or a growing peptide.
Puromycin resembles methyltyrosine sufficiently closely to accept the growing peptide chain. No futher change can take place because of the amide link at the 3′ position of the sugar ring in puromycin instead of the expected, and more easily broken, ester linkage.

Figure 3.2 The resemblance of puromycin to an aminoacyl adenosine.

the expected ester; the peptide chain consequently falls off the ribosome. All of the peptides so formed have the correct amino acid at the N terminus but puromycin at the C terminus.

The overall process is known as the puromycin fragmentation reaction and requires 50S subunits, a peptidyl- or aminoacyl-tRNA fragment, puromycin and K^+, Mg^{2+} and the presence of 33 per cent ethanol. This reaction is inhibited by other antibiotics that bind at or near the A-site of the ribosome, notably chloramphenicol with a K_i of approximately 10^{-5} M (Pestka, 1970).

Chloramphenicol was extremely useful in earlier decades for combatting a wide range of infections, but it soon became clear that the drug was responsible for haematological side-effects, notably bone marrow suppression and, more seriously, aplastic anaemia. One in 25 000 patients who take the drug succumb to the latter condition, which is frequently fatal (Kucers, 1982). Bone marrow suppression is a consequence of mitochondrial injury which appears to result from the inhibition of mitochondrial protein synthesis. Aplastic anaemia, however, may result from reduction of the nitro group of the drug to less chemically stable intermediates since thiamphenicol, which has an SO_2CH_3 group instead of the nitro, does not show the same toxicity (Yunis *et al.*, 1980).

The recognition, however, that various serious infections, notably with anaerobic bacteria such as *Bacteroides fragilis*, are sensitive to chloramphenicol has to some extent brought back the use of the drug - particularly when, as in this case, the safer β-lactams are rendered less effective by the bacterial production of β-lactamases. Nevertheless, chloramphenicol therapy should be limited to those infections for which the benefit of the drug outweighs the risks of the side-effects. If other drugs are available that are equally effective but are less toxic, they should be used instead (see Kucers, 1982).

One condition where chloramphenicol appears to be the antibiotic of choice is for the treatment of meningitis and brain abscesses caused by *Haemophilus influenzae* type b (Kucer, 1982) although, in the USA, chloramphenicol is recommended only for those strains of organism that are ampicillin resistant (AMA Drug Evaluation, 1983). *B. fragilis* infection of the central nervous system and systemic infections by *Salmonella* species (e.g. typhoid fever) are often treated by chloramphenicol (AMA Drug Evaluation, 1983).

3.4 Tetracyclines

Various Streptomycetes elaborate members of the tetracycline family of antibiotics, notably oxytetracycline from *S. rimosus* and chlortetracycline from *S. aureofaciens*. Later additions to the family, such as minocycline and doxycycline, have been synthesized chemically and have clinical advantages in being better absorbed by host tissues.

For drugs which have been used for so many years, there is still a surprising amount of disagreement as to which protein is the target of their action. Tetracycline binds to the 30S ribosomal subunit in a 1:1 fashion. Specific targets are primarily proteins S4 and S18 with S7, S13 and S14 as minor targets as shown by photoincorporation studies with 70S ribosomes (Goldman *et al.*, 1980). There is only one strong binding site for tetracyclines on the ribosome and it is likely that the occupancy of this site is sufficient to block protein synthesis. The binding constant is 2×10^{-5} M (Tritton, 1977). Tetracycline binding stabilizes the ribosome thermally and raises the melting temperature of the ribosomal RNA. This effect correlates moderately well with the inhibition of polyphenylalanine synthesis for a series of tetracyclines (Tritton, 1977).

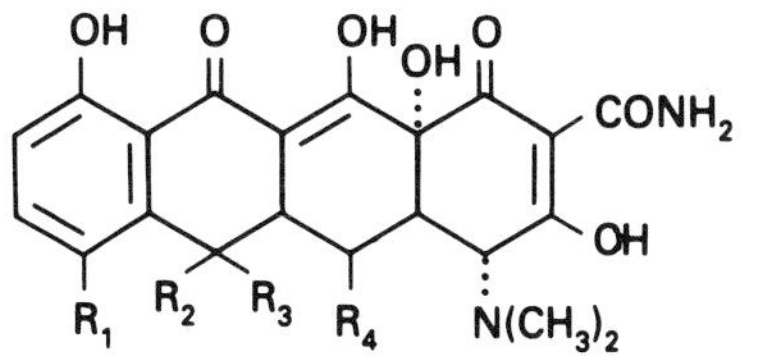

	R_1	R_2	R_3	R_4
Chlortetracycline	Cl	OH	CH_3	H
Oxytetracycline	H	OH	CH_3	OH
Doxycycline	H	H	CH_3	OH
Minocycline	$N(CH_3)_2$	H	H	H

As a consequence of this binding, aminoacyl-tRNA is prevented from binding to the A-site, and so no peptide bond can be formed. Elongation factor, EF_1, dependent hydrolysis of GTP is not affected by tetracycline (Modolell *et al.*, 1971). The interaction between the ribosome and EF_1 may be interrupted directly, or indirectly by interfering with the process whereby EF_1 recognizes the anti-codon on the tRNA (Smythies *et al.*, 1972).

Some interesting studies have been carried out on the process by which tetracyclines gain access to the cell (reviewed in Chopra and Howe, 1978). In Gram-negative species, passive diffusion of the non-ionized form of the drug takes place through hydrophilic pores in the outer cell membrane, specifically through protein 1A, one of three proteins situated there. In contrast, minocycline and doxycycline, which are more lipophilic than the original natural products, pass more readily through the lipid bilayer rather than the pore. The result is better uptake, as shown by studies on *E. coli* in which the uptake of minocycline was 10 to 20 times more rapid than that of tetracycline (McMurray *et al.*, 1982). The second process involves an energy-dependent active transport system that pumps tetracycline through the cytoplasmic membrane. Less is known about transport into Gram-positive bacteria but it also requires an energy-dependent process. Plasmids which confer resistance to tetracyclines in *E. coli*, and possibly other species, code for proteins which are located in the cell envelope and probably interfere with the uptake of the drug (Chopra and Howe, 1978).

The tetracyclines are active against a wide range of Gram-positive and -negative bacteria but a number of organisms such as *Pseudomonas aeruginosa*, and staphylococcal and streptococcal species have acquired resistance possibly as a result of indiscriminate use of these drugs. Tetracyclines are still used for the treatment of chlamydial infections, notably urethritis, but resistance limits their use in other infections (reviewed in Speer *et al.*, 1992).

3.5 Erythromycin

Erythromycin is a larger molecule than the other antibiotics considered in this chapter, with a molecular weight of 734. It consists of a 14-membered lactone ring with two sugar molecules attached, and was first detected as a product of *Streptomyces erythreus*. Erythromycin can be either bacteriostatic or bacteriocidal, depending on both the organism and the concentration of the drug. In general it is not active against most aerobic Gram-negative bacilli. It is active against some Gram-positive cocci such as *Streptococcus pneumoniae*, and is particularly useful in patients who show an allergy to penicillins. Erythromycin can also be helpful against staphylococcal strains that are resistant to penicillins. It is still the drug of choice in Legionnaire's disease caused by *Legionella pneumophila* and for the treatment of diphtheria (Kucers, 1982; Blagg and Gleckman, 1981). Unfortunately, erythromycin usage is not free from side-effects, notably liver toxicity, reversible hearing loss and pseudomembraneous colitis (Blagg and Gleckman, 1981) - see under clindamycin.

The binding site for erythromycin lies on the 50S subunit of the 70S ribo-

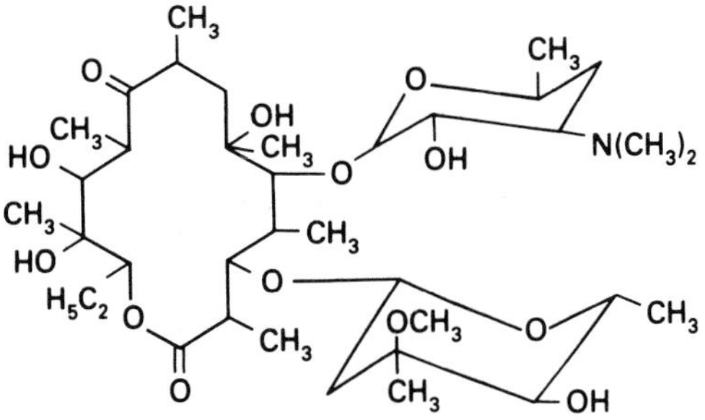

Erythromycin

some. The precise position remains uncertain, however, as both protein L16 and the ribosomal RNA itself have been implicated (Brisson-Noel *et al.*, 1988). The loss of protein L16 by washing with ammonium chloride parallels the loss of binding to the ribosome in one study (Bernabeu *et al.*, 1977), whereas methylation of adenine base in position 2058 of *E. coli* 23S RNA prevents the binding of various macrolides to the ribosome (Cundliffe, 1987). It is possible that the macrolide binding site is composed of both protein and RNA, particularly as free L16 or any other protein will not bind erythromycin.

The action of erythromycin is to cause premature release of the growing peptidyl-tRNA chain from the ribosome. This may occur either because the drug weakens the binding between ribosome and peptidyl-RNA or, alternatively, because it interferes with the translocation step when the newly synthesized peptidyl-tRNA moves from acceptor to donor site.

A recent addition to the antibacterial armoury is a close relative of erythromycin, azithromycin. Structurally, it contains a nitrogen atom in the macrolide ring as its name implies (azote is Latin for nitrogen) and acts by a similar mechanism. This drug is slightly less active against Gram-positive bacteria but rather more active against Gram-negative than its parent. It is also apparently less toxic (Drew and Gallis, 1992).

3.6 Clindamycin

Clindamycin is the 7-deoxy,7-chloro analogue of lincomycin, which was initially obtained from *Streptomyces lincolnensis*. The former is less toxic, however, and its use has superseded that of lincomycin. Broadly speaking, the spectrum of activity of clindamycin resembles that of erythromycin, and the former is of considerable use in the treatment of infections caused by anaerobic bacteria, notably *B. fragilis* (Dhawan and Thadepalli, 1982). Central nervous system infections are best managed by chloramphenicol which crosses the blood–brain barrier more effectively (Fass *et al.*, 1973). More recent work suggests that metronidazole and cefoxitin are at least as effective as clindamycin *in vivo* (Dubreil *et al.*, 1984).

Antibiotics of the lincomycin family inhibit protein synthesis in whole bacteria, and in cell-free systems, by interacting with the 50S subunit at a site very close to that occupied by erythromycin and chloramphenicol. This is shown by a parallel loss of binding for all three antibiotics when protein L16 is washed out of the ribosome by ammonium chloride and ethanol treatment (Bernabeu *et al.*, 1977). Furthermore, lincomycin inhibits the binding of chloramphenicol and erythromycin and *vice versa* (Fernandez-Munos *et al.*, 1971). One molecule of lincomycin binds per ribosome, preventing peptide bond formation by blocking binding of the 3′ terminal end of the substrate to the A-site on the ribosome (Pestka, 1970). The binding constant for lincomycin binding to *E. coli* ribosomes was $3{\cdot}4 \times 10^{-5}$ M as measured by direct binding of radiolabelled drug by equilibrium dialysis. Under conditions of the puromycin fragment assay with 33 per cent ethanol the binding was increased to give a figure of $1{\cdot}7 \times 10^{-6}$ M, but this is clearly not a physiological situation.

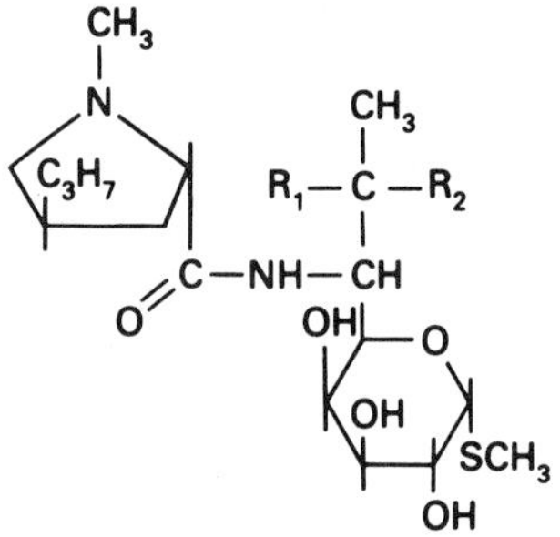

	R_1	R_2
Lincomycin	OH	H
Clindamycin	H	Cl

Clindamycin suffers from the drawback that diarrhoea frequently accompanies treatment, which can occasionally degenerate into pseudomembraneous colitis, a sometimes lethal disease which is caused by *Clostridium difficile* and characterized by diarrhoea, fever and abdominal pain (Dhawan and Thadepalli, 1982).

Questions

1. How does bacterial protein synthesis differ from mammalian?
2. What mechanisms of resistance occur against the aminoglycosides? Which is the most important clinically?
3. Which antibacterials block peptide bond formation? How do they do this?
4. Which antibacterial can be used for the treatment of infections caused by anaerobic bacteria? Why is this possible?

References

AMA Drug Evaluation, 1983, Prepared by A.M.A. Drug Division 5th edition (Philadelphia, U.S.A.: W.B. Saunders Co), p. 1255.

Amyes, S.G.B. and Gemmell, C.G., 1992, *J. Med. Microbiol*, **36**, 4-29.

Bailey, R.R., 1981, *Drugs*, **22**, 321-7.

Bernabeu, C., Vazquez, D. and Ballesta, J.P.G., 1977, *Eur. J. Biochem.*, **79**, 469-77.

Birge, E.A. and Kurland, C.G., 1969, *Science*, **166**, 1282-4.

Blagg, N.A. and Gleckman, R.A., 1981, *Postgrad. Med.*, **69**, 59-67.

Brisson-Noel, A., Trieu-Coot, P. and Courvalin, P., 1988, *J. Antimicrob. Chemother.*, **22**, Suppl. B, 13-23.

Bryan, L.E., 1988, *J. Antimicrob. Chemother.*, **22**, Suppl. A, 1-15.

Buckel, P., Buchberger, A., Bock, A. and Wittman, H.G., 1977, *Molec. Gen. Genetics*, **158**, 47-54.

Chopra, I. and Howe, T.G.B., 1978, *Microbiol. Rev.*, **42**, 707-24.

Cuchural, J.G. *et al.*, 1988, *Antimicrob. Ag. Chemother.*, **32**, 717-22.

Cundliffe, E., 1987, *Biochimie*, **69**, 863-9.

Davies, B.D., 1988, *J. Antimicrob. Chemother.*, **22**, 1-3.

Davies, J. and Courvalin, P., 1977, *Am. J. Med.*, **62**, 868-72.

DeTorres, O.H., 1981, *Clin. Ther.*, **3**, 399-412.

Dhawan, V.K. and Thadepalli, H., 1982, *Rev. Infect. Dis.*, **4**, 1133-53.

Drew, R.H. and Gallis, H.A., 1992, *Pharmacother.*, **12**, 161-73.

Dubreil, L., Devos, J., Neut, C. and Romond, C., 1984, *Antimicrob. Ag. Chemother.*, **25**, 764-6.

Fass, R.J., Scholand, J.F., Hodges, G.R. and Saslaw, S., 1973, *Ann. Intern. Med.*, **78**, 853-9.

Fernandez-Munos, R., Monro, R.E., Torres-Pinedo, R. and Vazquez, D., 1971, *Eur. J. Biochem.*, **23**, 185-93.

Goldman, R.A., Cooperman, B.S., Strycharz, W.A., Williams, B.A. and Tritton, T.R., 1980, *FEBS Letters*, **118**, 113-18.

Kaltschmidt, E. and Wittman, H.G., 1970, *Proc. Nat. Acad. Sci. (U.S.A.)*, **67**, 1276-82.

Kozak, M., 1983, *Microbiol. Rev.*, **47**, 1-45.

Kucers, A., 1982, *Lancet*, **ii**, 425-8.

Lando, D., Cousin, M.A., Ojasoo, T. and Raynaud, J.P., 1976, *Eur. J. Biochem.*, **66**, 597-606.

LeGoffic, F., Capmau, M.L., Tangy, F. and Baillarge, M., 1979, *Eur. J. Biochem.*, **102**, 73-81.

McMurray, L.M., Cullinane, J.C. and Levy, S.B., 1982, *Antimicrob. Ag. Chemother.*, **22**, 791-9.

Maitra, U., Stringer, E.A. and Chaudhuri, A., 1982, *Annu. Rev. Biochem.*, **51**, 869-900.

Merrick, W.C., 1992, *Microbiol. Rev.*, **56**, 291-315.

Modolell, J., Cabrer, B., Parmeggiani, A. and Vazquez, D., 1971, *Proc. Natl. Acad. Sci (U.S.A.)*, **68**, 1796-800.

Nierhaus, D. and Nierhaus, K.H., 1973, *Proc. Natl. Acad. Sci. (U.S.A.)*, **70**, 2224-8.

Pestka, S., 1970, *Arch. Biochem. Biophys.*, **136**, 80-8.

Pongs, O. and Messer, W., 1976, *J. Mol. Biol.*, **101**, 171-84.

Smythies, J.R., Benington, F. and Morin, R.D., 1972, *Experientia*, **28**, 1253-4.

Speer, B.S., Shoemaker, N.B. and Salyers, A.A., 1992, *Clin. Microbiol. Rev.*, **5**, 387-99.

Tai, P.-C., Wallace, B.J. and Davis, B.D., 1978, *Proc. Natl. Acad. Sci. (U.S.A.)*, **75**, 275-9.

Tritton, T.R., 1977, *Biochemistry*, **16**, 4133-8.

Turk, D.C., 1977, *J. Med. Microbiol.*, **10**, 127-31.

Wallace, B.J., Tai, P.-C., Herzog, E.L. and Davis, B.D., 1973, *Proc. Natl. Acad. Sci. (U.S.A.)*, **70**, 1234-7.

Yunis, A.A., Miller, A.M., Salem, Z. and Arimura, G.K., 1980, *Clin. Toxicol.*, **17**, 359-73.

Chapter 4

Carbohydrate metabolism

4.1 Introduction

Metronidazole is a drug which is used widely for the treatment of infections caused by anaerobic organisms, either (a) protozoa such as *Trichomonas vaginalis*, which provokes vaginal discharges, and *Entamoeba histolytica*, causing amoebic dysentery, or (b) bacteria, e.g. in abscesses or tissue necroses caused by *Bacteroides fragilis* and antibiotic-induced pseudomembraneous colitis (*Clostridium difficile*). For a general review on metronidazole, see Muller (1981). Although there is agreement that the drug has to be reduced to show activity (metronidazole is in fact a pro-drug) the ultimate target may be DNA in bacteria and, possibly, carbohydrate metabolism in protozoa. In view of this uncertainty, therefore, it has been placed in a chapter on its own.

Carbohydrate metabolism in anaerobes appears to be similar to that in aerobes - at least as far as the conversion of glucose to pyruvate is concerned. Although anaerobes subsequently transform pyruvate to acetyl-coenzyme A, they use a very different type of enzyme from the mammalian or bacterial pyruvate dehydrogenase. Pyruvate decarboxylation in anaerobes is linked to the reduction of hydrogen ion by ferredoxin (Fd), a protein which contains four iron–sulphur centres and is a very powerful reducing agent. This reaction is catalysed by hydrogenase, to give hydrogen gas, and the reduction of molecular oxygen to give hydrogen peroxide. Pyruvate dehydrogenase (pyruvate:ferredoxin oxidoreductase) appears to have been a very early product of evolution as it is also present in Archaebacteria which are thought to have existed initially in a highly reduced environment (Kerscher and Oesterhelt, 1982). The overall reaction is shown below:

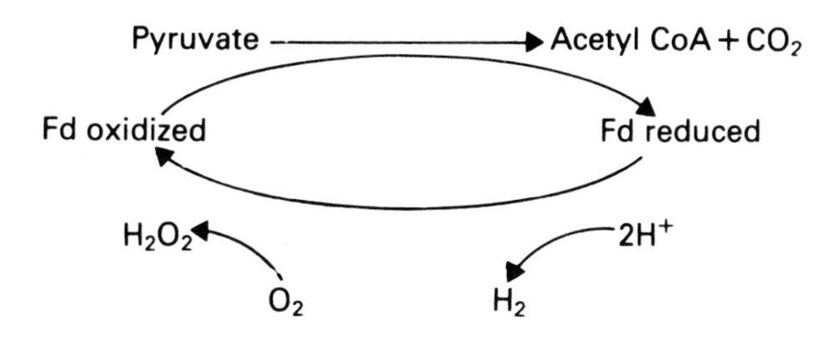

In trichomonads, acetyl-coenzyme A is converted to acetate directly by an enzyme which simultaneously transfers the coenzyme A moiety to succinate. Succinyl-coenzyme A is then hydrolysed to succinate and free coenzyme A coupled with phosphorylation of ADP to ATP (this is known as substrate level phosphorylation in contrast to oxidative phosphorylation in which an electron transport chain is required). In prokaryotes, acetate is still the end-product but an intervening step of acetylphosphate formation is included (sometimes referred to as the phosphoroclastic reaction). In trichomonads, but not in bacteria, pyruvate decarboxylase and the other enzymes involved in the later stages of carbohydrate metabolism are enclosed in a cytoplasmic organelle called the hydrogenosome (anaerobic protozoa do not have mitochondria and peroxisomes) (Lindmark *et al.*, 1975).

4.2 Metronidazole action - anaerobic protozoal and bacterial infections

Metronidazole is the original and most widely used of a family of alkyl-substituted 5-nitroimidazoles. It is inactive against organisms that grow aerobically and have reducing activity equivalent to a redox potential of −350 mV whereas anaerobic organisms (whether Gram-positive or -negative) possess a reducing system sufficiently powerful to reduce the nitro group (E_o = −470 mV; O'Brien and Morris, 1972). These products are likely to be the nitro group radical-ion for one-electron reduction, nitroso for two-electron reduction, reduction to a hydroxylamino group for four while a 6-electron reduction would give rise to products at the oxidation level of the amino group. Whether all or only some of these products occur in the intact cell is not established but the instability of some of them makes their detection extremely difficult. The chemical instability of one product at least allows it to react with DNA, causing strand breakage and cell death (reviewed in Edwards, 1993). Intracellular reduction of the drug causes more to be transported into the cell to maintain equilibrium. The net effect is that accumulation of the drug occurs in sensitive cells (Muller, 1981).

Metronidazole

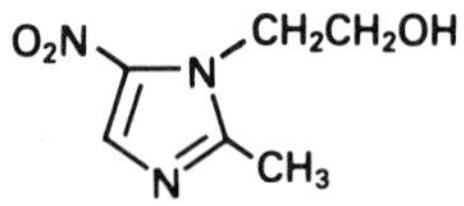

O_2N N CH_2CH_2OH N CH_3

The powerful reducing agent required to carry out this reduction is probably the iron–sulphur protein ferredoxin (E_0 = −640 mV; O'Brien and Morris, 1972), acting as a cofactor to the enzyme pyruvate:ferredoxin oxidoreductase. This enzyme catalyses the decarboxylation of pyruvate to acetylphosphate and carbon dioxide. There is a correlation in anaerobes between enzyme level, degree of susceptibility of the organism and rate of drug uptake (Narikawa, 1986). Ferredoxins contain equal numbers of iron and sulphur atoms at the active site (i.e. 4 Fe:4 S in mammalian systems), capable of acting as redox agents in a respiratory chain because of the ability of the iron to change valency. They are characterized by the presence of acid-labile sulphur atoms - so-called because treatment with acid liberates H_2S. Ferredoxin-linked reduction of metronidazole has been demonstrated in cell-free extracts of *Clostridium pasteurianum*, and it appears that the drug acts as an electron sink, siphoning reducing equivalents away from other cellular reductive processes. Methylviologen is less effective, while the pyridine nucleotide coenzymes are totally ineffective. In principle, aromatic nitro groups can be reduced in four stages:

$$R - NO_2 \xrightarrow{1e} R - \dot{N}O_2 - \xrightarrow{1e} R - NO \xrightarrow{2e} R - NHOH \xrightarrow{2e} R - NH_2$$

Although it is difficult to detect the intermediates in the metronidazole pathway because of instability, the antimicrobial action may involve a four-election reduction taking place to the oxidation level of a hydroxylamino derivative (Lockerby *et al.*, 1984; Kedderis *et al.*, 1988). Alternatively, the radical anion may take up a proton to give $R\text{-}NO_2H$, which removes electrons from (oxidizes) DNA causing strand breaks. In the process, the active agent is reduced to the nitroso form R–No (Edwards, 1993). In essence this is oxidation by the hydroxide radical $OH^{\cdot}$.

Experimentally, the first effects detectable after drug treatment are the cessation of first hydrogen emission and then carbon dioxide, as seen with *C. welchii* for example (Edwards *et al.*, 1973). The radical ion produced by metronidazole acting as an electron acceptor in place of hydrogen ion has been detected by electron spin resonance (e.s.r.) measurements both in hydrogenosomes from *T. vaginalis* (Lloyd and Kristensen, 1985) and in the intact organism (Chapman *et al.*, 1985). The radical ion is then reduced further to the active species that attacks crucial cell constituents. If the reduced product(s) were chemically stable it is likely that the effect of the drug would be reversible, in the sense that when all the drug had been reduced, hydrogen ion would continue as the electron acceptor and hydrogen evolution would restart.

In aerobic situations, oxygen can accept the electron from the radical ion because the redox potential of the O_2/O_2^- is greater than that of RNO_2/RNO_2^-. Superoxide ion and the uncharged drug molecule are produced - a futile metabolic cycle. Such a cycle has been observed in rat liver microsomes in the presence of NADPH (Perez-Reyes *et al.*, 1979), and in hydrogenosomes from *Tritrichomonas foetus*, the cattle pathogen, in the presence of oxygen (Moreno *et al.*, 1984). In the latter case, pyruvate and addition of coenzyme A stimulated the production of the radical ion, but NADPH was totally ineffective, as one would expect. Oxygen thereby acts to detoxify the drug and so protects aerobic organisms.

The precise nature of the protozoal target for the reduced active species is not known. Chapman *et al.* (1985) showed that progressive damage occurs to the radical generating system as a consequence of metronidazole reduction and considered that the radical ion is likely to be reactive enough not to travel very far in the whole cell. Furthermore, Lloyd and Kristensen (1985) proposed that a metabolite of the drug may irreversibly interfere with part of the system for hydrogen production, since the level of gas release falls progressively with time in the presence of metronidazole. The likelihood, therefore, of a reactive metabolite crossing both the hydrogenosomal and nuclear membranes to attack the DNA is not very high.

In bacteria, where there are neither nuclear nor hydrogenosomal membranes, DNA synthesis has been proposed as the target for the active species. Although protein and RNA synthesis in *B. fragilis* are unaffected by treatment with 10 μg/ml metronidazole, DNA synthesis, as measured by [^{3}H]thymidine incorporation, halted 20 minutes after treatment. The DNA remained structurally intact, as shown by subsequent extraction and gel electrophoresis, and it is unlikely that the level of supercoiling was affected, thus ruling out an effect on DNA gyrase (Chapter 2). The precise target may be in a multi-enzyme replication step since DNA polymerase was also not inhibited by drug in sonicated cells and DNA was clearly capable of serving as a template for RNA synthesis (Sigeti *et al.*, 1983). Earlier studies had shown that if metronidazole is reduced in the presence of DNA, the reduced products will bind covalently to the nucleic acid but without interfering with the sedimentation velocity or melting temperature (LaRusso *et al.*, 1978).

The development of resistance to 5-nitromidazoles does not occur frequently. Two possibilities that, in theory, can occur with an anti-infective agent are: (a) reduced transport into the cell, which would be very unlikely with such a small molecule as metronidazole, or (b) metabolism by enzymes which have not yet been discovered. Resistance modes normally give a clue as to the mode of action of a given drug, but in this case there is little evidence. Oxygen detoxification of metronidazole has been suggested as a possible method of inducing resistance (Morent *et al.*, 1984), since oxygen can react with the radical ion to reform uncharged drug. Indeed, if sensitivity tests to 5-nitroimidazoles are carried out under aerobic conditions, the microorganisms often appear more or less resistant. In studies with *T. vaginalis* isolates that markedly differed in their susceptibility to the drug *in vivo*, the availability of oxygen was shown to be an important factor in that NADH oxidase, which operates to keep oxygen tensions low, was lowered in one strain (Clackson and Coombs, 1982).

Mutations in pyruvate:ferredoxin oxidoreductase could be a possible mode of resistance, as shown for *Clostridium perfringens* by mutagenesis with *N*-methyl-*N'*-nitro-*N*-nitrosoguanidine (Sindar *et al.*, 1982). In these mutants, pyruvate and lactate levels rose whereas acetate and carbon dioxide levels fall during growth, and so the enzyme activity was not totally abolished. Muller and Gorrell (1983), however, discounted the importance of this enzyme since activity can be quite low even in susceptible strains. Furthermore, a number of clinically resistant isolates defied identification of their locus of resistance

(Muller and Gorrell, 1983). The overall difficulty in assigning the resistance to a given site is that the organism, by virtue of its ferredoxin or other powerful reducing agent, is almost certain to be able to reduce metronidazole to chemically unstable intermediate(s). The nature of the ultimate target could vary from species to species depending on the nature of the nearest sensitive molecule.

4.3 Medical use of metronidazole

Metronidazole was originally introduced on to the market in 1959 for the treatment of protozoal infections. The development of the drug for the treatment of infections caused by anaerobic bacteria coincided with, and played a major part in, the recognition of the importance of these organisms in infectious disease (Finegold, 1981; Bartlett, 1982). Metronidazole is regarded as the drug of choice in the treatment of these conditions, particularly those likely to be caused by *B. fragilis*. In one study on the suceptibility of anaerobic bacteria isolated from patients in several French hospitals, metronidazole and cefoxitin were found to be equally effective while clindamycin was less so. Metronidazole was considered to be the more suitable drug, partly because fewer resistant strains have been recorded and, in addition, a smaller dose is normally required than for cefoxitin (Dubreil *et al.*, 1984).

Metronidazole is also being used in conjunction with bismuth salts, histamine H_2 receptor blockers and other antibiotics, such as tetracycline, for the treatment of *Helicobacter pylori* associated with duodenal ulcers (Rauws, 1992).

The role of *H. pylori* in ulcer formation is not fully understood. Almost all patients with ulcers proven by endoscopy are colonized with the bacterium. In contrast, some 20 per cent of asymptomatic volunteers are also colonized, which implies that the aetiology of ulcers is multi-factorial and, indeed, acid secretion and pepsin are also strongly implicated.

Ulcers take longer to heal with an H_2 blocker such as ranitidine, alone than with a combination of ranitidine, bismuth and an antibiotic. Ulcers are also more likely to relapse if *H. pylori* is not eradicated. *H. pylori* possesses urease which converts urea to ammonia, thus raising the pH in the immediate vicinity and, in addition, various hydrolytic enzymes such as lipase, phospholipase and proteases which could cause damage to the mucosal lining (Rauws, 1992).

H. pylori can grow if the oxygen content of the atmosphere is low, i.e. 5 per cent, and is known as microaerophilic. The bacterium can survive, but probably not grow, in anaerobic conditions. Resistance to metronidazole, like the early reports of *B. fragilis* resistance, is associated with microaerophilic conditions where presumably the detoxicant effects of oxygen come into play. If *H. pylori* is incubated under anaerobic conditions before testing, it is susceptible to the drug (Cederbrand *et al.*, 1992).

Metronidazole is particularly effective *in vivo*, being readily absorbed from the gastrointestinal tract and widely disseminated in body fluids and tissues. Furthermore, in the considerable period of the use of the drug, very few genuinely resistant strains have been reported. Another potential problem with the use of the drug has been the possibility of mutagenicity as indicated by the Ames test (Ames *et al.*, 1975). This test measures the ability of a compound to reverse a mutation in *Salmonella typhimuria* that lacks an enzyme in the histidine biosynthetic pathway (phosphoribosyl adenosine triphosphate synthetase). The bacterium is unable to grow in histidine-deficient medium unless a reverse mutation occurs. Since the microsomal mixed-function oxidase system which involves cytochrome P-450 is often responsible for the metabolism of inactive precursors to rank carcinogens, liver microsomes and a NADPH generating system are also included. Metronidazole is positive in this test.

It should be noted that the Ames test is not the only test used for potential carcinogens and that some of its results are controversial. In practice over 30 years, the use of metronidazole apparently has not given rise to an abnormal incidence of tumours (Friedman and Selby, 1989). The drug is also very cheap, at least in oral tablet form.

4.4 α-Glucosidase inhibitors - acarbose as antidiabetic

Various approaches have been made to the treatment of non-insulin-dependent diabetes mellitus; diet to control carbohydrate uptake, the sulphonylureas which stimulate insulin secretion and sensitize the target tissues to insulin action, and the injection of insulin itself. Recently, a different approach has become possible by the development of α-glucosidase inhibitors, which delay the hydrolysis of complex polysaccharides to glucose, slow glucose absorption and reduce its plasma levels (Lebovitz, 1992).

The carbohydrate content of the diet in Western countries is composed of the polysaccharide starch (≈ 60 per cent) and the disaccharide sucrose (≈ 30 per cent). Salivary and pancreatic α-amylases hydrolyse starch to disaccharides (maltose), trisaccharides (maltotriose) and polysaccharides (dextrins). The brush border of the intestinal mucosa contains the α-glucosidases. The most important enzymes are maltase hydrolyzing maltose, sucrase hydrolyzing sucrose and isomaltase that catalyses the breakdown of maltotriose. In addition, glucoamylase removes one glucose residue at a time from the non-reducing end of a dextrin. With the exception of sucrase, which produces fructose as well as glucose, the product of all these enzymes is glucose (see William-Olsson, 1985; Clissold and Edwards, 1992).

Acarbose is one of the leading α-glucosidase inhibitors and is a modified tetrasaccharide with two α-D-glucose residues linked to acarviosine - a 4,6-dideoxyhexose attached to cyclohexitol through a secondary amino group. The K_i for sucrase is $1{\cdot}5 \times 10^{-6}$ M, the K_m for the substrate sucrose is $1{\cdot}9 \times 10^{-3}$ M so the drug has a 1500-fold greater affinity for the enzyme than does the substrate. The secondary amino group of acarbose, instead of the usual glycosidic oxygen bond of the substrate, accepts a proton from a carboxylate anion on the enzyme thus preventing the first step in hydrolysis (William-Olson, 1985).

Acarbose has lowered glucose levels in diabetics after a meal, but not necessarily insulin levels, irrespective of whether the patients were receiving sulphonylureas (see Clissold and Edwards, 1992). In patients receiving insulin, daily requirements were sometimes reduced. Acarbose may also be useful in preventing the diabetic complications that result from higher plasma glucose, but the evidence requires long-term trials which have not yet been carried out (Zimmerman, 1992).

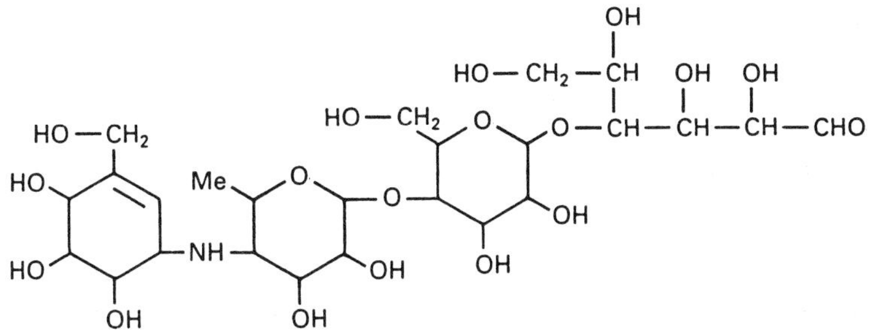

Acarbose

4.5 Sialidase inhibitor - 4-guanidino-Neu5Ac2en for influenza

In earlier years it was a dream of the medicinal scientist to design a drug on the basis of the X-ray structure and mechanism of an enzyme. This dream may be realized in a potential drug for influenza through inhibition of the viral enzyme sialidase (neuraminidase, N-acetylneuraminic acid hydrolase).

Sialidase has been known for a long time and used to be sold, moderately purified, as 'receptor-destroying enzyme'. It cleaves α-keto-linked sialic acids from terminal positions on oligosaccharide side-chains. Haemagglutinin, the other protein on the virus surface binds to sialic acid residues on target and other cells. Sialidase cuts off these false receptors and allows the virus to reach its target cell. In addition, after the virus has replicated inside the cell, sialidase cleaves the sialic acid residues from the debris of the cell and allows the virus to escape.

The enzyme reaction passes through a sialosyl cation intermediate with a proton donated by a water molecule. The sugar transition state contains a double bond in which the ring is flattened (Chung *et al.*, 1992). An unsaturated compound Neu5Ac2en that mimics this state was a moderate inhibitor of the enzyme (K_i near 10^{-6} M). The influenza virus, however, unlike the bacterial and mammalian analogues, has a pocket near the oxygen at the 4-position of the sugar ring. A glutamate residue is sited in this pocket and the suggestion was made that amino or guanidino side-chains would form salt-bridges with this residue producing very powerful inhibitors. This turned out to be the case with the amino compound K_i at 5×10^{-8} M and the guanidino at 2×10^{-10} M respectively (von Itzstein *et al.*, 1993). The latter may be a slow-binding inhibitor.

Both compounds block the replication of both types of influenza virus, A and B, in tissue culture and are active *in vivo* intranasally. Although these compounds are good leads there is much that can go wrong in the development process. Nevertheless, it is hoped that a compound will be available for use before the next worldwide influenza epidemic occurs.

R	NAME
OH	Neu-S Ac 2en
NH_2	4-Amino-Neu-S Ac 2en
NHC(=NH)NH_2	4-Guanidino-Neu-S Ac 2en

Questions

1. How many electrons are required to reduce metronidazole to (a) the radical ion, (b) the hydroxylamino derivative?
2. Why is metronidazole not active against *B. fragilis* grown in the presence of oxygen?
3. What part do anaerobic organisms play in (a) bacterial and (b) protozoal infection?
4. What enzymes hydrolyze oligosaccharides in the intestine?
5. How does acarbose antagonize the symptoms of diabetes?

References

Ames, B.N., McCann, J. and Yamasaki, E., 1975, *Mutat. Res.*, **31**, 347–64.

Bartlett, J.G., 1982, *Lancet*, **ii**, 478–81.

Cederbrand, G., Kahlmeter, G. and Ljungh, A., 1992, *J. Antimicrob. Chemother.*, **29**, 115–20.

Chapman, A., Cammack, R., Linstead, D. and Lloyd, D., 1985, *J. Gen. Microbiol.*, **131**, 2141–4.

Chung, K.J., Pegg, M.S., Taylor, N.R. and von Itzstein, M., 1992, *Eur. J. Biochem.*, **207**, 333–43.
Clackson, T.E. and Coombs, G.G., 1982, *J. Protozool.*, **29**, 636.
Clissold, S.P. and Edwards, C., 1988, *Drugs*, **35**, 214–43.
Dubreil, L., Devos, J., Neut, C. and Romond, C., 1984, *Antimicrob. Ag. Chemother.*, **25**, 764–6.
Edwards, D.I., 1993, *J. Antimicrob. Chemother.*, **31**, 9–20.
Edwards, D.I., Dye, M. and Carne, H., 1973, *J. Gen. Microbiol.*, **76**, 135–45.
Finegold, S.M., 1981, *Scand. J. Infect. Dis.*, Suppl. **26**, 9–13.
Friedman, G.D. and Selby, J.V., 1989, *J. Am. Med. Assoc.*, **261**, 866.
Kedderis, G.L., Argenbright, L.S. and Miwa, G.I., 1988, *Arch Biochem. Biophys.*, **262**, 40–8.
Kerscher, L. and Oesterhelt, D., 1982, *Trends in Biochem. Sci.*, **3**, 371–4.
LaRusso, N.F., Tomasz, M., Kaplan, D. and Muller, M., 1978, *Antimicrob. Ag. Chemother.*, **13**, 19–24.
Lebovitz, H.E., 1992, *Drugs*, **44**, Suppl. 3, 21–8.
Lindmark, D.G., Muller, M. and Shio, H., 1975, *J. Parasitol.*, **61**, 552–4.
Lindmark, D.G. and Muller, M., 1976, *Antimicrob. Ag. Chemother.*, **10**, 476–82.
Lloyd, D. and Kristensen, B., 1985, *J. Gen. Microbiol.*, **131**, 849–53.
Lockerby, D.L., Rabin, H.R., Bryan, L.E. and Laishley, F.J., 1984, *Antimicrob. Ag. Chemother.*, **26**, 665–9.
Moreno, S.N.J., Mason, R.P. and Decampo, R., 1984, *J. Biol. Chem.*, **259**, 8252–9.
Muller, M., 1981, *Scand. J. Infect. Dis.*, Suppl. **26**, 31–41.
Muller, M. and Gorrell, T.E., 1983, *Antimicrob. Ag. Chemother.*, **24**, 667–73.
Narikawa, S., 1986, *J. Antimicrob. Chemother.*, **18**, 565–74.
O'Brien, R.W. and Morris, J.G., 1972, *Arch. Mikrobiol.*, **84**, 225–33.
Perez-Reyes, E., Kalyanaraman, B. and Mason, R.P., 1979, *Molec. Pharmacol.*, **17**, 239–44.
Rauws, E.A.J., 1992, *Drugs*, **44**, 921–77.
Sigeti, J.S., Guiney, D.G. and Davis, C.E., 1983, *J. Infect. Dis.*, **148**, 1083–9.
Sindar, P., Britz, M.L. and Wilkinson, R.G., 1982, *J. Med. Microbiol.*, **15**, 503–9.
von Itzstein *et al.*, 1993, *Nature*, **363**, 418–23.
William-Olsson, T., 1985, *Acta Med. Scand.*, **706**, Suppl. 1–39.
Zimmerman, B.R., 1992, *Drugs*, **44**, Suppl. 3, 54–9.

Chapter 5

Cell wall biosynthesis

5.1 Introduction

Fungi and bacteria both require a structure external to their cell membrane to maintain the integrity of the organism - preventing a high internal osmotic pressure from rupturing the cell in a hypotonic external environment. Drugs that interfere with the biosynthesis of the cell wall might therefore be expected to be of value in combating infection by these organisms, particularly since mammalian cells do not require such a structure. The β-lactams (penicillins and cephalosporins) and vancomycin fall into this category. β-Lactams, indeed, have been on the market for many decades; the first observation that led to the identification of the precursor penicillin was made in 1929, although it took nearly 20 years for the drug to reach the market. In addition, cilofungin is known to interfere with the biosynthesis of β-glucan in the fungal cell wall and is effective in the treatment of human mycoses.

Peptidoglycan is a polymeric molecule which consists of parallel polysaccharide chains covalently joined by peptide cross-links. The whole structure can be regarded as a single molecule encompassing the entire cell in a sort of rigid bag. The glycan chain is a polymer consisting of alternating pyranose residues of *N*-acetylglucosamine and *N*-acetylmuramic acid connected in a β-(1-4) linkage (similar to the fungal cell wall polysaccharide chitin) with D-lactyl groups attached to alternate sugar residues. Attached via amide links through the carboxyl groups of the D-lactate are tetrapeptide chains (Fig. 5.1). The sequence of this peptide is L-Ala-γ-D-Glu-L-Lys (or *meso*-diaminopimelic acid)-D-Ala (*meso*-

Table 5.1 Drugs discussed in Chapter 5

Section	*Target*	*Drug*	*Use*
5.2	Peptidoglycan synthesis	β-Lactams	Bacterial infections
5.3	Peptidoglycan synthesis	Vancomycin	Bacterial infections
5.4	Mycolic acid synthesis	Isoniazid	Tuberculosis
5.5	β-Glucan synthesis	Cilofungin	Candida infections

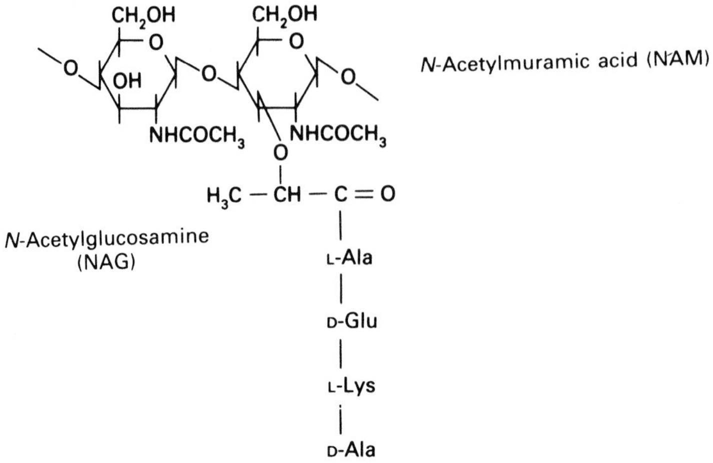

N-Acetylmuramic acid (NAM)

N-Acetylglucosamine (NAG)

L-Ala

D-Glu

L-Lys

D-Ala

The γ-carboxyl group of D-Glu is used for peptide bond formation.

Figure 5.1 Peptidoglycan repeating unit (*S. aureus*).

diaminopimelic acid is found in Gram-negative organisms, lysine in the Gram-positive *Staphylococcus aureus*).

The bifunctional amino acid in the chain (e.g. lysine) acts as a cross-link to accept the terminal carboxyl group of a D-Ala residue from an adjacent chain. This linkage may be either direct as in *Escherichia coli*, or through a short connecting peptide such as the pentaglycine found in *Staphylococcus aureus* (Fig. 5.2). This structure is resistant to peptidases which cannot attack peptides containing D-amino acids. The enzyme lysozyme, however (found in tears and egg-white) is able to split the polysaccharide back-bone at the β-(1–4) link, to yield disaccharides of the basic repeating unit to which peptide chains are still attached. This breaks the strength of the peptidoglycan so that it can no longer contain the osmotic pressure inside the cell and so the membrane ruptures with loss of cell contents.

The biosynthesis of the peptidoglycan involves the action of about 30 enzymes, starting with the formation of active precurors by soluble enzymes in the cytoplasm. The first major precursor or monomer is the pentapeptide attached to uridine diphosphate: UDP-MurNAc-L-Ala-γ-D-Glu-L-Lys-D-Ala-D-Ala (MurNAc is shorthand for *N*-acetylmuramic acid). Synthesis of the D-Ala-D-Ala fragment, which is added last, occurs by conversion of L-alanine to D-alanine (epimerization) followed by condensation of two D-alanine molecules. The monomer is then transferred to a C_{55}-terpene alcohol with a terminal phosphate group, known as bactoprenol, which acts as a membrane-bound carrier instead of UDP, and is linked to a phospholipid in the cell membrane via a pyrophosphate bridge. Condensation of UDP-*N*-acetylmuramyl-pentapeptide with UDP-*N*-acetylglucosamine is effected. In the case of the Gram-positive organism *Staphylococcus aureus*, five glycine residues are transferred from

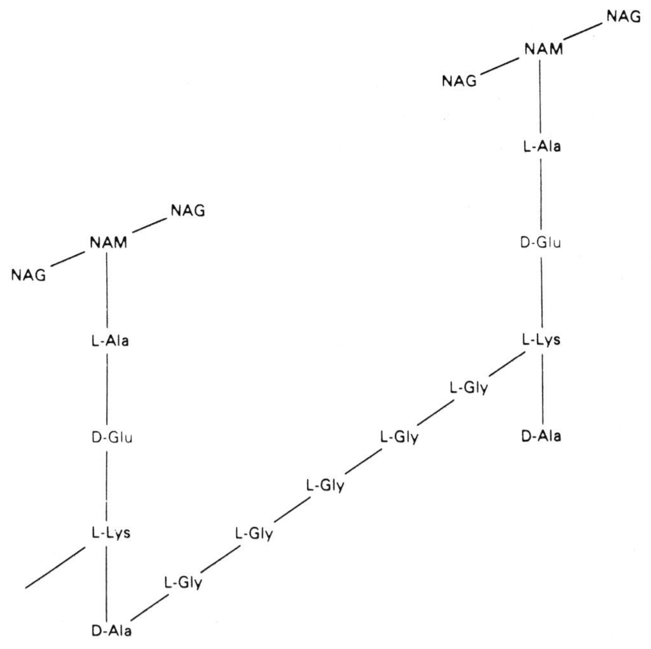

NAG, *N*-acetylglucosamine; NAM, *N*-acetylmuramic acid.
The γ-carboxyl group of D-Glu is used for peptide bond formation.

Figure 5.2 Mode of peptidoglycan cross-linking (*S. aureus*).

glycinyl-tRNA to form a chain attached at one end to the ϵ-amino group of the L-lysine residue in the side-chain. The unit is then inserted into the cell wall and simultaneously the bactoprenol is released as a pyrophosphate. The pyrophosphate then has to be converted to a monophosphate before being used again.

The third and final stage involves the formation of the cross-link. The terminal glycine is linked to the penultimate D-alanine residue of a neighbouring chain (a reaction catalysed by a transpeptidase) and at the same time releasing the terminal D-alanine. It is this last step in peptidoglycan biosynthesis that was originally thought to be the only target for the β-lactam antibiotics, probably because of their structural resemblance to the D-Ala-D-Ala dipeptide (Tipper and Strominger, 1965). The overall process is reviewed in more detail in Gale *et al.* (1981).

5.2 β-Lactams

β-Lactams comprise a number of different classes of antibiotic whose salient feature is a four-membered cyclic amide ring. This may be fused to a six-membered ring with or without sulphur, as in the cephalosporins and carbacephems, a five-membered ring in the penicillins and penems or no ring at all in the monobactams. The activity of penicillin was first observed in 1929 by Alexander Fleming. The antibiotic is elaborated by a fungus, *Penicillium notatum*, and the extraction and identification of the antibiotic from broth cultures was carried out by Chain, Florey and Abraham starting in 1939. The drug became available for human use during the Second World War. A multitude of semi-synthetic penicillins have been produced with a variety of substituents on the 6-amino group.

Cephalosporium acremonium was the first source of the cephalosporin family of antibiotics, and was isolated in 1948 from a sewer outlet off the Sardinian coast. One of the earliest cephalosporins identified was cephalosporin C which can be hydrolysed to yield 7-aminocephalosporanic acid; this has allowed the preparation of a vast range of semi-synthetic agents. The cephamycins are a closely related family of antibacterials which are derived from *Streptomyces* species, in particular *Streptomyces lactamdurans*. The latter produces cephamycin C, which has a methoxyl group at position 7 of the β-lactam ring of the 7-aminocephalosporanic acid nucleus. Recently, the monobactam series of structures has been developed of which one of the main compounds of interest is aztreonam (Neu, 1986).

Originally the mode of action of the penicillins was attributed to inhibition of the transpeptidase that catalyses the final cross-linking of the peptide side-chain of the nascent peptidoglycan (Tipper and Strominger, 1965). In *Staphylococcus aureus* treated with penicillin these authors detected large amounts of non-cross-linked peptidoglycan units, confirmed in later studies by the detection of soluble linear peptidoglycan in penicillin-treated cells (Waxman *et al.*, 1980).

More recent studies have shown, however, that the situation is not so clear-cut (reviewed in Spratt, 1980, 1983). In addition to the transpeptidase, both a D-alanine carboxypeptidase and a peptidoglycan endopeptidase have been found that are sensitive to penicillin, and it has become clear that bacteria possess a number of enzymes that are β-lactam sensitive. It appears, however, that the carboxypeptidase reaction at least is not crucial for the organism, as 6-aminopenicillanic acid inactivates at least 95 per cent of the enzyme at non-lethal concentrations in *Bacillus subtilis*, whereas cephalothin kills the bacterium at concentrations that do not affect the enzyme (Blumberg and Strominger, 1971).

5.2.1 Penicillin-binding proteins

In contrast, more recent approaches have been to study the pattern of binding when radiolabelled benzylpenicillin is incubated with bacterial cells, the membrane portion is isolated and solubilized, and the proteins are separated by gel electrophoresis (Curtis *et al.*, 1979a; Spratt and Pardee, 1975). A number of proteins to which the β-lactams will bind have been identified by this process and are known as penicillin-binding proteins (PBP) – numbered 1–6 in decreas-

Penicillins

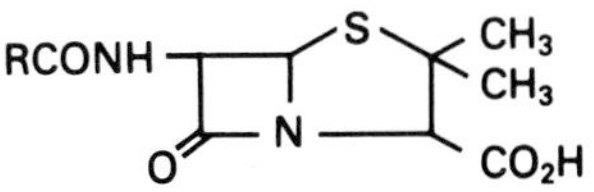

RCONH — (penicillin nucleus: S, CH_3, CH_3, N, O, CO_2H)

Benzylpenicillin : R = $C_6H_5-CH_2-$

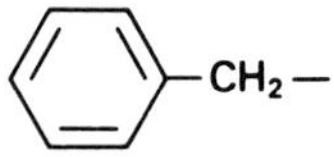

Amoxycillin : R = $HO-C_6H_4-CH(NH_2)-$

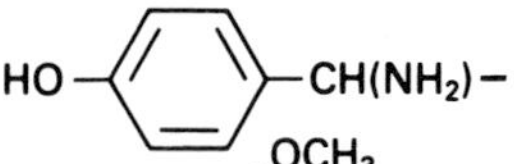

Methicillin : R = 2,6-(OCH_3)$_2C_6H_3-$

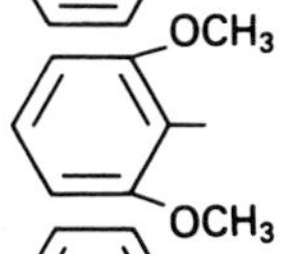

Ampicillin : R = $C_6H_5-CH(NH_2)-$

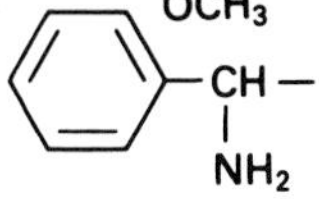

Cephalosporins

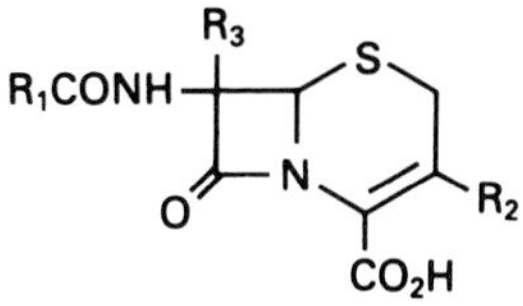

R_1CONH — (cephalosporin nucleus: R_3, S, N, O, R_2, CO_2H)

	R_1	R_2	R_3
Cefoxitin	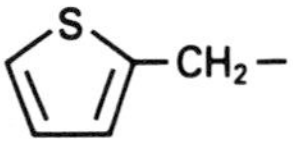 2-thienyl$-CH_2-$	$-CH_2OCONH_2$	$-OCH_3$
Cephalosporin C	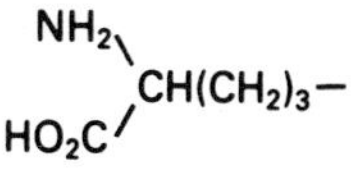 $NH_2(HO_2C)CH(CH_2)_3-$	$-CH_2OCOCH_3$	$-H$
Cefuroxime	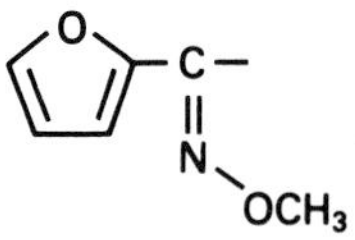 2-furyl$-C(=N-OCH_3)-$	$-CH_2OCONH_2$	$-H$
Ceftazidime	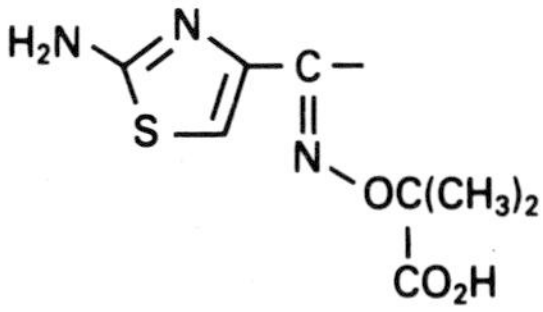 2-amino-4-thiazolyl (H_2N, N, S)$-C(=N-OC(CH_3)_2CO_2H)-$	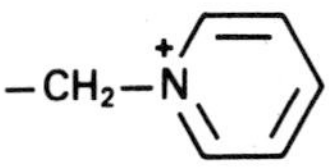 $-CH_2-N^+$ (pyridinium)	$-H$

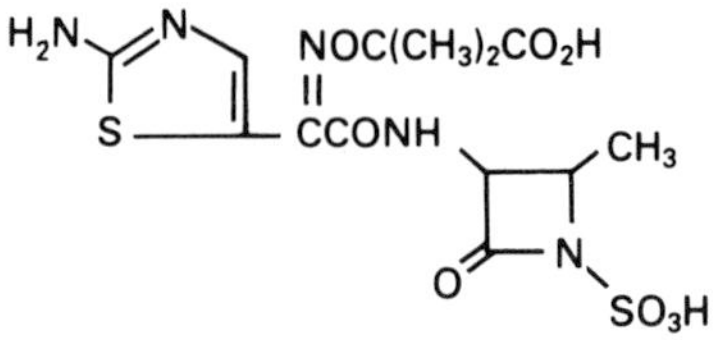

Aztreonam

ing order of apparent molecular weight. The enzymic reactions associated with these proteins are not fully understood, although they must carry out transpeptidation and endopeptidase reactions.

Antibiotics that bind to these proteins produce a variety of effects: in *E. coli*, blockade of PBP1a and 1b leads to cell lysis although it appears that the loss of either one alone is not lethal, presumably because the other protein can compensate; PBP2 inhibition leads to cessation of growth and change of shape from rod to sphere and eventual lysis; PBP3 appears to be responsible for the formation of the cross-wall to divide the two daughter cells at cell division (septation), and blockade of this protein results in the formation of long filaments and eventual cell death. Binding to the three other penicillin-binding proteins does not lead to any obvious growth defects (reviewed in Spratt, 1983).

Other distinctions that correlate with the separation of the penicillin-binding proteins into two groups depending on their importance to the bacterium are that PBPs 1–3 have molecular weights between 60 and 92 kDa in *E. coli*, for example, and exhibit transglycosylase and transpeptidase activity, while PBPs 4–6 have molecular weights 40–49 kDa and show D-alanine carboxypeptidase activity. This work has been carried out both with agents that bind specifically to one or other penicillin-binding protein, and also by the use of mutants that lack, or produce thermolabile varieties of, these proteins (Spratt, 1980, 1983).

It is clearly of importance to correlate the binding of an antibiotic to a given penicillin-binding protein with the effect on the whole cell. Cefoxitin, for example, exhibits an I_{50} of 0·1 μg/ml for PBP1a and 3·9 μg/ml for PBP1b from *E. coli* K12, values which are close to the minimum inhibitory concentration (MIC, the minimum concentration that totally inhibits growth; Chapter 1) (Curtis *et al.*, 1979b), and so it is likely that the penicillin-binding proteins are the targets, particularly as the drug causes the bacteria to lyse. Mecillinam is an example of β-lactam that binds particularly effectively to PBP2 ($I_{50} < 0{\cdot}25$ μg/ml), kills *E. coli* with an MIC of 0·05 μg/ml, producing in the process spherical forms of the organism as we would expect from a binding of PBP2 (see above). In sharp contrast, ceftazidime has a potent affinity to PBP3 in both *E. coli* and *Pseudomonas aeruginosa*, with an I_{50} of 0·06 μg/ml for PBP3 of the former bacterium and MIC of 0·2 μg/ml (Hayes and Orr, 1983). As a consequence, filamentation of the organism results.

Resistance to β-lactams among Gram-negative bacteria may result from a reduction in drug affinity by the penicillin-binding protein (Malouin and Bryan,

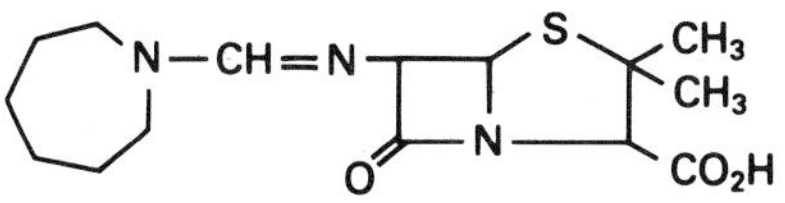

Mecillinam

1985). In Gram-positive bacteria, this can also happen and, in addition, the acquisition of a different PBP.

Staphylococcus aureus normally has five PBPs (1, 2, 3, 3′, 4) of which three (1, 2, 3) are essential for cell survival and growth. Some strains have become resistant to methicillin and other β-lactams by acquiring an altered PBP2 (PBP2A, molecular weight 78 kDa) which has a low affinity for these drugs. The PBP gene is chromosomal and seems to have been derived from the fusion of the genes for a β-lactamase and a PBP from another source. The gene is highly conserved in that limited proteolysis of the PBP from unrelated strains produces similar peptides. Although almost all cells within a population carry the gene, only a very few express it, so there must be other factors involved in methicillin resistance (Hackbarth and Chambers, 1989). The altered PBP does not reduce growth or virulence as the mutant strains are as virulent as the wild type at causing life-threatening infections, for example pneumonia or meningitis. The crucial site of action is probably the septum (the division formed from plasma membrane and cell wall that forms across the middle of the original cell, splitting it into two daughter cells); methicillin-sensitive cells lyse when growth at this point is blocked, whereas resistant cells do not. Infection is often transmitted within an institution (nosocomial), often a hospital and may reach epidemic proportions. The problem is that the resistant strains may not be recognized in time to be treated with an appropriate drug such as vancomycin (Chambers, 1988).

5.2.2 β-Lactamases

Bacterial resistance to the β-lactams may be derived, as with other antibiotics, by preventing the ingress of the antibiotic into the cell or by altering the enzyme target. Although the former possibility does not appear to play a major part, it is clear that the latter does (Malouin and Bryan, 1986). In addition, another major method for the development of bacterial resistance to the β-lactams is detoxification by enzymes which destroy the β-lactam ring, namely β-lactamases (see Amyes and Gemmell, 1992). The genetic coding for the enzyme may be found both on a chromosome or a plasmid. In Gram-positive bacterial β-lactamase production is the major method of resistance in that the bacteria produce a large quantity of the enzyme and secrete it extracellularly.

In Gram-negative bacteria the situation is more complex. The organism produces less β-lactamase and secretes it into the periplasmic space, i.e. the space between the peptidoglycan band around the cytoplasmic membrane and the bilayer outer membrane, which is very hydrophobic (construction of the

Gram-negative cell envelope is reviewed in Costerton and Cheng, 1975). The enzyme is thereby positioned at the site where the peptidoglycan is formed on the outer side of the cytoplasmic membrane, and it can therefore protect the organism to the greatest extent. Furthermore, the outer membrane presents a much more serious obstacle to the transport of the antibiotic in Gram-negative bacteria than it does in Gram-positive ones. As a consequence, the possession of a β-lactamase correlates better with bacterial resistance for a Gram-positive than for a Gram-negative bacterium (Sykes and Matthew, 1976).

β-Lactamases have been classified in various ways but probably the most widely accepted classification depends on their amino acid sequence and catalytic properties. Two groups of enzymes with serine at the active site have a preference for penicillins on the one hand (class A), and cephalosporins on the other (class C). The latter are usually chromosomally encoded enzymes derived from Gram-negative Enterobacteria (Jaurin and Grundstroem, 1981). Class B includes the Zn^{2+}-containing enzymes (see Frere and Joris, 1985). The enzymes are frequently extremely active, although the most useful way to describe the activity is by the ratio of catalytic rate constant to Michaelis constant (K_{cat} to K_m). For benzylpenicillin this figure is $9{\cdot}8 \times 10^5$ m/s, which is close to the diffusion limit, i.e. the substrate is hydrolysed almost as fast as it can diffuse to the active site of the enzyme. The enzyme splits most penicillins by breaking open the β-lactam ring to yield penicilloic acids (Fig. 5.3).

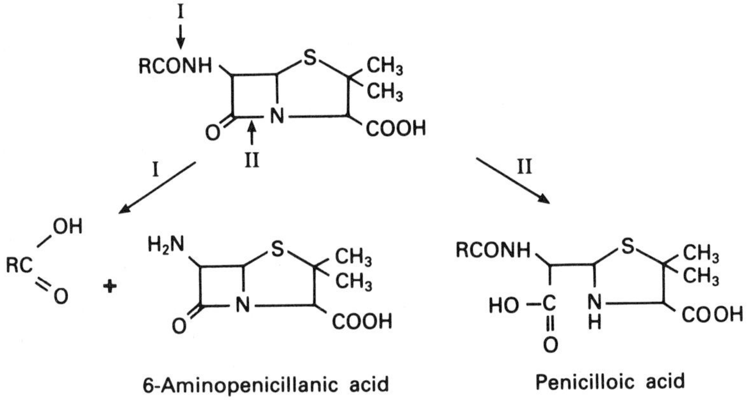

6-Aminopenicillanic acid

Penicilloic acid

I. Penicillin amidase breaks the amide link in the side-chain of the drug. This enzyme is not used by the microorganism for protection against the drug. Penicillin amidase has been used to prepare 6-aminopenicillanic acid from which new semi-synthetic derivatives can be made by linking the amino group to new acyl side-chains.

II. β-Lactam hydrolysis of the β-lactam ring proceeds through attack by a nucleophilic serine hydroxyl on the β-lactam carbonyl, yielding a transient acyl-enzyme intermediate. It is this enzyme that confers drug resistance upon the bacterium, and organic chemists have been at pains to synthesize less susceptible new compounds.

Figure 5.3 Pathways of β-lactam metabolism.

β-Lactamases vary markedly from each other in primary sequences having in some cases less than 40 per cent homology, and so the overall shape of the molecule is the key, rather like the G-protein family (see section 9.1.1). Indeed, some PBPs are closer in structure to β-lactamases than to other PBPs. The common feature both classes share is a serine at the active site, which forms an acyl-enzyme intermediate (reviewed in Ghuysen, 1991).

In kinetic terms, the anti-bacterial binds reversibly to the enzyme. The complex then isomerizes irreversibly into a complex with a covalent link between enzyme and anti-bacterial which breaks down into enzyme and product:

$$E + I \overset{K}{\leftrightarrow} EI \overset{k_2}{\rightarrow} EI^* \overset{k_3}{\rightarrow} E + P$$

where EI* represents inactivated enzyme, K is an equilibrium constant and k_2, k_3 are rate constants. Provided that k_3 remains small and k_2/K is high the enzyme remains inactivated as in the PBPs, whereas if k_3 is high we have a recipe for drug detoxification as in the β-lactamases.

Translated into molecular terms, the active site serine makes a nucleophilic attack on the carbon atom of the carbonyl group in the lactam ring. The leaving group is too bulky to diffuse away and remains attached to the serine. Only a water molecule can attack the complex and release the leaving group. In the PBPs the access is extremely slow or negligible, while in the β-lactamases it occurs rapidly (Ghuysen, 1991). This may be because the β-lactamases have a glutamate moiety able to act as a general base in the active site by abstracting a proton from the water molecule to allow hydroxyl attack on the carbonyl group. PBPs have a phenylalanine at the equivalent position which cannot do this. β-Lactamases may not need the type of charge relay system seen in the serine proteases, partly because the lactam as a cyclic amide is intrinsically more reactive than an open chain amide. Furthermore, the lactam carbonyl is likely to be further polarized by binding to two backbone amides (Escobar *et al.*, 1991; Strynadka *et al.*, 1992).

A number of suicide substrates have been described, i.e. drugs which the enzyme converts into products that destroy the enzyme's activity (Chapter 1). These are frequently derived from fermentation cultures, as in the case of clavulanic acid from *Streptomyces clavuligerus*. Clavulanic acid interacts with β-lactamases in a complex fashion but, in general, the first step involves the formation of a complex that is non-covalent and readily dissociable. Secondly, the ligand acts as a substrate and the β-lactam ring is opened with the active site serine forming an ester link. A true substrate would now dissociate from the enzyme, but the clavulanic acid moiety rearranges to produce a conjugated series of double bonds giving rise to an ultraviolet absorbance spectrum. Subsequently, in some cases, the enzyme–inhibitor complex is reactivated and the enzyme recovers its full activity. In others the complex is stable and the enzyme remains irreversibly inhibited (Labia *et al.*, 1985).

5.2.3 β-Lactam therapy

It has been of great importance to synthesize drugs which are resistant to β-lactamase detoxification. The introduction of a 2,6-dimethoxyphenacyl group to form an amide with the amino group at position 6 of the β-lactam ring sterically hindered β-lactam hydrolysis as in methicillin. Replacing the CH_2 between the ring and the carbonyl group with a methoxyimine side-chain, as in cefuroxime, also conferred stability to β-lactamases. Another approach is to insert a 7α-methoxy group in the β-lactam ring as in the family of cephamycins of which cefoxitin is an example, although larger substituents in this position produced a reduction in activity. Further substitution of the imino grouping, as in ceftazidime, was designed to increase intrinsic activity against Gram-negative organisms of clinical importance (such as *E. coli* and *Klebsiella pneumoniae*) by virtue of the aminothiazolyl side-chain, whilst retaining resistance to β-lactamases. Unfortunately, however, the series of plasmid-coded β-lactamases have mutated to produce enzymes that hydrolyse the drug (see Amyes and Gemmell, 1992). The introduction, however, of a positively charged quaternary nitrogen at the 3 position has produced drugs that are more active against enteric bacteria and *Pseudomonas aeruginosa*, largely through increased resistance to beta-lactamases (Hancock and Bellido, 1992).

Another approach has been to design structures that have a carbocyclic five- or six-membered ring attached to the lactam, known as carbapenems and carbacephems. The most important examples of these classes are imipenem and loracarbef. They have the great advantage that they are resistant to class A and C β-lactamases (imipenem has a *trans* side-chain) albeit sensitive to the class B zinc-containing enzymes. They also have a very wide spectrum of activity against a range of organisms.

Imipenem binds strongly to PBP2 of *E. coli* and *Pseudomonas aeruginosa* and to all four PBPs of *Staphylococcus aureus*. Imipenem appears to damage *Ps. aeruginosa* more severely than other β-lactams without allowing regrowth. Imipenem resistance results from a change in the permeability of the outer membrane, specifically in the porins that form channels to allow the inward passage of nutrients (reviewed in Buckley *et al.*, 1992). Imipenem resembles a substrate for the kidney dipeptidase, dehydropeptidase I, and is rapidly degraded *in vivo* by this enzyme. The drug is administered in a 1:1 mixture with cilastatin, a compound which inhibits dehydropeptidase (Farrell *et al.*, 1987). Loracarbef is given as a single drug and has the advantage of being orally active (Cooper, 1992).

An alternative approach is to use β-lactams that are potent β-lactamase inhibitors, but not necessarily anti-bacterial, to act in synergy with β-lactam antibiotics. Various compounds with these properties have been discovered in culture filtrates of Streptomycetes. In particular, clavulanic acid has been developed for this purpose. A combination of clavulanic acid with amoxicillin (trade name Augmentin) is effective in treating infections produced by organisms that produce β-lactamase in large quantities, notably *Bacteroides fragilis*.

It can be shown *in vitro* that clavulanic acid will enhance the activity of β-lactams against *B. fragilis* by lowering the MIC by two orders of magnitude (Bansal *et al.*, 1985).

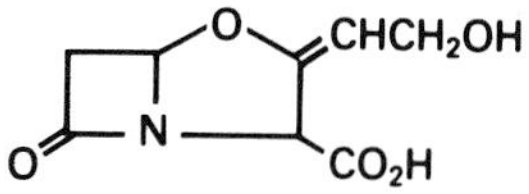

Clavulanic acid

The β-lactams are probably the most widely used anti-bacterials at the present time, partly because of their broad spectrum of action and partly because of their very low toxicity. They must be about the least toxic drugs available on the market.

Oral activity is clearly a great advantage for the use of drugs in the community, while injectable drugs are more easily given in a hospital setting. Although the original penicillin (G) required injection, some α-amino-β-lactams are orally active. They are structural analogues of tripeptides that are taken up by an oligopeptide transport system, used for small nutrient peptides, that operates across the brush border membrane of the enterocyte (Kramer *et al.*, 1992). β-Lactams are themselves synthesized from a tripeptide, aminoadipyl-cysteinyl-valine. The transport system is chirally sensitive in that D-cephalexin is taken up while the L-isomer is not (Kramer *et al.*, 1992).

A goal of β-lactam synthesis has been to make orally active drugs, either by synthesis of a structure that is orally active, such as ampicillin or cephalexin, or by modification of a structure that is only active by injection, i.e. by forming a pro-drug. Cefuroxime axetil is an example of a pro-drug where the orally inactive cefuroxime is converted at position 4 of the carboxyl to the acetoxyethyl ester. The pro-drug is absorbed because it is more lipophilic not because it is taken up through the peptide transport system. The pro-drug is absorbed intact and then hydrolyzed in the intestinal mucosa or portal blood (Williams and Harding, 1984).

The esterification of the carboxyl group introduces another chiral centre into the molecule (positions 6 and 7 are already chiral) and the hydrolysis of the isomeric esters *in vivo* may depend on the chirality of the 1′ position; 1′R,6R,7R is rapidly hydrolyzed by an esterase in rat stomach while 1′S,6R,7R is almost completely resistant (Mosher *et al.*, 1992).

The absorption of cefuroxime axetil was not straightforward, however, as the first available tablets dissolved at variable rates and so the dose was variable; a new test for measuring the rate of disintegration of a tablet had to be developed (Emmerson, 1988). Furthermore, 50 per cent of the drug is absorbed if taken after food but only 30 per cent after overnight fasting

(Williams and Harding, 1984). This is one practical aspect of the pharmaceutical industry that can be overlooked in the quest for new medicines.

5.3 Vancomycin

Vancomycin is a large glycopeptide with a molecular weight in the region of 1500, obtained from the culture filtrates of *Streptomyces orientalis*. It has a complex molecular structure (see Sheldrick *et al.*, 1978) containing an amino sugar called vancosamine linked to three aromatic rings; several amino acid residues and a bis-resorcinol system. Vancomycin is active primarily against Gram-positive bacteria. It is not a new antibiotic, having been discovered in the mid-1950s, but while it demonstrated undesirable side-effects it had no advantage over the β-lactams, especially with the advent of the β-lactamase-stable β-lactams.

Vancomycin

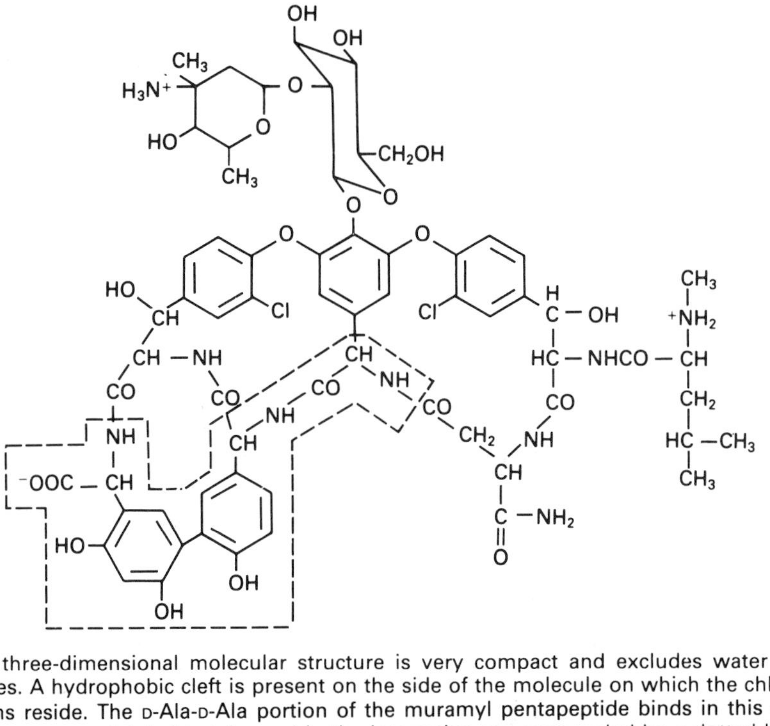

The three-dimensional molecular structure is very compact and excludes water molecules. A hydrophobic cleft is present on the side of the molecule on which the chlorine atoms reside. The D-Ala-D-Ala portion of the muramyl pentapeptide binds in this cleft. The groups postulated to take part in the interaction are surrounded by a dotted line. See Fig. 5.4 for the detailed interaction proposed by Sheldrick *et al.* (1978).

Latterly, however, the importance of methicillin-resistant *Staphylococcus aureus* infections has indicated the need for a tried and tested antibiotic such as vancomycin. In addition, pseudo-membraneous colitis responds well to vancomycin. This condition is induced by the action of other antibiotics which 'scour' the gut leaving it almost empty of bacteria and thus allowing *Clostridium difficile* to colonize. Furthermore, more modern preparations appear to be 'cleaner' and, therefore, the occurrence of side-effects is markedly reduced compared with impure preparations of earlier years (Kucers, 1984; Watanakunakorn, 1984).

The mode of action of vancomycin involves interference with the biosynthesis of cell wall peptidoglycan, but in a different way from the β-lactams. Vancomycin binds tightly to UDP-pentapeptides containing D-alanine-D-alanine at the free carboxyl end, thus causing UDP-*N*-acetylmuramyl-peptide precursors to accumulate. The tightest binding was observed with UDP-*N*-acetylmuramyl-glycyl-D-glutamylhomoseryl-D-alanyl-D-alanine, with a binding constant of 6×10^{-7} M, as measured by ultraviolet difference spectroscopy. The smallest fragment to which binding was detected was acetyl-D-alanine-D-alanine (Perkins, 1969). The molecule carries a marked hydrophobic cleft on one side to which the peptide is bound as shown by nuclear magnetic resonance studies. The binding is initiated by the interaction between the peptide carboxylate group and a protonated amine on the antibiotic (Williamson and Williams, 1984; Fig. 5.4).

As the target is a structural component of the cell wall, rather than an enzyme, resistance to vancomycin was expected to be very unlikely. Recently, however, resistant clinical isolates have turned up, where the pentapeptide terminates in D-lactate with greatly reduced affinity for vancomycin

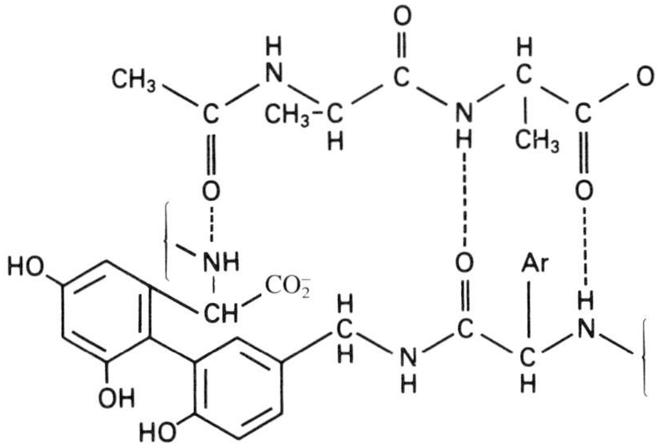

Interaction with acetyl-D-Ala-D-Ala is shown above. Ar, the trihydroxylated benzene ring in the middle of the structure. Dotted lines indicate hydrogen bonds which are postulated to be the major force in holding the complex together. The portion of the molecule that interacts with the dipeptide moiety is outlined in the structure on previous page.

Figure 5.4 Vancomycin interaction with D-ala-D-ala moiety of pentapeptide.

(Handwerger *et al.*, 1992). Figure 5.4 shows three of the five hydrogen bonds holding the drug-pentapeptide complex together - the other two arise from the carboxylate oxygen not involved in binding in the figure (Barna and Williams, 1984). Resistance arises from the substitution of D-lactate for D-alanine and so the NH in the amide link of the dipeptide is replaced by the oxygen of an ester link, thus removing one of the hydrogen bonds; a major contribution to the binding is thereby lost. The pentapeptide altered from amide to ester may act more efficiently as a substrate with the cross-linking enzymes to form the wild-type peptidoglycan.

The primary mode of action of vancomycin is inhibition of cell wall biosynthesis, and consequently results in cell lysis. Gram-positive organisms are primarily affected because vancomycin is too large to cross the outer membrane of Gram-negative organisms. The drug may also alter the permeability of the cell membrane and, in addition, interfere with ribonucleic acid synthesis (Watanakunakorn, 1984).

5.4 Mycolic acid synthesis - isoniazid for tuberculosis

Tuberculosis, known as consumption in earlier years, is the leading cause of death from infectious diseases (Collins, 1993). The disease has recently become much more frequent in developing countries and is also a serious complication of AIDS, where the patient is suffering from a weakened immune system. Tuberculosis is mainly caused by an infection with *Mycobacterium tuberculosis* and can be lethal. The lung is most often the site of infection, although the brain or skin can also be affected. The symptoms of pulmonary tuberculosis are cough, fever and, in the later stages, extreme loss of weight, severe night sweats and often haemorrhage. The disease gets its name from a granule composed of leukocytes, epithelial cells, a few giant cells and mycobacteria, called the tubercle. As the disease progresses, the tubercles fuse to form a soft mass and invade the healthy tissue. Eventually cavities form and the lung collapses.

Most front-line anti-bacterial agents are inactive, or only weakly active, against mycobacteria. Isoniazid (isonicotinic acid hydrazide) and rifampicin are

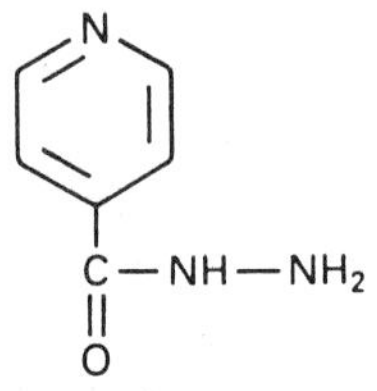

N

C—NH—NH_2

O

Isoniazid

still the mainstays of anti-tubercular chemotherapy. They are normally given in conjunction with a third drug, in case the microorganism is resistant to one or both. Rifampicin binds to bacterial RNA polymerase (see section 2.3.5) but isoniazid's mode of action is still in doubt, despite having been used clinically for over 40 years. Various proposals include reaction of the hydrazine with the carbonyl group of pyridoxal phosphate enzymes, which can happen even *in vitro* (Shoeb *et al.*, 1985) or conversion of isoniazid to the isonicotinic acid analogue of NAD. The evidence for these effects is, however, weak. The reason for the inclusion of isoniazid in this chapter is that one important hypothesis is inhibition of mycolic acid formation, a major component (≈ 40 per cent) of mycobacterial cell walls (Quemard *et al.*, 1991). This assignment is controversial as there is also good evidence that activation of isoniazid by peroxidase is involved. It may be, however, that these two mechanisms are not mutually exclusive, in that peroxidase may activate isoniazid to a product that may attack mycolic acid synthesis.

Mycolic acids, a family of closely related lipids found in mycobacteria and a few other bacterial genera, are α-alkyl-β-hydroxy long-chain fatty acids with between 60 and 90 carbon atoms. They contain branched methyl groups, double bonds and cyclopropane rings. The general formula of the major mycolic acid, α-mycolic acid, present in *M. tuberculosis* is shown below (from Qureshi *et al.*, 1978):

```
               CH2                      CH2                    OH          COOH
              /   \                    /   \                   |            |
CH3(CH2)a CH  -  CH (CH2)b CH  —  CH (CH2)c CH  —     CH    (CH2)d CH3
```

where a = 17, 18, 19; b = 10; c = 15, 17, 19, 21 and d = 21, 23.

The fatty acid synthetase which produces the C_{24} and C_{26} acids required for the straight α chain part of the mycolic acid has been purified from *M. tuberculosis* and is composed of two 500 kDa monomers. The enzyme is unusual in being a soluble cytoplasmic source of fatty acids with carbon chains longer than C_{18} or C_{20}. The enzyme also resembles the mammalian fatty acid synthetase complex in that both acetyl and malonyl-coenzyme are substrates and all the enzymic steps are sited on a single polypeptide. Fatty acids are released from the synthetase as coenzyme esters (Kikuchi *et al.*, 1992).

Mycolic acid synthesis has long been recognized as a possible target for isoniazid. Sensitivity to isoniazid paralleled inhibition of mycolic acid synthesis in two closely related strains of *M. aurum*. The concentration for mycolic acid inhibition in a cell-free system was five times higher than the MIC (I_{50} of $3{\cdot}5 \times 10^{-5}$ M compared with MIC of $7{\cdot}3 \times 10^{-6}$ M) but this relationship may be explained by the ability of the mycobacterium to concentrate isoniazid. Peroxidase activity was similar in the two strains (Quemard *et al.*, 1991).

There is also evidence that oxidation by peroxidase is the key target. Mycotubercular resistance to isoniazid is associated clinically with gene deletion of an enzyme which has peroxidase and low catalase activity. Hydrogen peroxide seems important for the sensitivity, because if the catalase activity is higher, as in some mycobacterial species, hydrogen peroxide is broken down to water and oxygen and the microorganism is resistant (Shoeb *et al.*, 1985). Hydrogen peroxide oxidizes isoniazid to a radical with peroxidase as the catalyst. Insertion of the gene into resistant *M. tuberculosis* or *E. coli* confers sensitivity to isoniazid. The suggestion is that the radical, whether drug-derived or superoxide, is responsible for damaging the cellular machinery (Zhang *et al.*, 1992).

It is likely that this uncertainty about the mode of action of isoniazid will soon be resolved now that tuberculosis is becoming a much more serious therapeutic problem.

5.5 β-Glucan synthetase

The fungal cell wall consists of three polysaccharides in varying proportions: chitin, mannan and β-glucan, held together by cross-links. Chitin is a homopolymer of *N*-acetylglucosamine, mannan of mannose complexed to a small protein, and β-glucan of glucose. β-Glucan provides the main structural polymer and is largely made of glucose 1,3-linked with a small proportion of 1,6-linked material to provide the cross-links (Cabib *et al.*, 1982).

β-Glucan, like most polysaccharides, is synthesized from a nucleotide sugar, UDP-glucose, by the cell membrane enzyme β-glucan synthetase, and is inhibited non-competitively by the semisynthetic lipopeptide, cilofungin, with an apparent K_i of $2{\cdot}5 \times 10^{-6}$ M. Both the peptide nucleus and the lipophilic side-chain are essential for the inhibition (Tang and Parr, 1991); the 3-hydroxy-4-methylproline residue enhances activity, but the L-homotyrosine moiety is crucial for both antifungal activity and β-glucan synthetase activity (Zambias *et al.*, 1992).

Cilofungin inhibits the growth of the pathogenic *Candida albicans* and *C. tropicalis in vitro* with minimum inhibitory concentrations (MIC) $<0{\cdot}31$ μg/ml, but other *Candida* species less powerfully. The growing organism forms protoplasts, a form without a cell wall in the presence of the drug, and may burst depending on the external osmotic pressure. To growing cells, therefore, cilofungin is fungicidal and to stationary cells it is fungistatic *in vitro*. Althoug β-glucan is widespread in the fungal kingdom, other fungi are resistant to the action of the drug, possibly because they contain less β-glucan (Hall *et al.*, 1988).

Cilofungin is under development for the treatment of systemic or topical infections by *Candida albicans*, particularly in immunosuppressed patients. The cell wall of *Pneumocystis carinii* also contains 1,3-β-glucan, and echinocandin analogues may also be useful for the treatment of this microorganism in AIDS patients.

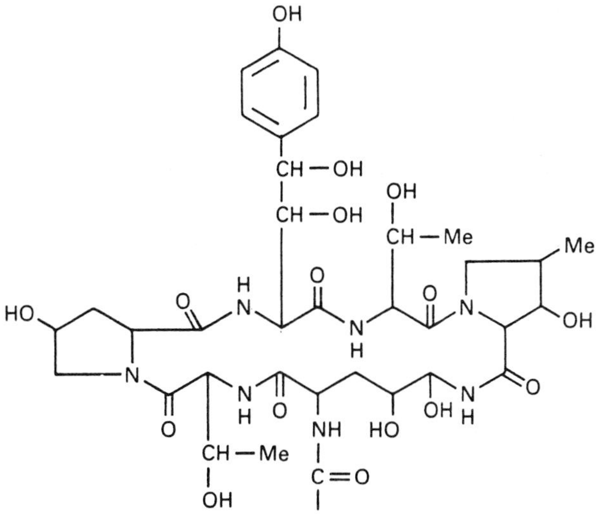

Cilofungin

Questions

1. What are the effects on *E. coli* of drug binding to penicillin-binding proteins (a) 2 and (b) 3?
2. What families of β-lactams are available to the clinician? How do they differ structurally?
3. What is the therapeutic advantage of an orally active β-lactam compared with one that has to be given intravenously?
4. How do β-lactamases and penicillin-binding proteins differ kinetically?
5. What methods do bacteria use to detoxify β-lactams? What is their relative importance clinically?
6. Vancomycin is particularly useful in the treatment of infections caused by which bacteria? Why is it effective?
7. How have bacteria developed vancomycin resistance?
8. Isoniazid blocks the synthesis of which component of the mycobacterial cell wall?
9. What are the main constituents of the fungal cell wall? With the synthesis of which component does cilofungin interfere?

References

Amyes, S.G.B. and Gemmell, C.G., 1992, *J. Med. Microbiol.,* **36**, 4–29.

Bansal, M.B., Church, S.K., Onjema-Lalobo, M. and Thadepalli, H., 1985, *Chemotherapy (Basel),* **31**, 173–7.

Barna, J.C.J. and Williams, D.H., 1984, *Annu. Rev. Microbiol.,* **38**, 339–57.

Blumberg, P.M. and Strominger, J.L., 1971, *Proc. Natl. Acad. Sci. (U.S.A.)*, **68**, 2814-17.
Buckley, M.M., Brogden, R.N., Barradell, L.B. and Goa, K., 1992, *Drugs*, **44**, 408-44.
Cabib, E., Roberts, R. and Bowers, B., 1982, *Ann. Rev. Biochem.*, **51**, 763-93.
Chambers, H.F., 1988, *Clin. Microbiol. Rev.*, **1**, 173-86.
Collins, F.M., 1993, *Crit. Rev. Microbiol.*, **19**, 1-16.
Cooper, R.D.G., 1992, *Amer. J. Med.*, **92**, Suppl 6A, 2-6.
Costerton, J.W. and Cheng, K.J., 1975, *Bact. Rev.*, **38**, 87-110.
Curtis, N.A.C., Brown, C., Boxall, M. and Boulton, M.G., 1979a, *Antimicrob. Ag. Chemother.*, **15**, 332-6.
Curtis, N.A.C., Orr, D., Ross, G.W. and Boulton, M.G., 1979b, *Antimicrob. Ag. Chemother.*, **16**, 533-9.
Emmerson, A.M., 1988, *J. Antimicrob. Chemother.*, **22**, 101-4.
Escobar, W.A., Tan, A.K. and Fink, A.L., 1991, *Biochemistry*, **30**, 10783-7.
Farrell, O.A., Allegretto, N.J. and Hitchcock, M.F., 1987, *Arch. Biochem. Biophys.*, **256**, 253-9.
Frere, J.M. and Joris, B., 1985, *CRC Crit. Rev. Microbiol.*, **11**, 299-396.
Gale, E.F., Cundliffe, E., Reynolds, P.E., Richmond, M.H. and Waring, M.J., 1981, Eds. *The Molecular Basis of Antibiotic Action*, 2nd Ed. (London: John Wiley) Ch. 3.
Ghuysen, J.-M., 1991, *Annu. Rev. Microbiol.*, **45**, 37-67.
Gooday, G.W., 1979, *J. Gen. Microbiol.*, **99**, 1-11.
Hackbarth, C.J. and Chambers, H.F., 1989, *Antimicrob. Ag. Chemother.*, **33**, 991-4.
Hall, G.S., Myles, C., Pratt, K.J. and Washington, J.A., 1988, *Antimicrob. Ag. Chemother.*, **32**, 1331-5.
Hancock, R.E.W. and Bellido, F., 1992, *J. Antimicrob. Chemother.*, **29**, Suppl. A, 1-6.
Handwerger, S., Pucci, M.J., Volk, K.J., Liu, J. and Lee, M.S., 1992, *J. Bacteriol.*, **174**, 5982-4.
Hayes, M.V. and Orr, D.C., 1983, *J. Antimicrob. Chemother.*, **12**, 119-26.
Jaurin, B. and Grundstroem, T., 1981, *Revs. Infect. Dis.*, **8**, Suppl. 3, S237-59.
Kikuchi, S., Rainwater, D.L. and Kolattukudy, P.E., 1992, *Arch. Biochem. Biophys.*, **295**, 318-26.
Kramer, W., Girbig, F., Gutjahr, U., Kowalewski, S., Adam, F. and Schiebler, W., 1992, *Eur. J. Biochem.*, **204**, 923-30.
Kucers, A., 1984, *J. Antimicrob. Chemother.*, **14**, 564-7.
Labia, H., Barthelemy, M. and Peduzzi, J., 1985, *Drugs Exptl. Clin. Res.*, **11**, 765-70.
Malouin, F. and Bryan, L.E., 1986, *Antimicrob. Ag. Chemother.*, **30**, 1-6.
Mosher, G.L., McBee, J. and Shaw, D.B., 1992, *Pharm. Res.*, **9**, 687-90.
Neu, H.C., 1986, *Revs. Infect. Dis.*, **8**, Suppl. 3, S237-59.
Perkins, H.R., 1969, *Biochem. J.*, **111**, 195-205.
Quemard, A., Lacave, C. and Laneelle, G., 1991, *Antimicrob. Ag. Chemother.*, **35**, 1035-9.
Qureshi, N., Takayama, K., Jordi, H.C. and Schnoes, H.K., 1978, *J. Biol. Chem.*, **253**, 5411-17.
Samraoui, B., Sutton, B.J., Todd, R.J., Artymiuk, P.J., Waley, S.G. and Phillips, D.C., 1986, *Nature*, **320**, 378-80.
Sheldrick, G.M., Jones, P.G., Kennard, O., Williams, D.H. and Smith, G.A., 1978, *Nature*, **271**, 223-5.
Shoeb, H.A., Bowman, B.U., Ottolenghi, A.C. and Merola, A.J., 1985, *Antimicrob. Ag. Chemother.*, **27**, 399-403.
Spratt, B.G. and Pardee, A.B., 1975, *Nature*, **254**, 516-17.
Spratt, B.G., 1980, *Phil. Trans. R. Soc. Lond. B.*, **289**, 273-83.
Spratt, B.G., 1983, *J. Gen. Microbiol.*, **129**, 1247-60.
Strynadka, N.C.J., Adachi, H., Jensen, S.E., Johns, K., Sielecki, A., Betzel, C., Sutoh, K. and James, M.N.G., 1992, *Nature*, **359**, 700-705.
Sykes, R.B. and Matthew, M., 1976, *J. Antimicrob. Chemother.*, **2**, 115-57.

Tang, J. and Parr, T.R., 1991, *Antimicrob. Ag. Chemother.,* **35**, 99-103.
Tipper, D.J. and Strominger, J.L., 1965, *Proc. Natl. Acad. Sci. (U.S.A.),* **54**, 1133-41.
Watanakunakorn, C., 1984, *J. Antimicrob. Chemother.,* **14**, Suppl. D, 7-18.
Waxman, D.J., Yu, W. and Strominger, J.L., 1980, *J. Biol. Chem.,* **255**, 11577-87.
Williams, P.E.O. and Harding, S., 1984, *J. Antimicrob. Chemother.,* **13**, 191-6.
Williamson, M.P. and Williams, D.G., 1984, *Eur. J. Biochem.,* **138**, 345-8.
Zambias, R.A., Hammond, M.L., Heck, J.V., Bartizal, K., Trainor, C., Abruzzo, G., Schmatz, D.M. and Nollstadt, K.M., 1992, *J. Med. Chem.,* **35**, 2843-55.
Zhang, Y., Heym, B., Allen, B., Young, D. and Cole, S., 1992, *Nature,* **358**, 591-3.

Chapter 6

Steroid biosynthesis and action

6.1 Introduction

The major sterols to be discussed in this chapter are ergosterol and cholesterol, which are essential constituents of cell membranes of yeasts and fungi, and of mammals respectively. Both sterols are synthesized in a similar fashion start-

Table 6.1 Drugs discussed in Chapter 6

Section	*Target*	*Drug*	*Use*
6.2 Sterol biosynthesis			
6.2.1	β-Hydroxy-β-methyl glutaryl		Hypercholesterolaemia
	CoA reductase	Lovastatin	Hypercholesterolaemia
6.2.2	Squalene epoxidase	Naftifine	Dermatophyte infections
		Terbinafine	Dermatophyte infections
		Tolnaftate	Dermatophyte infections
6.2.3	Lanosterol 14α-demethylase	Ketoconazole	Fungal infection
		Itraconazole	
		Fluconazole	
6.3 Steroid biosynthesis			
6.3.1	Steroid 17,20-lyase	Ketoconazole	Hormone-dependent prostate tumours
6.3.2	Steroid 11β-hydroxylase	Metyrapone	Hypercortisolaemia
6.3.3	Aromatase	Aminoglutethimide	Hormone-dependent breast cancer
6.3.4	5α-reductase	Finasteride	Prostate hyperplasia
6.4 Steroid receptor ligands			
6.4.1	Oestrogen receptor agonists	Ethinyloestradiol	Oral contraceptive
		Mestranol	Oral contraceptive
	Progesterone receptor agonists	Norethisterone	Oral contraceptive
		Norgestrel	Oral contraceptive
6.4.2	Oestrogen receptor antagonist	Tamoxifen	Hormone-dependent breast cancer
	Oestrogen receptor antagonist	Clomiphene	Fertility stimulant
6.4.3	Progesterone receptor antagonist	Mifepristone	Pregnancy terminator
6.4.4	Androgen antagonist	Flutamide	Prostate cancer
		Nilutamide	
6.4.5	Aldosterone receptor antagonist	Spironolactone	Diuretic

ing from acetyl-coenzyme A (Fig. 6.1). The synthesis takes place in different parts of the cell: acetate to mevalonate is carried out in the microsomal fraction while mevalonate to squalene is a cytoplasmic function; squalene conversion to ergosterol and cholesterol is performed in the microsomes and this is the first part of the pathway to require oxygen. The pathways to ergosterol and cholesterol are common up to lanosterol. Furthermore, the removal of the methyl group attached to the carbon atom at position 14 of lanosterol is also common, although in yeasts and fungi it does not necessarily occur next in sequence (Fig. 6.3). Not all the enzymes have been purified, but it appears that, although yeast and mammalian enzymes catalyse the same interconversions, they are not necessarily identical, and thus lend themselves to selective inhibition.

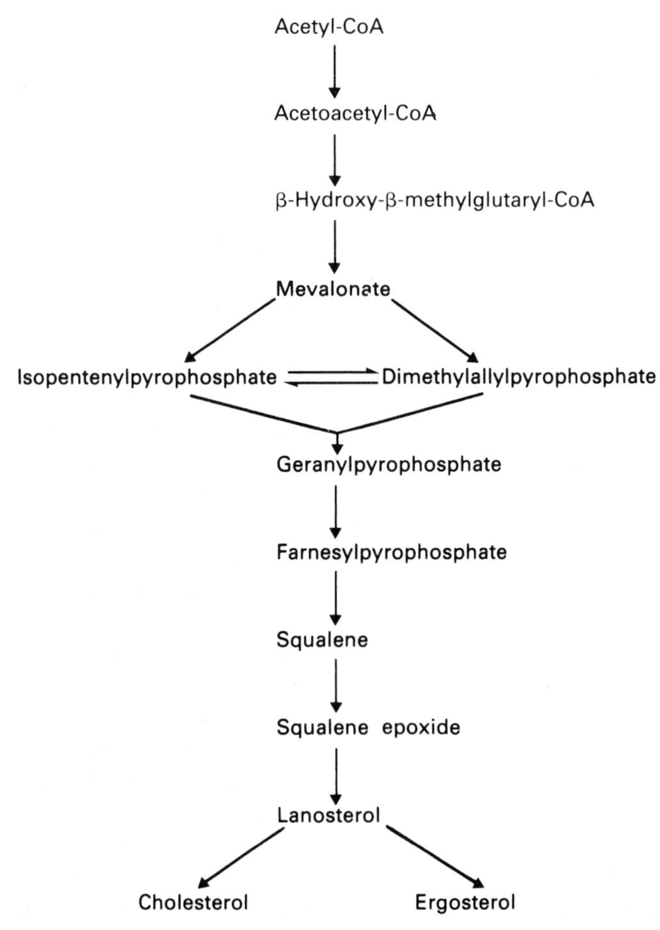

Figure 6.1 Sterol biosynthetic pathway.

Cholesterol can undergo a variety of reactions. Part of the side-chain is removed as the first step in the formation of the steroid hormones (Fig. 6.2). This leads to (a) *glucocorticoids*, such as cortisol which, amongst other actions, increase gluconeogenesis and stimulate lipolysis, (b) *mineralocorticoids*, such as aldosterone that cause sodium to be reabsorbed in the kidneys thus ensuring that sufficient water is retained to maintain the body's osmotic balance and (c) the final removal of the side-chain yields the *sex steroids*, such as testosterone in the male and oestradiol in the female. Although the throughput down these pathways is fairly small, the effects of the hormones produced is very marked.

The greatest proportion of cholesterol is converted into bile acids which are released into the gut to emulsify triglycerides so that they can be transported into the circulation. Cholesterol can also be esterified at the 3-hydroxy position with long-chain fatty acids donated from either triacyl glycerols such as lecithin (catalysed by the plasma enzyme lecithin cholesterol acyl-transferase: LCAT) or their respective coenzyme-A esters (tissue-bound acyl-CoA acyltransferase: ACAT). The esters so formed act as a storage depot for cholesterol, and also in the former case to keep cholesterol within the circulating lipoprotein and prevent it from transferring into the tissues.

The cholesterol nucleus cannot be broken down directly to carbon dioxide and water, and so if cholesterol levels rise there is no simple metabolic pathway that can remove it. Elevated cholesterol, when attached to very low density lipoproteins, gives rise to one of the major diseases facing the Western world today (section 6.2.1).

Ergosterol, strangely enough, is only known to undergo (a) esterification and (b) conversion to ergocalciferol (vitamin D_2) under the influence of ultraviolet light. At present there are no other metabolic transformations described, although by analogy with cholesterol we might expect some conversions, perhaps to a yeast hormone. On the other hand, like cholesterol, the ergosterol nucleus cannot be broken down directly to carbon dioxide and water.

The sterol biosynthetic pathway has not, until recently, attracted much attention from the pharmaceutical point of view, except perhaps from those seeking to lower cholesterol levels. Much effort went into this direction in the 1960s, after the important connection between cholesterol and coronary heart disease was found. There was considerable success in reducing the flow of metabolites through this pathway, but the agents led instead to the formation of toxic intermediates and were of no value as drugs. Recently, however, the discovery of non-toxic agents that interfere with the rate-limiting step in the early part of the pathway, β-hydroxy-β-methylglutaryl-CoA reduction to mevalonate, has greatly improved the prospects for the treatment of hypercholesterolaemia and, hopefully, coronary heart disease.

More recently, the part of the pathway leading from lanosterol to ergosterol has been spotlighted as the target of various anti-fungal agents, some useful as agricultural fungicides and others in human medicine. This interest is likely to turn what was previously rather a backwater, inhabited only by organic

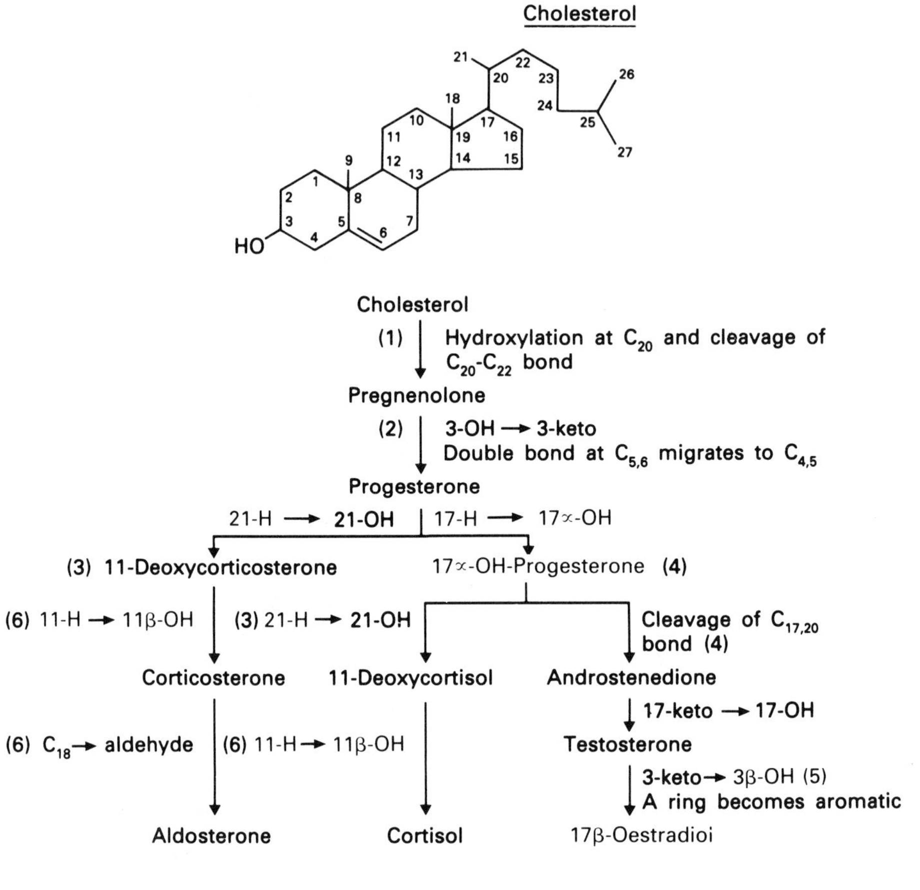

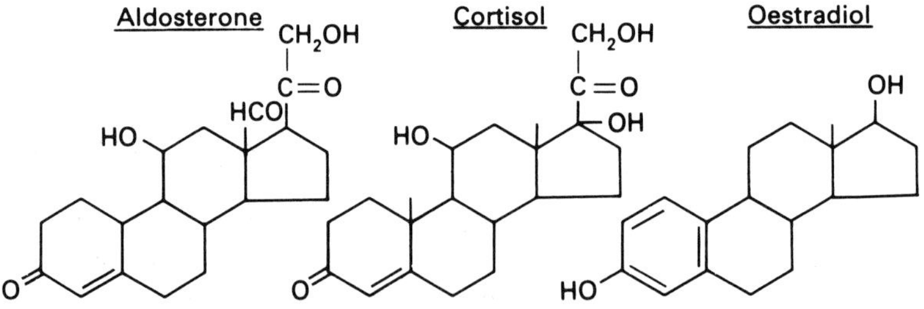

(1) Cholesterol 20,22-lyase, mitochondrial.
(2) 3-β-Hydroxysteroid dehydrogenase, mitochondrial.
(3) 21-Steroid hydroxylase, microsomal.
(4) 17-α-Steroid hydroxylase and 17,20-lyase, microsomal.
(5) Aromatase, microsomal.
(6) 11β-Steroid hydroxylase, mitochondrial.
All the enzymes except 3 are cytochrome *P*-450s, similar to the liver mixed-function oxidases, and thus require NADPH and molecular oxygen to function. Unlike the latter, however, they are selective for their particular substrates.

Figure 6.2 Main pathways of steroid hormone biosynthesis.

Lanosterol

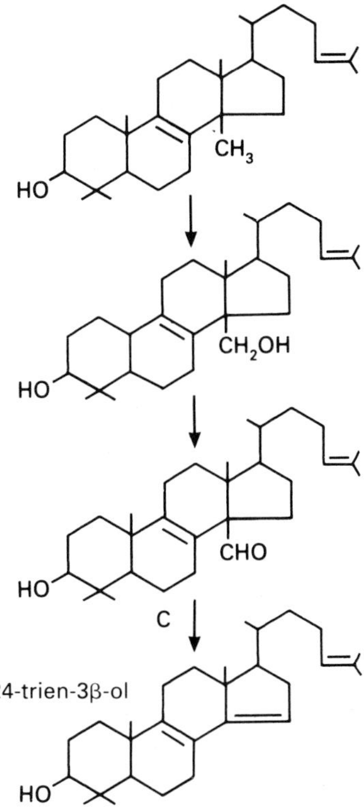

CH_3

HO

CH_2OH

HO

CHO

HO

C

4,4-Dimethylcholesta-8,14,24-trien-3β-ol

HO

The 14α-methyl group of lanosterol is sequentially hydroxylated in the presence of a cytochrome *P*-450-linked enzyme, the flavoprotein cytochrome *P*-450 reductase, NADPH and molecular oxygen. The final oxidation yields formic acid, together with a sterol double-bonded at position 14–15.

Figure 6.3 Lanosterol 14α-demethylation.

chemists interested in the mechanism of squalene epoxide cyclization, for example, into one of the most fruitful areas in biochemistry.

The conversion of cholesterol into various steroid hormones has been of major pharmaceutical importance for many years because of the development of oral contraceptives. Although the use of oestrogens and progestins to prevent conception is not to combat disease, it is nonetheless one of the major uses of drugs. Possibly because the use of drugs has greatly increased the life expectancy of the population of a large part of the world, the use of drugs to restrain population growth may be regarded as a moral imperative for the industry. Alternatively, it may be argued that the industry is merely responding to a need, in that women increasingly desire to regulate their reproductive function. Whichever view is taken, there is still considerable effort being expended to find other means of achieving the same end, particularly in view of the uncertainty as to whether the pill can cause breast cancer (Anon, 1986).

Another pharmaceutical outlet for inhibitors of steroid biosynthesis, which are often hydroxylations carried out by cytochrome *P*-450 dedicated to a particular reaction, has been to lower the levels of sex hormones, glucocorticoids etc., in conditions where these pathways are hyperactive, and in situations where tumours are particularly dependent on sex steroids. Antagonists of mineralocorticoid action are also useful as diuretics.

6.2 Sterol biosynthesis

6.2.1 β-Hydroxy-β-methylglutaryl-CoA reductase - lovastatin as hypocholesterolaemic

The lowering of plasma cholesterol levels, whether by inhibition of biosynthesis or reduction in gastrointestinal absorption, has been a major pharmaceutical target for many years, since the establishment of a strong causative connection between plasma lipid, the deposition of fat and other material in the arteries, or atherosclerosis and coronary heart disease. In outline, atherosclerosis results in the narrowing of the arteries in the following way. Cholesterol is transported via the plasma in three types of lipoprotein complexes characterized by their densities: very low density (VLDL) which has the highest lipid content, low density (LDL) which is intermediate and high density (HDL) with the least lipid. These complexes are of high molecular weight, with a core of cholesteryl esters, triglycerides and free cholesterol surrounded by phospholipid and lipoprotein and they provide the transport mechanism for hydrophobic lipids in an aqueous environment.

At the surface of the cells in the lining of the arteries (e.g. the aortic intimal cells) are receptors for the protein component of LDL, which is the lipoprotein whose levels correlate most closely with the development of atherosclerosis. LDL binds and the complex is taken up by the cell. There, any remaining free cholesterol is converted to esters by acyl-CoA cholesterol acyltransferase (ACAT). These esters can be laid down intracellularly to produce what is known as a 'fatty streak', and at this stage the deposition can be reversed. Subsequently, lipids begin to accumulate extracellularly, particularly at points where previous injury to the lining has occurred. Fibrin and collagen deposit around the growing plaque. In addition, the smooth muscle cells proliferate at that point to form at least 50 per cent of the advanced plaque, and the artery begins to be blocked. At this point, the lesion is no longer reversible. High plasma cholesterol levels, particularly associated with LDL, are correlated with plaque formation, and with the development of coronary heart disease (Wissler, 1991).

The rate-limiting step in the earlier part of the pathway is β-hydroxy-β-methylglutaryl-CoA reductase, which is subject to feedback inhibition by dietary cholesterol. A powerful inhibitor of this enzyme, isolated from broths of the fungus *Aspergillus terreus*, is lovastatin (originally known as mevinolin -

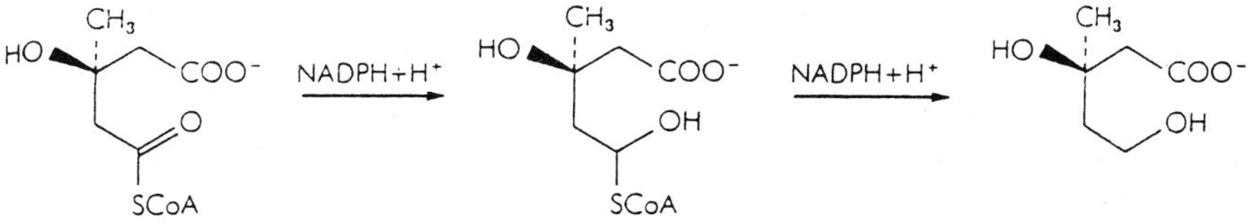

HMG-CoA NADPH+H⁺ → Half-reduced Intermediate NADPH+H⁺ → Mevalonate

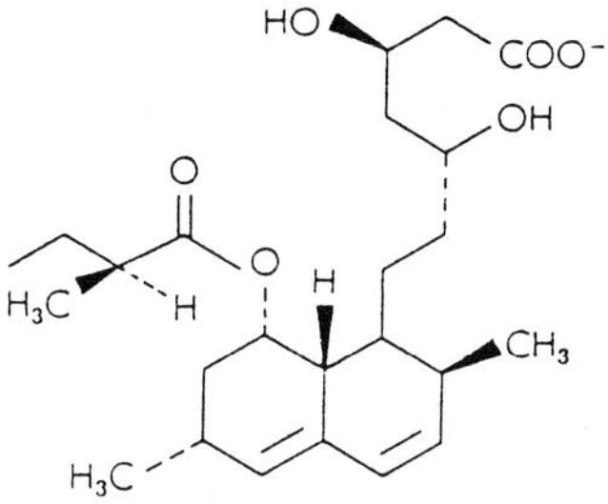

Active form of lovastatin

HMG-CoA is reduced in two stages by two molecules of reduced nicotinamide adenine dinucleotide phosphate (NADPH) to form mevalonate. The half-reduced intermediate is a structural analogue of the active form of lovastatin.

Figure 6.4

Alberts *et al.*, 1980). The drug is isolated as the lactone form but the active principle is the open chain hydroxy acid. Lovastatin inhibits the enzyme competitively with a K_i of 6×10^{-10} M; the upper part of the drug in Fig. 6.4 closely resembles the half-reduced reaction intermediate (Grundy, 1988).

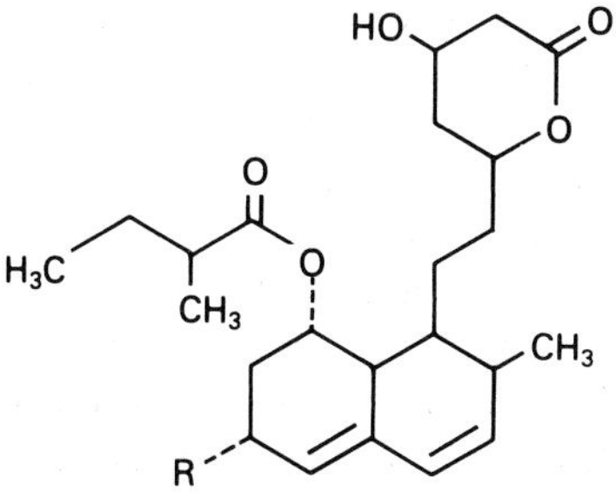

Lovastatin

Lovastatin does not work, however, by sharply lowering the rate of cholesterol synthesis, as plasma cholesterol levels are only slightly reduced in man. It is the intracellular levels of cholesterol that are important. Cholesterol suppresses the synthesis of hydroxymethylglutaryl-CoA (HMG-CoA) synthase,

HMG-CoA reductase and the LDL receptor by binding to a DNA sequence in the promoter of the genes for all three proteins (Smith *et al.*, 1988). When lovastatin reduces cholesterol levels in the cell, the synthesis of all three proteins increases. Although raised HMG-CoA synthase and reductase can partially offset lovastatin inhibition of cholesterol synthesis, the net effect is still to increase the concentration of LDL receptors, thereby stimulating the removal of LDL from the circulation. In people who have a genetic deficiency of LDL receptors, this mechanism cannot work and there is no change in LDL levels or rates of production or breakdown (Uauy *et al.*, 1988).

Clearly, it is necessary to show that lowering LDL levels does in fact reduce heart attacks, and a number of trials have been carried out with cholesterol-lowering drugs. There is considerable controversy about the results of these trials. There appears to be no doubt that the drugs reduce the frequency of what is euphemistically called 'coronary events' (i.e. heart attacks). There does not, however, appear to be an overall reduction in mortality from all causes (Ravnskov, 1992; Smith and Pekkanen, 1992). Even the largest trial to date with lovastatin (EXCEL) seems to be giving equivocal results, although it is perhaps too early to tell (Bradford *et al.*, 1991). Improving the diet may be at least as effective as drug treatment (Smith and Pekkanen, 1992).

6.2.2 Squalene epoxidase - naftifine as anti-fungal

The intermediates immediately before squalene in the sterol biosynthetic pathway are rendered soluble by the presence of pyrophosphate groups. Squalene is the first substrate that is insoluble in aqueous media - indeed it must be one of the most, if not the most, insoluble compounds in the cell. It appears that a sterol carrier protein has evolved to handle this situation. In addition, squalene epoxidase is of considerable interest because it is responsible for the first oxidative step in the pathway.

The enzyme is present in the microsomes of eukaryotic cells, and requires the presence of cytoplasmic fraction to show activity, presumably because the latter contains the sterol carrier protein. The cofactors are NADPH or NADH and FAD, but the enzyme is not a cytochrome P-450. Although the enzyme catalyses the placement of an oxygen atom across positions 2 and 3 of the squalene ring to form an epoxide, it may also be able to put a second one across positions 22 and 23, because conditions that induce an accumulation of the mono-epoxide, such as inhibition of the next step in the pathway (squalene epoxide cyclase) by chloroquine (Chen and Leonhardt, 1984) yield a proportion of the di-epoxide as well. The enzyme has been purified from rat liver (Ono *et al.*, 1982) but not so far from any fungal or yeast source, although the *Candida albicans* enzyme has requirements for activity similar to the rat liver enzyme (Ryder and Dupont, 1984).

Recently a new class of antifungal agents, the allylamines, has been found to exert its antifungal effect by virtue of powerful inhibition at this step (reviewed by Balfour and Faulds, 1992). Naftifine was the first member of this series to be synthesized but a more recent one, terbinafine, is more potent at inhibiting both enzyme activity and fungal growth. Terbinafine shows non-competitive inhibition of the enzyme from *Candida albicans* with a K_i of 3×10^{-8} M (Petranyi *et al.*, 1984). The enzyme from rat liver is some 2000-times less sensitive to the drug, and this selectivity appears to be reflected *in vivo* (Petranyi *et al.*, 1984). The inhibition is apparently reversible, despite the system of conjugated double and triple bonds which might be expected to interact covalently with the enzyme.

Squalene epoxidase inhibition will lead to rising levels of squalene as the substrate and falling levels of ergosterol, the final end-product (Fig. 6.1). Against *Trichophyton* species, that cause skin infections such as athlete's foot, inhibition of fungal growth parallels rising levels of squalene. Terbinafine is fungicidal for these fungi. Fungistatic action against *Candida* species apparently correlates with falling levels of ergosterol (Balfour and Faulds, 1992).

Candida albicans is a dimorphic fungus, i.e. it can exist either as a yeast or as a mycelial form; the latter is believed to be the pathogenic form as it can force its way through tissues. The change from one form to the other, known as the dimorphic switch, has been the subject of much research. Terbinafine is almost inactive against the yeast form while very active against the mycelia, and thus blocks the switch. This could also explain fungistatic activity against *C. albicans* as the fungus could revert to yeast form in the presence of the drug and resume growth in its absence.

Terbinafine is orally and topically active in humans against a wide variety of fungal skin infections and may become a first-line therapy in the future (Balfour and Faulds, 1992).

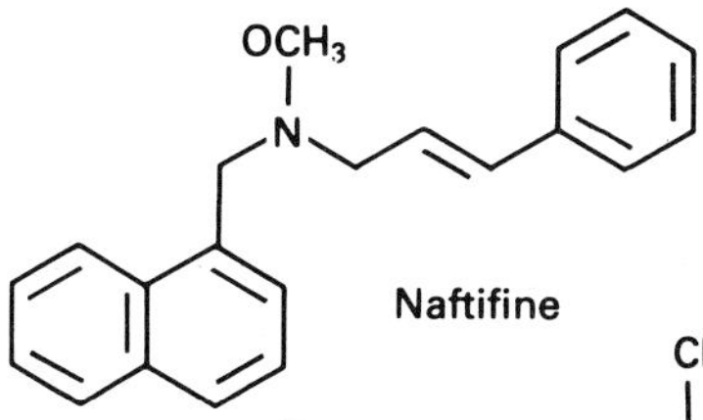

Naftifine

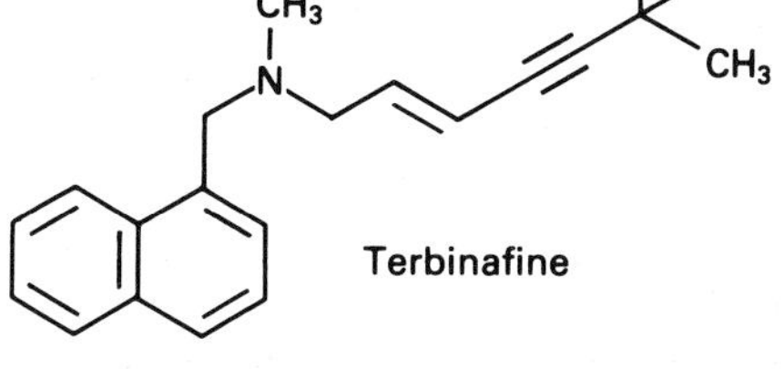

Terbinafine

Another family of compounds which have been in use for some time against topical infections is exemplified by tolnaftate. Recent studies have shown that it also inhibits squalene epoxidation. Although active against the *Candida* enzyme, the drug apparently cannot reach its site of action in these yeasts (Barrett-Bee *et al.*, 1986). At concentrations that inhibit ergosterol biosynthesis, both families of drugs appear to sensitize the *C. albicans* membrane to further damage (Georgopapadakou and Bertasso, 1992).

6.2.3 Lanosterol demethylation - ketoconazole as anti-fungal

One of the crucial rate-limiting steps in the later part of the pathway of cholesterol biosynthesis is the removal of the C_{32} methyl group from the 14α position of lanosterol with a concomitant introduction of a double bond at the 14–15 position (Fig. 6.3). These transformations appear to be carried out by a single microsomal enzyme that is cytochrome *P*-450 dependent, and requires NADPH, molecular oxygen and cytochrome *P*-450 reductase, a flavoprotein, for activity (reviewed by Coulson *et al.*, 1984). The methyl group is sequentially hydroxylated to yield —CH_2OH and —CHO, and then removed as formic acid as the double bond is inserted at the 14–15 position. Carbon monoxide which binds to the haem iron of cytochrome *P*-450 and interferes with a number of *P*-450 linked reactions, inhibits the first step but not the latter two (Gibbons *et al.*, 1979), but that may merely reflect tighter binding of the oxidized intermediates to the enzyme, i.e. carbon monoxide may be able to displace lanosterol, but not the oxidized intermediates, at the binding site. More recent work has confirmed that the same cytochrome *P*-450 will catalyze all three steps in the process (Aoyama *et al.*, 1987).

In fungi, the demethylation of lanosterol is not rate-limiting as it is in liver, but it is nevertheless essential for the formation of ergosterol and other sterols that are acceptable materials for the fungal cell membrane. Lanosterol and its analogues which retain the methyl group do not fit properly into the membrane, causing it to become permeable to protons and eventually burst (Nes *et al.*, 1978; Thomas *et al.*, 1983). The 14α-methyl group of lanosterol is axial, protruding from the otherwise planar face. Van der Waals interactions of the sterol underside with the fatty acyl groups in the phospholipid bilayer of the membrane are therefore very much less favourable (Bloch, 1979).

A number of medical and agrochemical anti-fungal agents in use today owe their activity to potent inhibition at this step. The majority are substituted nitrogen heterocycles. The presence of a lone pair of electrons which can be donated from the nitrogen to the iron of the haem appears to be mandatory. The binding of ligand to enzyme gives rise to a shift in the ultraviolet spectrum whereby the absorbance of the haem is shifted from 410 nm to 420–430 nm (Fig. 6.5; Van den Bossche *et al.*, 1984). This results in what is known as a Type II difference spectrum characterized by a maximum in the region of 420–430 nm and a minimum around 390–400 nm. Substrates of cytochrome *P*-450-linked reactions such as lanosterol give a totally reversed type of difference spectrum, called Type 1, where a peak is found at 380–390 nm and a trough at 410–420 nm.

Inhibitors of lanosterol demethylation in medical use include substituted imidazoles, such as ketoconazole, and the newer triazoles, fluconazole and itraconazole. The inhibitory effect of the 'azoles' is usually measured by the

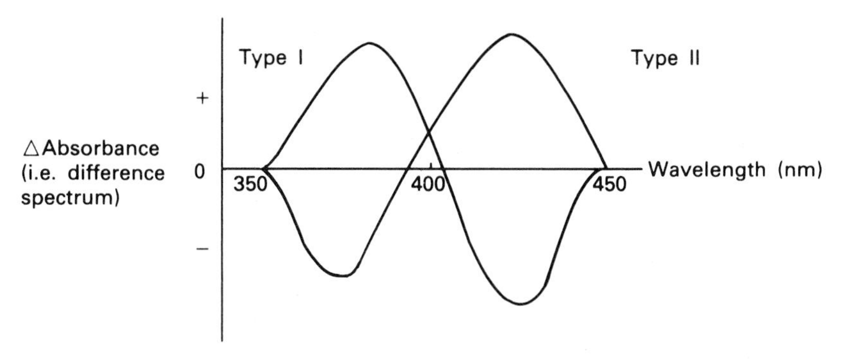

A difference spectrum is used because the changes in ultraviolet spectrum during interaction of cytochrome *P*-450 with a ligand are often too small to be detected by reading the direct spectrum. The spectrum of cytochrome and ligand uncomplexed is subtracted from the spectrum of the complex. This can be done electronically by subtracting spectra stored in memory or, experimentally, by the use of split compartment cells.
A type I difference spectrum is characteristic of a substrate binding to a cytochrome *P*-450 whereas type II is given by a poorer substrate or an inhibitor. The absorbance maximum of the cytochrome is shifted to longer wavelength from about 410 nm by such inhibitors to give the type II spectrum, while a substrate causes the wavelength maximum to move to shorter wavelength. Ketoconazole gives a type II difference spectra.
This subject is also discussed in Chapter 7, section 7.5.

Figure 6.5 Cytochrome *P*-450 difference spectra (diagrammatic representation).

Fluconazole

inhibition of incorporation of ^{14}C-acetate or -mevalonate into ergosterol, which is more convenient to use than a specific assay for lanosterol demethylation and is more flexible in that inhibition of various steps in the pathway may be identified. It is quite clear in this case from the pattern of inhibition obtained, namely a build-up of lanosterol and a decrease in ergosterol, where the inhibited step lies. Ketoconazole and fluconazole give I_{50} values in the region of 10^{-8} M to 10^{-9} M in this system (Grant and Clissold, 1990).

Ketoconazole has two asymmetric carbon atoms and thus has four stereoisomers, all of which show some activity against lanosterol demethylase. The most active is the 2*S*,4*R* isomer (Rotstein *et al.*, 1992). There is no information available about the stereoisomers of itraconazole, and fluconazole is not optically active as the central carbon atom has symmetrical substituents.

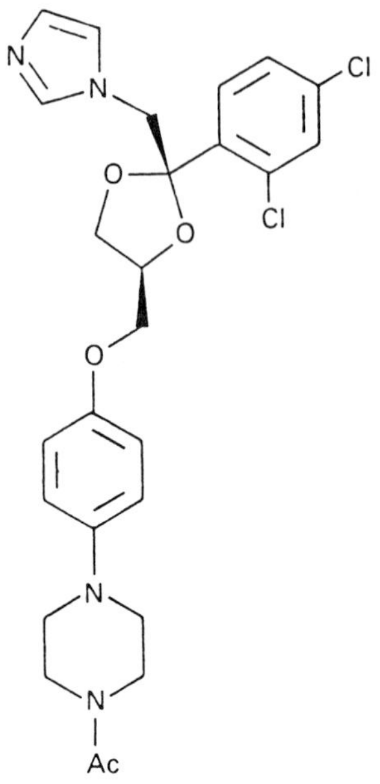

Ketoconazole

The main action of this class of drug is to disrupt the supply of ergosterol for membrane synthesis. This effect is likely to be fungistatic not fungicidal because the inhibitor will be effective only when the organism is growing and dividing.

The advent of the azoles has greatly improved the treatment of a variety of fungal infections. They are orally active and can be used for infections that are carried throughout the body (systemic) and dermatophyte infections that are confined to the surface of the skin (reviewed in Grant and Clissold, 1990; Hay, 1991).

Ketoconazole is relatively specific in inhibiting the fungal lanosterol demethylase as opposed to the same enzyme from rat liver (Van den Bossche *et al.*, 1980, 1984). Nevertheless, a number of other cytochrome *P*-450 conversions, particularly those in the interconversion of steroid hormones, are sensitive to inhibition by the drug. Two of the most obvious of these are (a) the rate-limiting step in the production of glucocorticoids and sex steroids, i.e. cholesterol 22,24 side-chain cleavage (Loose *et al.*, 1983) and (b) 17,20-lyase (Sikka *et al.*, 1985; reaction 4 in Fig. 6.2). These effects are probably responsible for the occurrence of undesirable side-effects such as the growing of breast tissue in men (gynaecomastia), probably through lowered testosterone levels (Pont *et al.*, 1982). Other problems include teratogenicity and some incidence of serious, indeed fatal, liver toxicity (Lewis *et al.*, 1984).

These inhibitions of cytochrome *P*-450 linked enzymes may be put to good effect, however, in conditions in which there is overproduction of ACTH leading to overstimulation of the adrenals (Cushing's syndrome); ketoconazole has been found effective in this instance (Angeli and Frairia, 1985). Furthermore, some tumours are dependent on sex steroids and respond to the drug (Amery *et al.*, 1986). In this case, the observation of side-effects has led to the development of the drug for another purpose.

6.3 Steroid biosynthesis

6.3.1 Steroid 17,20 lyase - ketoconazole for steroid-dependent tumours

The conversion of progesterone into 17α-hydroxyprogesterone followed by the removal of the acetyl side-chain to yield androstenedione (reaction 4 in Fig. 6.2) is believed to be carried out by one enzyme - but there is still a considerable lack of knowledge about some of the finer details of steroid interconversions (see Takemori and Kominami, 1984, for review). The enzyme is sited in the endoplasmic reticulum of adrenals and testes. It is a cytochrome *P*-450 and, therefore, requires molecular oxygen and NADPH for activity. The molecular weight of the purified enzyme is 52 000.

The inhibition is presumably effected by the donation of the imidazole nitrogen lone pair of electrons to the haem iron (as in lanosterol 14α-demethylase, Section 6.2.3). Sikka *et al.* (1985) find an I_{50} of 1×10^{-6} M for ketoconazole inhibition of the 17,20-lyase reaction. Ketoconazole appears to be relatively unselective in its inhibition of dedicated sterol and steroid *P*-450s. Other drugs are being developed to be selective for this enzyme system.

As mentioned in the previous section, the anti-fungal drug ketoconazole produces an occasional side-effect in a few men taking the drug in that it causes them to grow breast tissue (gynaecomastia). Testosterone levels are reduced reversibly by the drug in a dose-related fashion, while at the same time progesterone and 17α-hydroxyprogesterone accumulate. Since androstenedione is reduced to form testosterone, these results can be explained by an inhibition of the steroid 17,20-lyase (reviewed in Amery *et al.*, 1986). It appears that the gynaecomastia occurs in men whose 17β-oestradiol levels are high.

This unfortunate side-effect was, however, noted and put to good use in a male condition where the reduction of androgen levels is the goal of therapy, namely prostate cancer. The hormone stimulates the growth of the tumour. Any reduction in androgen level, and possibly antagonism by increasing levels of progesterone, is an advance on one of the previously used treatments - castration!

6.3.2 11β-Steroid hydroxylase - metyrapone for hypercortisolism

This enzyme catalyses the conversion of deoxycorticosterone to corticosterone, and deoxycortisol to cortisol (reaction 6 in Fig. 6.2). Like cholesterol

20,22-lyase, the enzyme is found in adrenal mitochondria and requires a powerful reducing agent containing iron–sulphur clusters like ferredoxin (Chapter 4), known as adrenodoxin, an enzyme to reduce it called adrenodoxin reductase, NADPH and molecular oxygen to form an electron transport chain. Metyrapone inhibits the enzyme competitively with a K_i of 1×10^{-7} M, and shows a type II ultraviolet difference spectrum with adrenal mitochondria (Williamson and O'Donnell, 1969). 11-β-Deoxycorticosterone shows a type I difference spectrum characteristic of a substrate. Both substrate and metyrapone stabilize the enzyme. Furthermore, metyrapone induces a shift in the electron paramagnetic resonance spectrum suggesting that the haem is involved in the interaction (Wilson *et al.*, 1969).

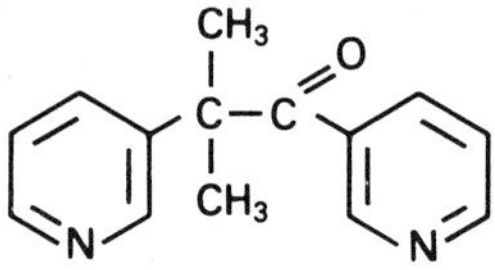

Metyrapone

Metyrapone has been used to treat the hypersecretion of cortisol by adrenal tumours (reviewed by Temple and Liddell, 1970), but on long-term administration the level of ACTH may rise to counteract the effect of the drug. The use of metyrapone alone for this indication is no longer recommended, and aminoglutethimide may be given in conjunction (Gold, 1979).

6.3.3 Aromatase - aminoglutethimide for hormone-dependent cancers

Another major area of interest concerning the synthesis of steroid hormones is the dependence of certain tumours on steroids for the maintenance of growth. Testosterone appears to be essential for the growth of some tumours of the prostate gland, while certain breast cancers are dependent on oestradiol. At least one third of human breast cancers are hormone-dependent, and regress when the sources of oestrogen are eliminated (Kiang *et al.*, 1978). Inhibition of steroid-producing enzymes is one way of achieving this.

Aminoglutethimide is an inhibitor of both cholesterol 20,22-lyase and aromatase - the latter so-called because it introduces an aromatic ring into the steroid nucleus (ring A, reaction 5 in Fig. 6.2) in catalysing the transformation of testosterone and androstenedione to 17-β-oestradiol and oestrone respectively.

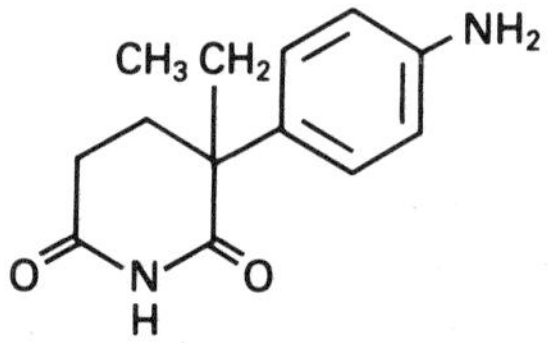

Aminoglutethimide

The mechanism of removal of the C_9 methyl group shows a formal similarity in chemical terms to the demethylation of lanosterol in that the reaction takes place in three stages; first to $—CH_2OH$, next —CHO, followed by total removal of the fragment as formic acid and the introduction of a double bond into the sterol skeleton. Conversion of the 3-keto to hydroxy produces a third double bond to give aromaticity. This has prompted the synthesis of some mechanism-based irreversible (suicide) inhibitors (Chapter 1; Covey *et al.*, 1981).

It is not clear whether aromatase or cholesterol side-chain cleavage is the key target for aminoglutethimide, although it inhibits aromatase at a lower concentration than cholesterol 20,22-lyase (K_i for aromatase 6×10^{-7} M; testosterone concentration $1{\cdot}5 \times 10^{-6}$ M; K_i for 20,22-lyase $4{\cdot}5 \times 10^{-6}$ M, cholesterol concentration as normally found in bovine adrenal mitochondria). Aromatase may be the enzyme that is the more clinically relevant since the fall in oestrogens after dosing is immediate, unlike progesterone levels which initially rise (Foster *et al.*, 1983).

These inhibitory effects have promoted the study of aminoglutethimide as a treatment for metastatic breast cancer, against which it is moderately effective (Smith *et al.*, 1978). Aminoglutethimide shows a number of undesirable side-effects and attempts to improve on the original structure have led to a number of compounds being tested in the clinic. These structures contain imidazole and are competitive inhibitors by virtue of binding to the haem iron. One leading compound is fadrozole (CGS16949A) which has a K_i for aromatase of $1{\cdot}9 \times 10^{-10}$ M (Santen, 1991). Non-steroidal aromatase inhibitors are very unlikely to have oestrogen agonist activity unlike tamoxifen and the other oestrogen antagonists, and thus can increase the options for treatment for breast cancer.

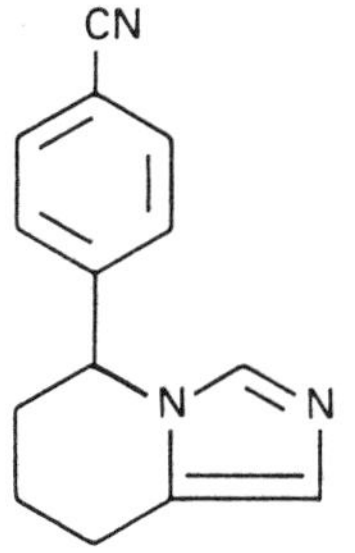

Fadrozole

6.3.4 5α-Reductase

Androgens control the development of male sexual characteristics. Testosterone is the main hormone responsible for these changes but inside the cell it is reduced to dihydrotestosterone which acts as the intracellular mediator. The prostate gland in older men can increase in tissue size and interfere with the flow of urine (benign hyperplasia - the cells are normal not cancerous hence benign). This condition is hormone-dependent. Blockade of the reducing enzyme, 5α-reductase, has been attempted to reduce the size of the prostate.

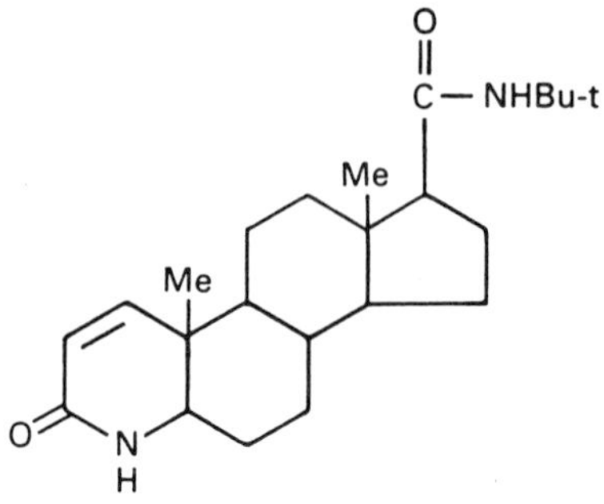

Finasteride

Finasteride is an 4-azasteroid which is a potent competitive inhibitor of the rat isoenzyme 1 with a K_i of 4×10^{-9} M. It inhibits the human isoenzyme 1 much less powerfully with a K_i of 3×10^{-7} M. Cloning and expression of the two enzymes in human embryonic kidney cells has shown that there is a tetrapeptide sequence (Val-Ser-Ile-Val) in the rat enzyme binding site (residues 22–25 at the amino terminus) that is responsible for drug sensitivity. The analogous sequence in man is Ala-Val-Phe-Ala (residues 26–29) which is partially resistant (Thigpen and Russell, 1992). Clearly there will be further attempts to design a more potent human enzyme inhibitor.

Despite the weaker inhibition in man, finasteride reduces the size of the prostate and increases the flow of urine. Levels of dihydrotestosterone in the serum were greatly reduced but testosterone levels were unchanged (testosterone can be metabolized to other products such as oestradiol) (McConnell *et al.*, 1992). There were side-effects such as decreased libido and impotence which could be attributed to hormone interference but testosterone itself seems to be able to substitute for dihydrotestosterone in many other tissues, thus limiting the side-effects experienced (Gormley *et al.*, 1992).

6.4 Steroid receptor ligands

6.4.1 Oestrogen/progesterone receptor agonists - oral contraceptives

Oestrogen and progesterone receptors are present in the nucleus - unlike the G-protein-linked receptors found in the cell membrane (Chapter 9). Corticosteroid receptors may be cytoplasmic with a connection to the nucleus (Chapter 7). Ligands cross the cell membrane by passive diffusion and bind to the steroid receptors; the complex acts as a transcription factor by activating particular sections of the genome to produce specific mRNA. General RNA synthesis occurs later followed by DNA biosynthesis (Murad and Kuret, 1990). Recent cloning studies have greatly increased our understanding of the structure of these receptors but we still do not know entirely how these processes function (reviewed in Gronemeyer, 1991; Bailieu *et al.*, 1990).

There are seven regions in steroid receptors, labelled A to F except the progesterone receptor which does not have F. Regions C and E are highly conserved and contain the DNA and hormone-binding domains respectively. At rest these receptors are bound by a 90 kDa protein called a heat-shock protein, so-called because its synthesis is greatly stimulated when the cell's temperature is raised to about 42°C. The heat-shock protein binds to the DNA binding domain and produces an oligomer with sedimentation coefficient of 8S. Another protein of molecular weight 59 kDa is also found in the 8S receptor–heat-shock protein complex.

When the ligand binds, the oligomer dissociates as a result of a conformational change and the drug-receptor complex is now 4S. There are two molecules of oestrogen receptor but probably only one of progesterone receptor complexed with the heat-shock protein. The heat-shock protein thus acts to repress the steroid signal and probably also to stabilize the receptor protein while not needed (Bailieu *et al.*, 1990).

The oestrogen and progesterone receptors bind to DNA as dimers. In the C region of the receptor the protein folds into two zinc fingers, so-called because each is produced by a zinc coordinated to four cysteine residues. The zinc fingers are essential for DNA binding, while the hormone binding domain and other areas are required to stabilize the binding to DNA. The promoter region of the hormone-inducible gene contains a sequence some 13 or 15 nucleotides long depending on the specific receptor; the sequence is palindromic i.e. the five bases at each end are complementary. The sequence of the oestrogen-responsive element is 5′-GGTCAXXXTGACC, where X can be any base. Transcription also requires the presence of two regions in the receptor known as transcription activation factors, one of which is in the N-terminal region, but the precise site of these areas is not known (Gronemeyer, 1991).

The progesterone receptor exists in two isoforms, named A and B, which result from the initiation of transcription from the same gene but two different sites. The two forms differ by some 130 amino acids in length at the N-terminus, with the A form being the smaller (Gronemeyer, 1991). There is another short eight amino acid stretch with more than four cationic residues in the centre of the receptor which appears to control nuclear siting.

One of the major uses of pharmaceutical products is of a prophylactic nature - the avoidance of the condition of pregnancy. Both oestrogen and progesterone receptor agonists are used for this, usually in conjunction, although progestins can be used alone if the patient is believed likely to suffer side-effects from oestrogen treatment. The rationale for the use of such agents is that they inhibit ovulation, with oestrogen mainly responsible for suppressing follicle-stimulating hormone (FSH) secretion, while the progestin suppresses luteinizing hormone (LH) secretion (Briggs *et al.*, 1970) (FSH and LH are known as gonadotrophins). Even if ovulation and subsequent fertilization did occur, it is very unlikely that the fertilized egg could implant into the lining of the womb. Oestrogen produces a cornified or keratinized layer on the surface lining of the womb to promote adhesion of the egg. Progesterone antagonizes this and, in addition, produces a very viscous mucus layer through which the egg would find it difficult to penetrate.

The oral contraceptives inhibit the production of FSH and LH by binding to receptors in the cytosol of hypothalamus and pituitary (Fig. 6.6). The effect on the former site suppresses the release of gonadotrophin releasing hormone (GnRH) while the effect on the latter is to reduce gonadotrophin release. This is a negative feedback loop that the endogeneous hormone controls, and is a necessary part of pregnancy (Schally, 1978). It is interesting in this context that the oral contraceptives do not trigger the positive feedback at the level of the pituitary whereby oestrogen together with GnRH stimulates the preovulatory surge of LH (Asch *et al.*, 1983 - see Fig. 6.6). The negative feedback closes the cycle since FSH stimulates the granulosa cells of the follicle to produce increasing amounts of oestrogen as the follicle develops. Following ovulation the granulosa cells differentiate into luteal cells which secrete progesterone under the influence of LH in the later part of the cycle (known as the luteal phase) until the corpus luteum begins to degenerate (Murad and Kuret, 1990). Both peptide hormones, LH and FSH, act through adenylate cyclase,

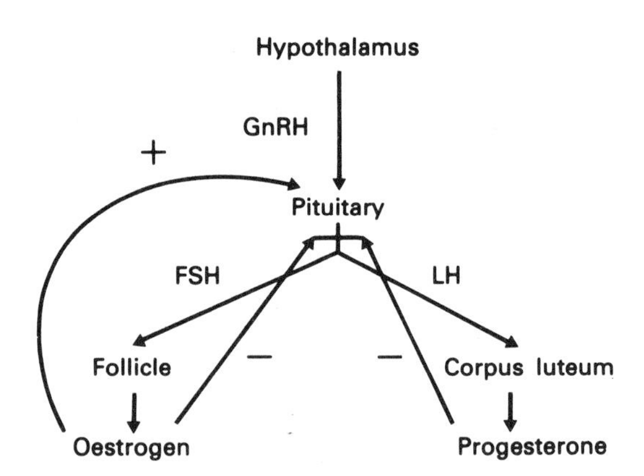

+ indicates a positive feedback
− indicates a negative feedback

The hypothalamus releases GnRH which acts on the pituitary to produce both FSH and LH. FSH stimulates follicles to grow and produce oestrogen which has a negative feedback on LH and FSH release. Eventually, however, an oestrogen level is reached that is sufficient, acting in concert with GnRH, to release LH and FSH in a surge from the pituitary. This converts one follicle to a corpus luteum, synthesizing progesterone that inhibits LH production. Only the negative feedback to the pituitary is shown for the sake of clarity.

Figure 6.6 Hypothalamus - Pituitary - Gonad axis.

after binding to their specific receptors, in a similar fashion to ACTH to stimulate cholesterol 20,22-lyase, cyclic-AMP levels and steroidogenesis (Funkenstein *et al.*, 1983; and references therein).

Two of the most frequently used oestrogens are ethinyloestradiol and its *O*-methyl analogue, mestranol. The former binds to rat anterior pituitary and hypothalamic cytosol with a potency similar to that of oestradiol, with a K_d from 1 to $1{\cdot}3 \times 10^{-10}$ M compared with 2·2 to $3{\cdot}2 \times 10^{-10}$ M for oestradiol. Mestranol, on the other hand, is bound much less tightly, with a K_d of 2·9 to $5{\cdot}5 \times 10^{-8}$ M, as shown by competitive binding with oestradiol. It has been suggested that mestranol is dealkylated to the more active ethinyloestradiol by liver mixed-function oxidases – as occurs in the rat (Eisenfeld, 1974).

R = CH_3, Mestranol
R = H, Ethinyloestradiol

After the drug-receptor complex is formed it is converted to a different species by a conformational change and translocated into the nucleus. Specific mRNA and certain proteins are synthesized initially, followed by a general increase in the synthesis of RNA. Specific DNA biosynthesis is a much later event (Murad and Kuret, 1990). How this process actually interferes with the secretion of GnRH is not known at present.

Norethisterone and norgestrel are typical examples of the most widely used progestins. They bind to receptors in the cytosol of human uterus cells with binding constants of $6{\cdot}8 \times 10^{-9}$ M and $2{\cdot}4 \times 10^{-9}$ M respectively (Kasid *et al.*, 1978; Srivastava, 1978). Both compounds compete with progesterone for these binding sites.

Diethylstilboestrol

R = CH_3, Norethisterone
R = C_2H_5, Norgestrel

If an oral contraceptive is required after intercourse, diethylstilboestrol, a non-steroid oestrogen is often used. Oestrogens have uses other than oral contraceptives including hormone replacement therapy during the menopause (Murad and Kuret, 1990).

6.4.2 Oestrogen antagonists - tamoxifen for oestrogen-dependent cancer, clomiphene to stimulate ovulation

Breast cancers may be divided into two distinct types: oestrogen-receptor-rich and receptor-poor. The receptor-rich group (about 50 per cent of pre-menopausal women and 75 per cent of post-menopausal women) depends on the hormone for growth. Oestrogen antagonism has been found to be very effective in a large proportion, particularly post-menopausal, of this group. Surgery is the treatment of choice for the pre-menopausal group. In the case of the receptor-poor group, hormone therapy alone is unlikely to be of much value, but the use of oestrogen antagonists as part of a cocktail of drugs including cytotoxic agents has recently been developed (Jordan, 1992).

The oestrogen antagonist of choice is tamoxifen, which is a potent ligand of oestrogen receptors, with a K_d of cytosolic receptors from human mammary tumours of $3{\cdot}7 \times 10^{-9}$ M compared with $1{\cdot}2 \times 10^{-10}$ M for 17β-oestradiol by direct measurement (Nicholson *et al.*, 1979). A figure of one to two orders of magnitude less activity is obtained in competitive binding studies, possibly due to a breakdown of the drug-receptor complex. Using a cell line of human breast cancer cells (MCF 7), Horwitz *et al.* (1978) found that the action of tamoxifen was biphasic in that at low extracellular levels (10^{-7} M) the drug showed agonist properties in stimulating growth and inducing progesterone receptors. Higher doses (10^{-6} M) were antagonistic. Interference with oestradiol may take the form of blocking the binding of oestradiol to its receptors, and of reducing the number of receptors available for the natural hormone. Consequently, the natural hormone is unable to maintain tumour growth.

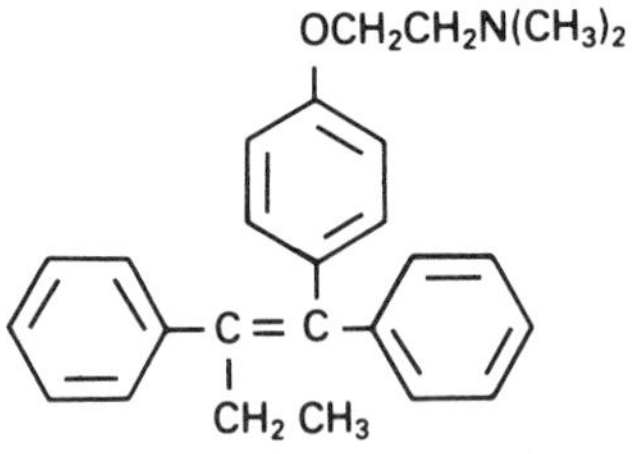

$OCH_2CH_2N(CH_3)_2$ — C=C — CH_2CH_3

Tamoxifen

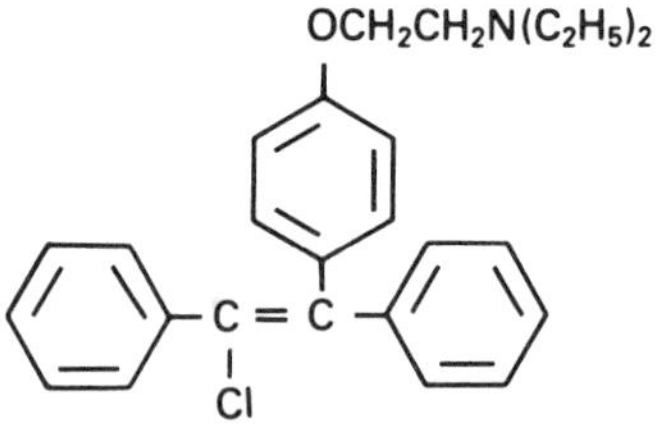

$OCH_2CH_2N(C_2H_5)_2$ — C=C — Cl

Clomiphene

A close analogue of tamoxifen, clomiphene, has found favour as an agent for the induction of fertility in women who either have no menstrual cycle or cycles in which they do not ovulate (Marshall, 1978). The most obvious physiological effect of this drug is the enlargement of the ovaries, and this hyperstimulation does on occasion give rise to multiple births - approximately 6 to 8 per cent compared with 25 per cent of births when gonadotrophins are given to induce fertility. Clomiphene at high dose is effective for the palliation of breast cancer but tamoxifen is preferred for this indication because of lower toxicity (Marshall, 1978).

Clomiphene acts as an anti-oestrogen showing side-effects consistent with this action, including hot flushes. The major pharmacological effect is to prevent the normal feedback inhibition of oestrogen on the release of GnRH from the hypothalamus (Fig. 6.6). Higher GnRH levels then release more luteinizing hormone and follicle-stimulating hormone from the pituitary and ovulation is stimulated.

The binding constant of clomiphene to various rat brain and uterine receptors is in the region of 10^{-8} M compared with 10^{-10} M for 17β-oestradiol under the same conditions. Unlike the studies with tamoxifen, direct binding and competitive experiments gave similar results (Ginsberg *et al.*, 1977). This family of anti-oestrogens show activity as antagonists in the *trans* isomer and as agonists in the *cis*. Tamoxifen is marketed as the *trans* isomer while clomiphene is a racemic mixture.

6.4.3 Progesterone antagonist - mifepristone as abortifacient

In addition to inhibition of progesterone biosynthesis as an approach to contraception and abortion, receptor antagonism has been found to be effective. Mifepristone (RU-486 or RU-38486) has been developed as a ligand of progesterone receptors and is apparently considerably more potent than progesterone itself (Schreiber *et al.*, 1983). Mifepristone also binds to glucocorticoid receptors but at markedly higher concentration. The drug also binds to

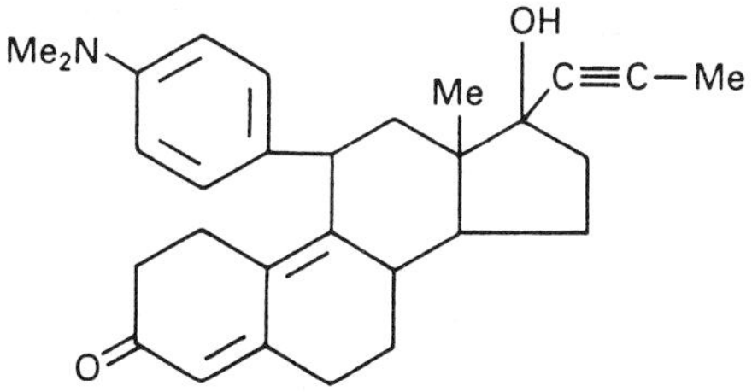

Mifepristone

receptors in the hypothalamus which leads to a fall in GnRH secretion and subsequently to luteolysis through a fall in LH levels. A marked fall in progesterone secretion is the result.

Mifepristone appears to antagonize progesterone in two ways. The drug binds very tightly to the B isoform of the progesterone receptor in the 8S complex and, by slowing the release of the heat-shock protein, mifepristone prevents progesterone binding. In addition, the drug receptor complex will still bind to DNA but fails to activate transcription (Gronemeyer *et al.*, 1992).

Information about the mifepristone binding site has come from a study of chicken progesterone receptors which cannot bind mifepristone because there is a cysteine instead of a glycine at the critical position 575. The region containing the glycine lies in a so-called 11β-pocket of the receptor, i.e. it lies near progesterone's 11β position. The bulky 11β-dimethylaminophenyl group of mifepristone fits very well into this pocket provided position 575 does not carry a blocking side-chain.

Mifepristone can be used either as a contraceptive if taken 72 hours after intercourse, although it may also work effectively after a longer time. In addition, the drug is licensed for use in conjunction with a prostaglandin to achieve abortion in up to nine weeks of pregnancy and is proving a realistic alternative to surgery, particularly in developing countries where surgical abortion carries a much greater risk (Henshaw and Templeton, 1992).

6.4.4 Androgen antagonist

Like the majority of breast cancers, prostate cancer is dependent on hormones, viz. androgens such as testosterone or androsterone, to maintain its growth. As noted earlier for oestrogen antagonism, the antagonism of androgens might be expected to inhibit the growth of prostate cancer.

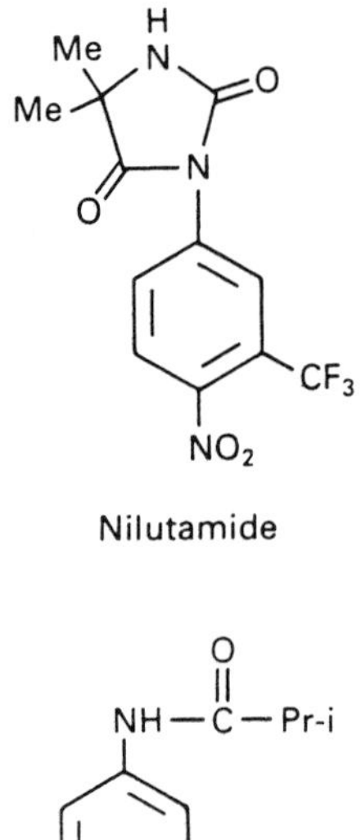

Nilutamide

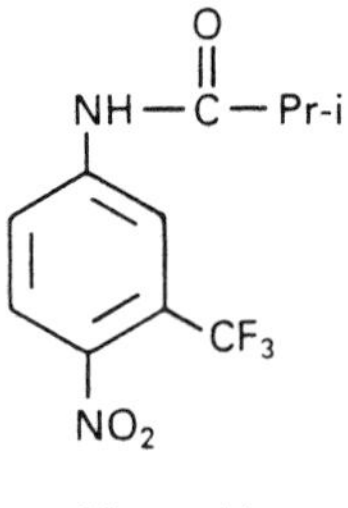

Flutamide

Flutamide and nilutamide have been developed for this indication. The former is a pro-drug for the 2-hydroxylated metabolite which has binding constants in the range of 0·8 to 2·05 × 10^{-8} M to androgen receptors from various genital organs including human prostate cancer (Simard *et al.*, 1986). Flutamide is often used in conjunction with luteinizing hormone releasing hormone agonists such as leuprolide, because flutamide blocks the transient increase of androgen release by leuprolide before down-regulation desensitizes the pituitary (Chapter 12). The efficacy of such combinations over single agent therapy is not yet proven (Brogden and Clissold, 1989).

It is interesting that both drugs contain an aromatic nitro group which in nilutamide may be reduced by NADPH and microsomal fraction, first by one-electron reduction to a radical ion, and then by further reduction to reactive products which become covalently attached to cellular protein (Berger *et al.*, 1992). The nitroimidazole, metronidazole, reacts in a similar fashion (Chapter 4).

6.4.5 Aldosterone antagonist - spironolactone as diuretic

Aldosterone is the steroid (known as a mineralocorticoid because it helps to control the levels of minerals in the blood plasma) which controls reabsorption of sodium ion in the distal portion of the kidney tubule (Fig. 6.7). Up to 90

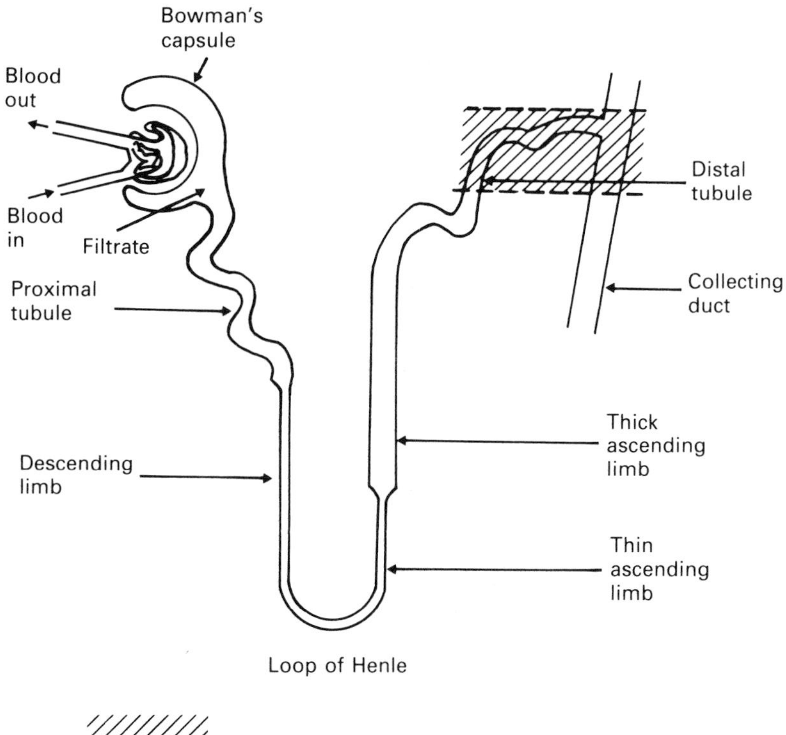

Figure 6.7 Site of aldosterone and spironolactone action.

per cent of the sodium is reabsorbed with chloride as the counter-ion, but a portion appears to be linked to the excretion of potassium or hydrogen ion thereby maintaining electrical neutrality. Water is reabsorbed with the sodium ion; the net effect is an increase in plasma volume and fall in plasma potassium.

Aldosterone is synthesized in the zona glomerulosa of the adrenal cortex. Like the other steroids mentioned in this chapter, the hormone functions by crossing the cell membrane and binding to a cytoplasmic receptor with a K_d of 2×10^{-9} M (Funder *et al.*, 1974). Two sets of binding sites in the cortex have been proposed by Farman *et al.* (1982), who obtained a biphasic Scatchard plot (see Appendix) for the binding of hormone to preparations from rabbit kidney cortex. Although such a plot does not always imply two sets of sites, and insufficient data points were obtained to reach a definite conclusion about the absolute values of the binding constant, these authors proposed that the higher affinity binding sites were for mineralocorticoids and the lower for glucocorticoids.

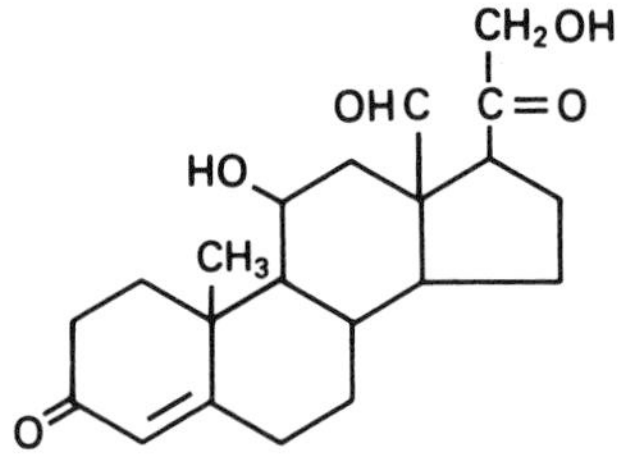

Aldosterone

The steroid–receptor complex undergoes a conformational change before it is translocated into the cell nucleus, where it binds to the DNA thus causing specific mRNAs to be synthesized (Corvol *et al.*, 1981). Several proteins in the cytosol and plasma membrane of the kidney cell are synthesized in response to aldosterone action – a process which is blocked by cycloheximide – but the precise role each plays in the facilitation of sodium reabsorption is not yet known (Scott *et al.*, 1981).

Steroid antagonists of aldosterone, of which the best known is spironolactone, are valuable diuretics in that not only do they antagonize the anti-diuretic action of the steroid by preventing sodium reabsorption, but also they do not cause a loss of potassium (this is referred to as potassium sparing action), an action not shared by the other types of diuretic. This is because aldosterone induces a low but measurable exchange of potassium for sodium noted above. Spironolactone and its analogues bind to the aldosterone receptor; the complex so formed does not rearrange into the active conformation and therefore cannot translocate into the nucleus (Funder *et al.*, 1974; Sakauye and Feldman, 1976). These drugs thus act in a fashion different from tamoxifen and its complex with the oestrogen receptor (section 6.4.3).

The binding constant for spironolactone is approximately 4×10^{-8} M when

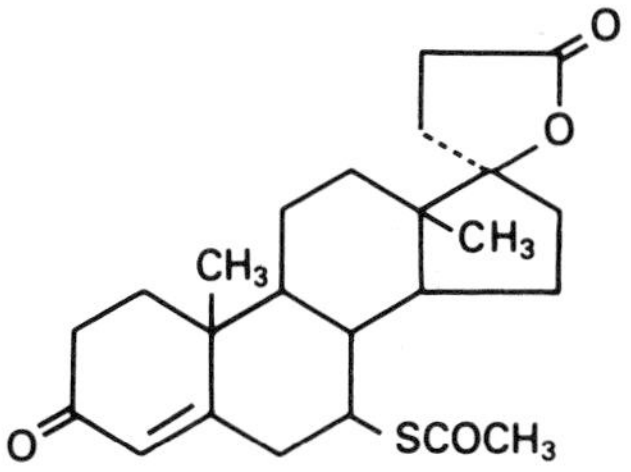

Spironolactone

measured competitively against ^{3}H-aldosterone binding to rat kidney (Farman *et al.*, 1974). The correlation of binding to the receptor and physiological effect has been studied by Rossier *et al.* (1983), who showed that receptor binding in toad bladder correlated with inhibition of sodium transport across the membrane induced by aldosterone. Aldosterone has to be present for spironolactone to be active, i.e. the latter is not a partial agonist, and the two compounds compete for the receptor (Rossier *et al.*, 1983; Horisberger and Giebisch, 1987).

Spirolactones, as a group, are also able to inhibit the biosynthesis of aldosterone but at rather higher concentrations than are needed to interfere with aldosterone receptor binding. The conclusion is, therefore, that the inhibition of biosynthesis plays little or no part in the antidiuretic action of the drugs (Corvol *et al.*, 1981).

Spironolactone is not, however, selective for aldosterone receptors, and is well able to block androgen receptors, albeit at higher doses. This is clearly a disadvantage when a diuretic effect only is required, because it can lead to the growth of breasts in men (gynaecomastia) and to menstrual disturbances in women. Nevertheless, in some conditions, notably hirsutism in women caused by high androgen levels, an anti-androgen effect is a positive advantage. The drug has been of clinical value in this condition (Shapiro and Evron, 1980).

Questions

1. Which protein is complexed with the oestrogen receptor in its inactive state?
2. Name three enzymes described in this chapter that use cytochrome *P*-450 as a prosthetic group.
3. Which enzymes are rate-limiting in the synthesis of cholesterol?
4. The plasma level of which lipoprotein is linked with heart disease? How does lovastatin interfere with its action?
5. Name one tumour that is hormone-dependent. Name a drug that can be used to treat this condition. How does the drug prevent the tumour growth?
6. Which enzyme does fluconazole block? What condition is fluconazole used to treat?

7. Why is terbinafine more effective against *Trichophyton* than *Candida* species?

References

Alberts, A.W., Chen, J., Kuron, G., Hunt, V., Huff, V., Hoffman, C., Rothbrook, J., Lopez, M., Joshua, H., Harris, E., Pitchett, A., Monaghan, R., Currie, S., Stapley, E., Albers-Schonberg, G., Hersens, O., Hirshfield, J., Hoogsteen, K., Liesch, J. and Springer, J., 1980., *Proc. Nat. Acad. Sci. (U.S.A.)*, **77**, 3957-61.

Amery, W.K., De Coster, R. and Caers, I., 1986, *Drug. Dev. Res.*, **8**, 299-307.

Angeli, A. and Frairia, R., 1985, *Lancet*, **i**, 821.

Anon, 1986, *Lancet*, **ii**, 665-6.

Aoyama, Y., Yoshida, Y., Sonoda, Y. and Sato, R., 1987, *J. Biol. Chem.*, **262**, 1239-43.

Asch, R.H., Balmaceda, J.P., Borghi, K.R., Niesvisky, R., Coy, D.H. and Schally, A.V., 1983, *J. Clin. Endocrinol. Metab.*, **57**, 367-72.

Bailieu, E.E., Binart, N., Cadepond, F., Catelli, M.G., Chambraud, B., Garnier, J., Gasc, J.M., Groyer-Schweizer, G., Oblin, M.E., Radanyi, C., Redeuilh, G., Renoir, J.M. and Sabbah, M., 1990, *Ann. N. Y. Acad. Sci.*, **595**, 300-15.

Balfour, J.A. and Faulds, D., 1992, *Drugs*, **43**, 259-84.

Barrett-Bee, K.J., Lane, A.C. and Turner, R.W., 1986, *J. Med. Vet. Mycol.*, **24**, 155-60.

Berger, V., Barson, A., Wolf, C., Chachaty, C., Fan, D., Fromenty, B. and Pessayre, D., 1992, *Biochem. Pharmacol.*, **43**, 654-7.

Bloch, K.E., 1979, *C.R.C. Crit. Rev. Biochem.*, **7**, 1-5.

Bradford, R.H., *et al.*, *Amer. J. Med.*, **91**, Suppl 1B, 18S-23S.

Briggs, M.H., Pitchford, A.G., Staniford, M., Barker, H.M. and Taylor, D., 1970, *Adv. Steroid. Biochem. Pharmacol.*, **2**, 111-222.

Brogden, R.N. and Clissold, S.P., 1989, *Drugs*, **38**, 185-203.

Chen, H.W. and Leonhardt, D.A., 1984, *J. Biol. Chem.*, **259**, 8156-62.

Corvol, P., Claire, M., Oblin, M.E., Geering, K. and Rossier, B., 1981, *Kidney Int.*, **20**, 1-6.

Coulson, C.J., King, D.J. and Wiseman, A., 1984, *Trends in Biochem. Sci.*, **9**, 446-9.

Covey, D.F., Hood, W.F. and Parikh, V.D., 1981, *J. Biol. Chem.*, **256**, 1076-9.

Eisenfeld, A., 1974, *Endocrinology*, **94**, 803-7.

Farman, N., Vandewalle, A. and Bonvallet, J.P., 1982, *Am J. Physiol.*, **242**, F69-77.

Foster, A.B., Jarman, M., Leung, C.-S., Rowlands, M.G. and Taylor, G.N., 1983, *J. Med. Chem.*, **26**, 50-4.

Funder, J.W., Feldman, D., Highland, E. and Edelman, T.S., 1974, *Biochem. Pharmacol.*, **23**, 1493-501.

Funkenstein, B., Waterman, M.R., Masters, B.S.S. and Simpson, E.R., 1983, *J. Biol. Chem.*, **243**, 10187-9.

Garnier, J., Gasc, J.M., Groyer-Schweizer, G., Oblin, M.E., Radanyi, C., Redeuilh, G., Renoir, J.M. and Sabbah, M., 1990, *Ann N. Y. Acad. Sci.*, **595**, 300-15.

Georgopapadakou, N.H. and Bertasso, A., 1992, *Antimicrob. Ag. Chemother.*, **36**, 1779-81.

Gibbons, G.F., Pullinger, C.R. and Mitropoulos, K.A., 1979, *Biochem. J.*, **183**, 309-15.

Ginsburg, M., Maclusky, N.J., Morris, I.D. and Thomas, P.J., 1977, *Brit. J. Pharmacol.*, **59**, 397-402.

Gold, E.M., 1979, *Ann. Intern. Med.*, **90**, 829-44.

Gormley, G.J., Stoner, E. and Bruskewitz, R.C., *et al.*, 1992, *N. Engl. J. Med.*, **327**, 1185-91.

Grant, S.M. and Clissold, S.P., 1990, *Drugs*, **39**, 877-916.

Gronemeyer, H., 1991, *Annu. Rev. Genet.*, **25**, 89-123.

Gronemeyer, H., Benhamou, B., Berry, M., Bocquel, M.T., Gofflo, D., Garcia, T., Lerouge, T., Metzger, D., Meyer, M.E., Tora, L., Vergezac, A. and Chambon, P., 1992, *J. Steroid Biochem. Molec. Biol.*, **3-8**, 217-21.
Grundy, S.M., 1988, *New Engl. J. Med.*, **319**, 24-33.
Hay, R.J., 1991, *J. Antimicrob. Chemother.*, **28**, Suppl A, 35-46.
Henshaw, R.C. and Templeton, A.A., 1992, *Drugs*, **44**, 531-6.
Horisberger, J.-D. and Giebisch, G., 1987, *Renal Physiol.*, **10**, 198-220.
Horwitz, K.B., Koseki, Y. and McGuire, W.L., 1978, *Endocrinology*, **103**, 1742-51.
Jordan, V.C., 1992, *Cancer*, **70**, 977-82.
Kasid, A., Buckshee, K., Hingorami, V. and Laumas, K.R., 1978, *Biochem. J.*, **176**, 531-9.
Kiang, D.T., Frenning, D.H., Goldman, A.I., Ascensao, V.F. and Kennedy, B.J., 1978, *New Engl. J. Med.*, **299**, 1330-4.
Lewis, J.H., Zimmerman, H.J., Benson, G.D. and Ishak, K.G., 1984, *Gastroenterology*, **86**, 503-13.
Loose, D., Kan, P.B., Hirst, A., Marcus, R.A. and Feldman, D., 1983, *J. Clin. Invest.*, **71**, 1495-9.
Marshall, J.R., 1978, *Clin. Obstet. Gynaecol.*, **21**, 147-62.
McConnell, J.D., Wilson, J.D., George, F.W., Geller, J., Pappas, R. and Stoner, E., 1992, *J. Clin. Endocrinol. Metab.*, **74**, 505-8.
Murad, F. and Kuret, J.A., 1990, Ch. 58 in the *Pharmacological Basis of Therapeutics*, (eds A.G. Gilman, T.W. Rall, A.S. Nies and P. Taylor), Pergamon, New York, 8th edn.
Nes, W.R., Sekula, B.C., Nes, W.D. and Adler, J.H., 1978, *J. Biol. Chem.*, **253**, 6218-25.
Nicholson, R.I., Syne, J.S., Daniel, C.P. and Griffiths, K., 1979, *Eur. J. Cancer*, **15**, 317-29.
Ono, T., Nakazone, K. and Kosaka, H., 1982, *Biochim. Biophys. Acta.*, **709**, 84-90.
Petranyi, G., Ryder, N.S. and Stutz, A., 1984, *Science*, **224**, 1239-41.
Pont, A., Williams, P.L., Azher, S., Reitz, R.E., Rochra, C., Smith, E.R. and Stevens, D.A., 1982, *Arch. Intern. Med.*, **142**, 2137.
Ravnskov, U., 1992, *Brit. Med. J.*, **305**, 15-19.
Rossier, B.C., Claire, M., Rafestin-Oblin, M.E., Geering, K., Gaggeler, H.P. and Corvol, P., 1983, *Am. J. Physiol.*, **244**, C24-31.
Rotstein, D.M., Kertesz, D.J., Walker, A.M. and Swinney, D.C., 1992, *J. Med. Chem.*, **35**, 2818-25.
Ryder, N.S. and Dupont, M.-C., 1984, *Biochim. Biophys. Acta*, **794**, 466-71.
Sakauye, C. and Feldman, D., 1976, *Am. J. Physiol.*, **231**, 93-7.
Santen, R.J., 1991, *J. Steroid Biochem. Molec. Biol.*, **40**, 247-53.
Schally, A.V., 1978, *Science*, **202**, 18-28.
Schreiber, J.R., Hsueh, A.J.W. and Barlieu, E.B., 1983, *Contraception*, **28**, 77-85.
Scott, W.N., Yang, C.P., Skipski, I.A., Cobb, M.H., Reich, I.M. and Terry, P.M., 1981, *Ann. N. Y. Acad. Sci.*, **372**, 15-28.
Shapiro, G. and Evron, S., 1980, *J. Clin. Endocrinol. Metab.*, **51**, 429-32.
Sikka, S.C., Swerdloff, R.S. and Rajfer, J., 1985, *Endocrinology*, **116**, 1920-5.
Smith, G.D. and Pekkanen, J., 1992, *Brit. Med. J.*, **304**, 431-4.
Smith, I.E., Fitzharris, B.M., McKinna, J.A., Fahmy, D.R., Nash, A.G., Neville, A.M., Gazet, J.-C., Ford, H.T. and Powles, T.J., 1978, *Lancet*, **ii**, 646-8.
Smith, J.R., Osborne, T.F., Brown, M.S., Goldstein, J.L. and Gil, G., 1988, *J. Biol. Chem.*, **263**, 18480-7.
Srivastava, A.K., 1978, *J. Steroid Biochem.*, **9**, 1241-4.
Takemori, S. and Kominami, S., 1984, *Trends in Biochem. Sci.*, **9**, 393-6.
Temple, T.E. and Liddle, G.W., 1970, *Annu. Rev. Pharmacol. Toxicol.*, **10**, 199-218.
Thigpen, A.E. and Russell, D.W., 1992, *J. Biol. Chem.*, **267**, 8577-83.
Thomas, P.G., Haslam, J.M. and Baldwin, B.C., 1983, *Biochem. Soc. Trans.*, **11**, 713.

Uauy, R., Vega, G.L., Grundy, S.M. and Bilheimer, D.M., 1988, *J. Pediatr.,* **113**, 387-92.
Van den Bossche, H., Willemsens, G., Cools, W., Cornelissen, F., Lauwers, W.F. and Van Cutsem, J.M., 1980, *Antimicrob. Ag. Chemother.,* **17**, 922-8.
Van den Bossche, H., Lauwers, W., Willemsens, G., Marischal, P., Cornelissen, P. and Cools, W., 1984, *Pesticide Sci.,* **15**, 188-98.
Williamson, D.G. and O'Donnell, V.J., 1969, *Biochemistry,* **8**, 1311-16.
Wilson, L.D., Oldham, S.B. and Harding, B.W., 1969, *Biochemistry,* **8**, 2975-80.
Wissler, R.W., 1991, *Amer. J. U. Med.,* **91**, Suppl 1B, 3S-9S.

Chapter 7

Prostaglandin and leukotriene biosynthesis and action

7.1 Introduction

The transformation of arachidonic acid into a multiplicity of lipid intermediates, many of which are extremely active biologically, has been an area of great interest in recent years. The prostaglandins themselves were first isolated and identified by Samuelson and co-workers in the 1960s, but the other agents with quaint, albeit accurate, names like 'rabbit aorta contracting substance' (thromboxane A_2) and 'slow reacting substance of anaphylaxis' (leukotrienes C_4 and D_4) defied identification until the early 1980s, largely because of their instability and low concentrations in biological fluids. The development of sensitive mass spectroscopic and gas chromatographic methods for their isolation was crucial in their identification.

The details of arachidonic acid metabolism have been elucidated at a remarkable rate over recent years. The present outline of the main pathways is shown in Fig. 7.1. The metabolism of arachidonic acid can give rise to four distinct pathways: the cyclooxygenase reaction leads to the prostaglandins D_2, E_2, F_2, G_2 and H_2, the thromboxanes and prostacyclin; 5-lipoxygenase leads to the leukotrienes (Fig. 7.2); a cytochrome *P*-450 linked reaction gives rise to the eicosatrienoic acids, and 12-lipoxygenase yields another family of hydroxyeicosatrienoic acids (Taylor and Ritter, 1986). It is the first two pathways with which this chapter is primarily concerned.

Table 7.1 Drugs discussed in Chapter 7

Target	*Drug*	*Major uses*
Phospholipase A_2	Corticosteroids	Asthma, arthritis
Prostaglandin synthetase	Aspirin, Indomethacin	Pain, inflammation in arthritis

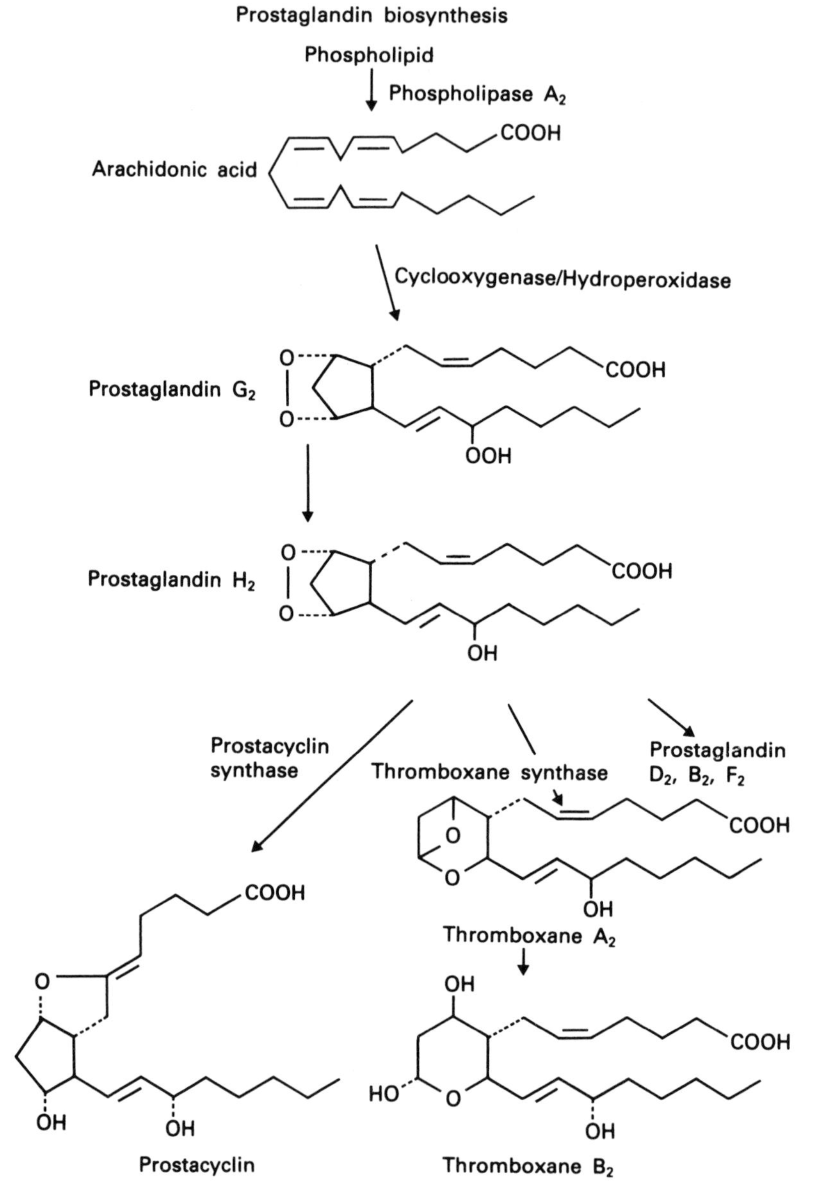

Figure 7.1 Prostaglandin biosynthesis.

7.2 Phospholipase inhibition - glucocorticoids for asthma and arthritis

Corticosteroids have long been the main therapy for the treatment of severe asthma and other allergic conditions. With the advent of aerosol systems that

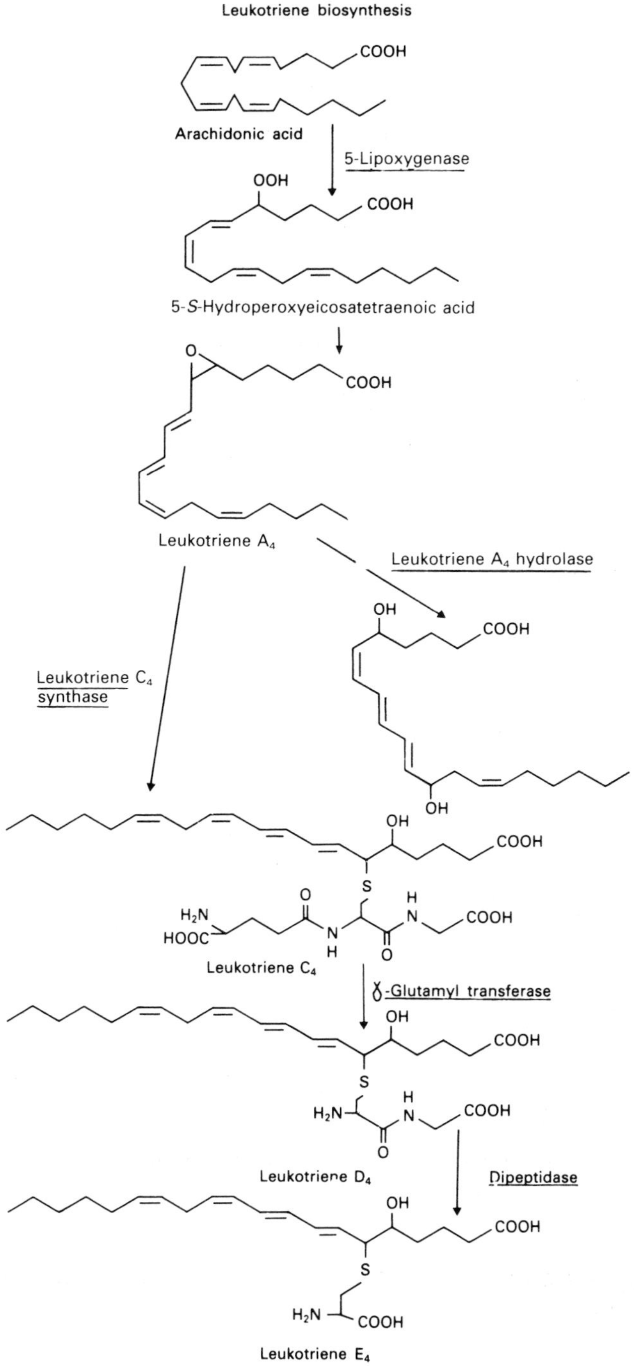

Leukotriene biosynthesis

COOH

Arachidonic acid

5-Lipoxygenase

OOH

COOH

5-*S*-Hydroperoxyeicosatetraenoic acid

O

COOH

Leukotriene A_4

Leukotriene A_4 hydrolase

OH

COOH

OH

Leukotriene C_4 synthase

OH

COOH

S

O

H

H_2N

N

COOH

HOOC

N

H

O

Leukotriene C_4

γ-Glutamyl transferase

OH

COOH

S

H

H_2N

N

COOH

O

Leukotriene D_4

Dipeptidase

OH

COOH

S

H_2N

COOH

Leukotriene E_4

Figure 7.2 Leukotriene biosynthesis.

deliver the drug direct to the lung, however, steroids can also be used for less severe asthma without the risk of the side-effects, sometimes severe, that occur with oral treatment. Control of symptoms can often be more effective than with cromoglycate or the β-agonists.

The initial step in prostaglandin and leukotriene biosynthesis is the release of arachidonic acid from phospholipids catalysed mainly by phospholipase A_2. The enzyme hydrolyses the fatty acyl group from position 2 of the phospholipid. Phospholipase A_2 is located on the inside of the plasma membrane and requires calcium ion for activity - but not calmodulin (Withnall *et al.*, 1984).

Corticosteroids inhibit phospholipase A_2 and thus block the synthesis of pro-inflammatory prostaglandin metabolites deriving from the prostaglandin synthetase and lipoxygenase pathways. Steroids switch on the gene that controls the synthesis of a protein called lipocortin, a protein of molecular mass 15 to 40 kDa depending on the cell type, that completely inhibits phospholipase A_2. Accordingly, the anti-inflammatory effects of steroids can be blocked by inhibitors of nucleic acid synthesis (see Sautebin *et al.*, 1992).

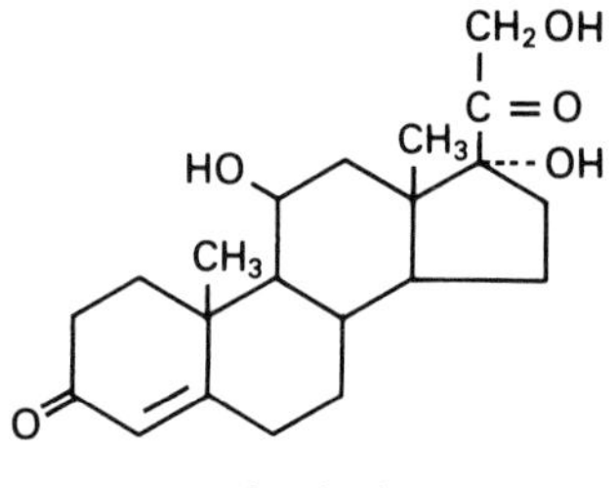

Cortisol

Lipocortin is now recognized as a member of a larger family of calcium binding proteins known as annexins. These proteins when activated by calcium can bind lipids and, consequently, draw two membranes together, suggesting that they may be involved in exocytosis (reviewed in Creutz, 1992). The N-terminal side is on the cytoplasmic face which contains tyrosine and serine residues as potential phosphorylation sites, the tyrosine being the target for protein kinase C. Phosphorylation reduces the ability of lipocortin to aggregate chromaffin granules (Wang and Creutz, 1992). In lipocortin there are four stretches or folds of 70 repeated amino acids, 17 of which are highly conserved between lipocortins. The calcium binding sites, three of high and two of low affinity, are on one side of the approximately planar molecule, which is also the membrane binding face. Each high affinity site involves the first few amino acids of the fold and an acidic residue in a second loop downstream.

The human lipocortin gene has been cloned and sequenced, and a glucocorticoid response element detected in the first intron similar to the hormone response elements in other steroid-induced genes. These hormone response elements respond to the transactivating function of the steroid receptor and initiate transcription of the gene (Chapter 6; Kovacic *et al.*, 1991).

The glucocorticoid receptor is a multimeric complex located in the cytoplasm containing heat-shock protein. In unstressed cells heat-shock proteins exist as monomers with no DNA binding ability. When the cell is stressed by heat or other stress, they assemble in the nucleus in the form of trimers that activate responsive genes (reviewed in Morimoto 1993). The receptor molecule is bound to heat-shock proteins of molecular weight 90, 70 and 56 kDa in a complex that does not bind to DNA. When the steroid binds, a dimer of the 90 kDa protein dissociates from the receptor and the receptor has a great affinity for DNA in that the DNA binding site is now unmasked. The activated receptor dimerizes either before or at the same time as it binds to DNA. The receptor molecule can be viewed as having three sections; an N-terminal region which is involved in transactivation and dimerization, a DNA binding section and a third domain carrying the steroid and 90 kDa protein binding sites and also involved in transactivation (reviewed in Dahlman-Wright *et al.*, 1992).

The receptor initiates glucocorticoid responses *in vivo* by turning on a number of genes through binding at the response element:

AGAACAnnnTGTTCT

where A, G, T, C are the respective nucleotide bases and n can be any of the four. This sequence is sometimes referred to as a partial palindrome, i.e. the two stretches of six amino acids produce an imperfect mirror-image in DNA base-pairing terms.

General DNA binding is mediated by the middle domain of the receptor containing a section 66 to 68 amino acids long, highly conserved between tissues. There are nine conserved cysteines, of which eight are involved in the tetrahedral coordination of the two zinc atoms consistent with cysteine ligands for structural zinc (see Chapter 8). The zinc atoms are located at the base of loops or 'zinc fingers' that protrude from the main structure of the protein. The zincs are apparently essential for structural integrity and DNA binding capacity.

7.3 Prostaglandin synthetase inhibitors - aspirin and other non-steroidal anti-inflammatories

One of the oldest medications still in use is aspirin or acetylsalicylic acid. The stimulus for the derivation of aspirin was the finding that the bark of the willow tree had a soothing effect on headaches and fevers. The active ingredient turned out to be a glycoside of salicyl alcohol, and subsequently salicylic acid was also found to be effective in reducing temperature (antipyretic action). Acetylation of the phenolic hydroxyl was the next step to yield aspirin (Flower *et al.*, 1985).

A very interesting advance in our understanding of pharmacology arose from the realization that aspirin and other similar drugs, classified as the non-steroidal anti-inflammatory agents, inhibited the synthesis of prostaglandins at the step catalysed by prostaglandin endoperoxide synthetase (Vane, 1971). There are a considerable number of these drugs which can be broadly grouped into structural families: salicylates such as aspirin, substituted propionic acids like naproxen and flurbiprofen, pyrazolones such as phenylbutazone, fenamates like meclofenamic acid, together with indomethacin which does not readily fall into any of the above structural groupings. Nevertheless, they all show similar pharmacological activity: anti-pyretic, analgesic and anti-inflammatory properties.

Prostaglandin synthetase is a homodimer with a molecular weight of

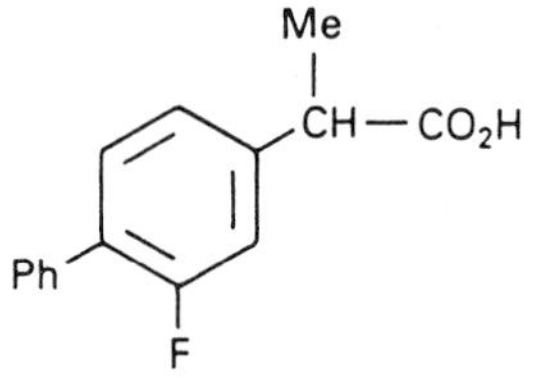

Flurbiprofen

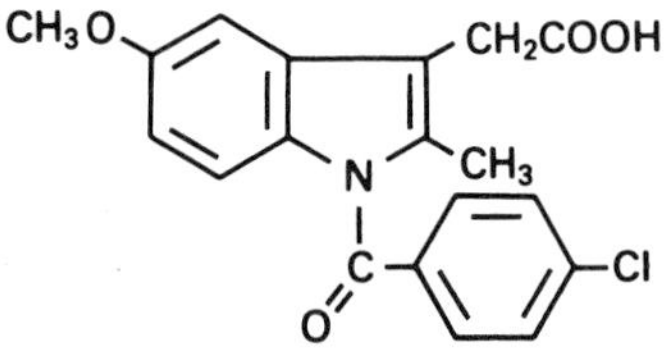

Indomethacin

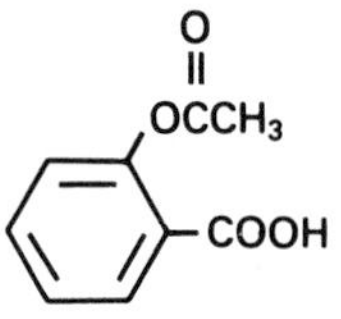

Acetylsalicylic acid

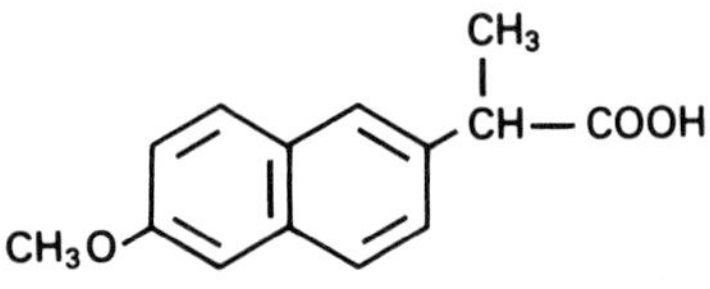

Naproxen

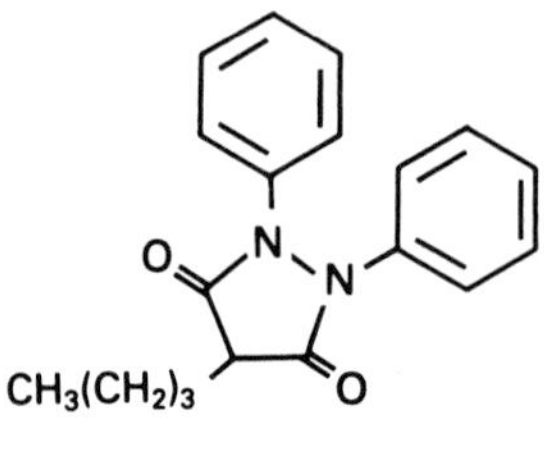

Phenylbutazone

144 kDa; it is a dioxygenase containing haem as an essential cofactor, membrane-bound in the microsomes and particularly rich in the vesicular gland. It exhibits two activities: cyclooxygenase and hydroperoxidase. Cyclooxygenase adds both atoms of a molecule of oxygen to the carbon atoms at positions 11 and 15 of the arachidonic acid molecule to form a 15-hydroperoxy-9,11-endoperoxide (prostaglandin G_2) followed by reduction of the hydroperoxy group to hydroxy by a hydroperoxidase activity yielding prostaglandin H_2 (Fig. 7.1). The two activities appear to operate separately, although the haem is shared between the two active sites. Aspirin, for example, inactivated cyclooxygenase activity but not hydroperoxidase by acetylating serine-530 at the amino terminus of the protein (Smith and Marnett, 1991; Smith *et al.*, 1992). In addition, iron can be replaced with manganese to give an enzyme which retains the cyclooxygenase but no hydroperoxidase activity.

The reaction mechanism proceeds via a protein free radical, although the precise details of the mechanism may differ from another iron enzyme, ribonucleotide reductase. Although the details are not fully understood, the reaction requires oxidization of the iron by a peroxide from the four- to five-valent state and the abstraction of a hydrogen atom from the 13-pro-*S* position of arachidonic acid by a radical. A tyrosyl radical (probably Tyr-385) is formed when the enzyme is incubated with arachidonate but may not be involved in the cycloxygenase reaction. The native enzyme loses activity as a result of catalysis - a sort of suicide inactivation (this is obviously an advantage for the cell to close down the production of such reactive intermediates but it bedevils efforts to unravel the mechanism). It is possible that the tyrosyl radical is somehow involved in the inactivation (Smith *et al.*, 1992). The anti-inflammatory agents first inhibit the cyclooxygenase activity reversibly and then a time-dependent inactivation follows with a stoichiometry of 1 mol drug:1 mol synthetase dimer. The enzyme retains some 4–10 per cent of activity. Dissociation constants calculated for the initial binding are $1{\cdot}7 \times 10^{-5}$ M for indomethacin, $2{\cdot}0 \times 10^{-7}$ M for flurbiprofen and 8×10^{-8} M for meclofenamic acid. There is a stereochemical basis to this action in that only the *S*(+) (dextrorotatory) isomer of flurbiprofen is active (Kulmacz and Lands, 1985), and the D-isomer of naproxen is 70 times more active than the L-isomer (Flower, 1974). Since the absorbance at 410 nm due to the haem group did not change, it is not likely to be involved in anti-inflammatory agent binding (Kulmacz and Lands, 1985).

The time-dependent change is likely to be conformational in the synthetase

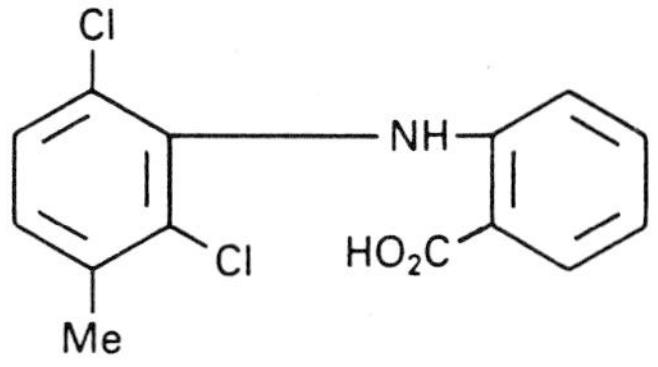

Cl
NH
Cl
HO_2C
Me

Meclofenamic acid

because indomethacin can be recovered intact after inhibiting the enzyme, so covalent reaction has not taken place. Esters of these agents, however, do not show irreversible inhibition, so the charge on the carboxyl is important, perhaps for binding to the arachidonic acid site.

The involvement of prostaglandin endoperoxide synthetase in the pathogenesis of inflammation rests on two types of evidence. The first is that the anti-inflammatory agents inhibit the enzyme in a rank order which correlates with their anti-inflammatory activity (Vane and Botting, 1987). The second lies in the discovery that prostaglandins play a major part in the oedema, inflammation, pain and fever that accompany rheumatoid arthritis. Other important mediators include leukotriene B_4 which attracts lymphocytes into the joint, platelet-activating factor and interleukin 1 (Harris, 1986).

The use of these aspirin-like drugs can result in the development of stomach ulcers. This side-effect has led to a greater understanding of the role of prostaglandins in protecting the stomach wall against acid by activating the secretion of mucus in the intestine. In addition, the anti-inflammatory drugs can suppress the release of gastric acid in the stomach. The drugs themselves may also cause tissue damage by diffusing into the cells in the neutral form. They are weak acids with pK_a in the region of 5, so they will be uncharged at the pH of the stomach. Inside the cells they will ionize at neutral pH and thus become trapped (Price and Fletcher, 1990).

7.4 Rheumatoid arthritis

Rheumatoid arthritis is characterized by inflammation of the synovium of particular joints associated with destructive effects, leading to the dissolution of cartilage and bone. Rheumatoid arthritis is not to be confused with osteoarthritis which contains less of an inflammatory component, in addition to cartilage breakdown. Both types of arthritis include an element of proliferative growth in different tissues. In osteoarthritis, this can reach the point of excruciating pain as the nodules grind in the joint and require removal, as in the operation for hip replacement for example. As well as the hip, the knee is commonly affected in osteoarthritis. It is a condition more common in later life, although not necessarily a part of the ageing process. Both types of arthritis respond to non-steroidal anti-inflammatory agents.

Rheumatoid arthritis is classified as an auto-immune disease in that antibodies of the immunoglobulin M class (IgMs) found are against the body's own constituents, notably the Fc fragment of immunoglobulin G (IgG). These are called rheumatoid factors which may be found in the serum as well as the joint, although the level does not necessarily parallel the severity of the disease. Indeed, the relevance of possible damaging effects of the autoantibodies to disease progression is not at all clear. The disease is complex involving both T and B cells of the immune system (see Insel, 1990).

The process seems to be initiated by an antigen that is attracted to the joint.

Antibodies made within the joint space complex with the antigen and activate the complement system. Antigen-presenting cells present the complex to T cells which produce cytokines, including interleukins, that recruit more T cells. The T cells differentiate and produce factors that induce both tissue destruction and inflammation. More lymphocytes are recruited, the cells of the lining of the synovium proliferate and synthesize connective tissue (pannus) eroding cartilage, bone, tendon and ligaments. At some stage the process becomes perpetuated, but the reasons for the persistence and fluctuation of inflammation are poorly understood. The initiating antigen may reappear or there may be a cycle of positive and negative immune responses.

Some view the T cell as driving the reaction, whereas others see the T cell as being held in check as the body tries to control the inflammation. Whichever view is correct, the T cell clearly plays a major role in the process. T cells need to have antigens presented to them in a way that they can be recognized - a function carried out by antigen-presenting cells. The latter express molecules on their surface encoded by the major histocompatibility complex (MHC). These molecules bind antigen which, in the case of the macrophage, may be derived from intracellular degradation of pathogens. The antigen receptor of T cells will only bind to antigen if it is complexed to an MHC molecule. MHC 1a is expressed on the lymphocytes in the synovium probably as a result of activation by γ-interferon. Lymphocytes associated with MHC 1a rich cells can produce lymphokines that promote synovial growth and B-cell differentiation to antibody secreting cells - both events occurring in the arthritic joint (Insel, 1990; Zvaifler, 1988). The persistent observation of an immune response in rheumatoid arthritis has led to frequent searches for an infective agent. Mycoplasma were suspected to be the culprits in earlier years, indeed one species of mycoplasma, *M. arthritidis*, can cause a transient condition which resembles arthritis, as its name suggests. More recently, however, attention has switched to viruses for the aetiology of the disease, but it is not clear whether the virus is the causative agent or merely a passenger in a damaged synovium.

The role of prostaglandins in this process is confined to pain sensitization, oedema formation, increase in blood flow and fever. Chemotactic agents that recruit more T cells are likely to be products of the lipoxygenase pathway such as leukotriene B_4. The non-steroidal anti-inflammatory agents can reduce or eliminate the pain and inflammation in the short term, but the progressive deterioration of the synovium and pannus formation continues, resulting in osteoarthritis at least in the formation of painful and irregular nodules (Baker and Rabinovitz, 1986).

7.5 Thromboxane synthesis

The next step in the prostaglandin metabolic pathway is conversion of prostaglandin H_2 to thromboxane A_2 (Fig. 7.1). This material is extremely unstable with a half-life of approximately 30 seconds in tissue fluids, being converted by non-enzymic hydrolysis of the epoxide bridge to thromboxane B_2, which is almost biologically inactive. Much interest has centred on thromboxane A_2 in view of its vasoconstrictor activity and its irreversible induction of platelet aggregation which can lead to blockage of blood vessels particularly in the heart. Thromboxane A_2 is generated in vascular smooth muscle and cardiac cells as well as platelets, among other tissues, thus being implicated in angina and acute myocardial infarction.

One major therapeutic objective has been to blockade the action of thromboxane A_2, either by inhibiting its synthesis or blocking its receptors. Thromboxane A_2 synthesis is rate-limited in the platelet by the availability of thromboxane synthetase. Prostaglandin H_2 diffuses out of the platelet and is taken up by the epithelial cells in the vessel wall (aortic intima). Microsomal prostacyclin synthetase catalyzes the formation of prostaglandin I_2 (prostacyclin), a vasodilator and extremely powerful inhibitor of platelet aggregation at between 1 and 10×10^{-9} M. Most of the prostacyclin is apparently synthesized this way. This diversionary process may be an important regulatory mechanism for limiting clot (thrombus) formation in damaged vessels (reviewed in Gresele *et al.*, 1991).

Thromboxane synthetase inhibitors were synthesized and tested in the clinic for the treatment of cardiovascular conditions including angina and hypertension. These drugs had, however, little consistent therapeutic effect (Fiddler and Lumley, 1990), probably because they caused a build-up of prostaglandin H_2 which binds more tightly to thromboxane receptors than thromboxane A_2 itself, and is a more powerful aggregating agent. Attempts are now being made to synthesize compounds which both inhibit thromboxane synthetase and are antagonists at thromboxane receptors. Receptor antagonists on their own are not expected to be effective, and also cannot redirect prostaglandin equivalents from platelet to vessel wall (Gresele *et al.*, 1991).

7.6 Leukotriene biosynthesis

Although no drug has yet reached the market that owes its activity to inhibition of leukotriene biosynthesis, a great deal of effort is going into the search for such a compound. This is largely because of the expectation that such a compound would be useful in the treatment of asthma, since the 'slow reacting substance of anaphylaxis', that is released in asthma and is responsible for much of the bronchoconstriction of the condition, is now recognized to be a mixture of leukotrienes. In addition, a leukotriene biosynthesis inhibitor might be useful in arthritis as leukotriene B_4 is a potent chemotactic agent for polymorphonuclear leukocytes.

The first committed step is peroxidation of arachidonic acid catalyzed by 5-lipoxygenase (Fig. 7.2) which contains a non-haem iron at the active site, producing 5-hydroperoxy-7,9,11,14-eicosatetraenoic acid which can then be converted into leukotriene A_4. This is the branch-point from where leukotriene B_4 may be obtained by hydrolase action or C_4 by the action of a glutathione *S*-transferase. Leukotrienes D_4 and E_4 result from the sequential action of γ-glutamyl peptidase and a dipeptidase.

Inhibition of the first enzyme in the pathway, 5-lipoxygenase, has turned out to be the most fruitful. Inhibitors, that either bind to the active site iron or compete with arachidonic acid, have shown promise in the clinic for allergic and inflammatory disorders (reviewed in McMillan and Walker, 1992).

Questions

1. What separate catalytic activities are carried out by prostaglandin endoperoxide synthetase?
2. Are the non-steroidal anti-inflammatory agents irreversible inhibitors of prostaglandin endoperoxide synthetase?
3. What therapeutic advantage would an inhibitor of thromboxane synthetase be expected to have over a thromboxane A_2 receptor antagonist?
4. What features of the lipocortin/annexin molecule are unusual? Can you list any cellular processes in which they might be involved?

References

Baker, D.G. and Rabinowitz, J.L., 1986, *J. Clin. Pharmacol.,* **26**, 2-21.
Creutz, C.E., 1992, *Science,* **258**, 924-30.
Dahlman-Wright, K., Wright, A., Carlstedt-Duke, J. and Gustaffson, J.-A., 1992, *J. Steroid Biochem. Molec. Biol.,* **41**, 249-72.
Fiddler, G.I. and Lumley, F., 1990, *Circulation,* **81**, Suppl 1, 69-78.
Flower, R.J., 1974, *Pharmacol. Rev.,* **26**, 33-67.
Flower, R.J., Moncada, S. and Vane, J.R., 1985, In: *The Pharmacological Basis of Therapeutics,* 7th Edn., Eds. A.G. Gilman, L.S. Goodman, T.W. Rall and F. Murad (New York: Macmillan), Ch. 29.
Gresele, P., Deckmyn, H., Nenci, G.G. and Vermylen, J., 1991, *Trends in Pharmacol. Sci.,* **12**, 158-63.
Harris, E.D., 1986, *Amer. J. Med.,* **80**, Suppl. 4B, 4-10.
Insel, P., 1990, Ch. 26 in the *Pharmacological Basis of Therapeutics*, 8th edn (eds A.G. Goodman, T.W. Rall, A. Nies and P. Taylor) Pergamon, New York.
Kovacic, R.T., Tizard, R., Cate, R.L., Frey, A.Z. and Wallner, B.P., 1991, *Biochemistry,* **30**, 9015-21.
Kulmacz, R.J. and Lands, W.E.M., 1985, *J. Biol. Chem.,* **260**, 12572-8.
McMillan, R.M. and Walker, E.R.H., 1992, *Trends in Pharmacol. Sci.,* **13**, 323-30.
Morimoto, R.I., 1993, *Science,* **259**, 1409-10.
Price, A.H. and Fletcher, M., 1990, *Drugs,* **40**, Suppl 5, 1-11.
Sautebin, L., Carnuccio, R., Ialenti, A. and Di Rosa, M., 1992, *Pharmacol. Res.,* **25**, 1-12.
Skorodin, M.S., 1993, *Arch. Intern. Med.,* **153**, 814-28.
Smith, W.L. and Marnett, L.J., 1991, *Biochem. Biophys. Acta,* **1083**, 1-17.
Smith, W.L., Eling, T.E., Kulmacz, R.J., Marnett, L.J. and Tsai, A.-L., 1992, *Biochemistry,* **31**, 3-7.

Taylor, G.W. and Ritter, J., 1986, *Trends in Pharmacol. Sci.*, **7**, March centrepage.
Vane, J.R., 1971, *Nature New Biology*, **231**, 232-6.
Vane, J.R. and Botting, R., 1987, *FASEB J.*, **1**, 89-96.
Wang, W. and Creutz, C.E., 1992, *Biochemistry* **31**, 9934-9.
Withnall, M.T., Brown, T.J. and Diocee, B.K., 1984, *Biochem. Biophys. Res. Commun.*, **121**, 507-13.
Zvaifler, N.T., 1988, *Amer. J. Med.*, **85**(4A), 12-17.

Chapter 8

Zinc metalloenzymes

8.1 Introduction

Approximately 200 enzymes have so far been discovered that contain at least one zinc atom as part of the integral functioning protein. Vallee and Galdes (1984) have classified the roles that zinc can play in metalloproteins.

1. Catalytic, in which the metal plays an active part in the catalysis. Notable examples of this are carbonic anhydrase, angiotensin-converting enzyme and alcohol dehydrogenase from liver.
2. Structural, where the zinc is required to maintain the structural integrity of the protein as in β-amylase and the glucocorticoid receptor (see Chapter 7).
3. Regulatory, either inhibitory or activatory, as in leucine aminopeptidase.
4. Other unspecified role(s) which may be regarded as non-catalytic for want of better knowledge.

The complete complement of 10 electrons in the 3d shell of the zinc atom has a screening effect on the positive charge of the nucleus, compared with calcium which lacks this screen. Consequently, zinc is able to form covalent coordinate complexes in addition to the ionic complexes characteristic of calcium. Ligands can, therefore, easily enter its coordination sphere and donate a pair of electrons to form a coordinate or dative bond. Zinc may have either four or six such ligands in simple inorganic compounds.

Furthermore, the external shell of zinc's electrons is polarizable and therefore distorts in an asymmetric environment. Characteristic features of the catalytic type of zinc protein are (a) a highly asymmetric environment of the zinc such that the atom can be regarded as either 4- or 5-coordinated and (b) three ligands to the protein amino acid side-chains, while water occupies the fourth position. Non-catalytic and structural zinc is normally bound to the protein by four ligands and is not as distorted as catalytic zinc. The polarizability plays a practical part in the catalysis as it has a higher energy than a non-distorted state and thus the activation energy for the reaction is correspondingly lowered (Vallee and Auld, 1990).

Covalent coordinate interactions require a closer inter-atomic approach than is seen in ionic compounds. Furthermore, zinc will accept a variety of ligands based on oxygen, nitrogen or sulphur, as exemplified by the formation of sodium zincate (Na_2ZnO_2) and zinc diamine chloride $[Zn(NH_3)_2]Cl_2$. In this way a variety of reaction types can be catalyzed, with cysteine, histidine or glutamic acid side-chains acting as anchors for the binding of zinc to the protein.

Zinc, as an electrophilic cation, can activate the substrate chemically by behaving as a Lewis acid (meaning that it can readily accept a share in a pair of electrons from a ligand, with the concomitant formation of a coordinate covalent bond). This has the effect of enhancing ligand nucleophilicity, as for example by ionizing a water molecule and binding the hydroxide ion formed, which can then participate in hydrolysis reactions that the initial water molecule itself could not achieve. In addition, zinc can facilitate catalysis by orientating reactants through coordination on the metal as a template (Vallee and Galdes, 1984).

Another feature of catalytic zinc proteins is the occurrence of acyl-enzyme intermediates in the catalytic process, as exemplified by acyl-phosphates in alkaline phosphatase. The nature of the nucleophile which breaks the acyl-enzyme intermediate is not certain, but in a number of hydrolytic metallo-enzymes the proposition of a zinc attached hydroxide ion has been made (Coleman, 1984). If the variation of enzyme activity with pH is measured (pH profile) there is a requirement for catalysis of a group with a pK_a of 7 to 7·5 which nuclear magnetic resonance (n.m.r.) studies have shown is not an imidazole. In effect the zinc is lowering the pK_a of the hydroxide moiety of the water molecule so that it can act as a nucleophile (see Vallee and Galdes, 1984).

It is the catalytic type of zinc atom with which we are concerned in this chapter, particularly as the inhibitors of the enzyme actually form complexes which involve direct attachment of a ligand to the zinc. Inhibitors of two enzymes containing catalytic zinc have found pharmaceutical use. Carbonic anhydrase inhibitors are useful both in situations where there is an excess of fluid (oedema) in various tissues and, therefore, in the treatment of glaucoma, and also when urine needs to be 'alkalinized' because acids like uric acid might otherwise precipitate. Secondly, angiotensin-converting enzyme (ACE) inhibitors are now playing a major part in the reduction of blood pressure. Potentially, a third application lies in inhibiting an enzyme that can degrade enkephalin. Such inhibitors may be approaching clinical trial as painkillers.

8.2 Carbonic anhydrase - methazolamide for glaucoma

An enzyme exists in a large number of tissues including kidney, erythrocyte and pancreas catalyzing the hydration of carbon dioxide to form carbonic acid, which immediately ionizes to form a bicarbonate ion:

$$CO_2 + H_2O \rightleftharpoons H_2CO_3 \rightleftharpoons HCO_3^- + H^+$$

This enzyme (carbonic anhydrase) can also catalyse the hydration of aldehydes and ketones, such as pyruvate, but it is unlikely that any of these reactions have physiological significance. It may seem unnecessary for an enzyme to carry out this reaction - estimates of the ration of enzyme-catalysed to uncatalysed rate (catalytic rate enhancement) range from 8000 (Maren, 1984) to 10^5 (Coleman, 1984), depending on the source of the enzyme. Nevertheless, the ubiquity and high level of the enzyme in, for example, the kidney suggest that the rate enhancement of this reaction plays a very important role in the metabolism of the cell.

Carbonic anhydrase normally exists in at least three forms (or isoenzymes) designated CA1, CA2 and CA3, which are distinguished mainly by a considerable difference in catalytic rate with CA2 being the most active. The isoenzymes also differ to some extent in primary sequence but not in overall length of some 260 amino acid residues, resulting in a molecular weight of approximately 30 kDa. There are 105 differences between human erythrocyte CA1 and CA2 which represents 59·6 per cent homology - the similarity is particularly close about the active site as would be expected. Other species show greater homology between the two forms (Deutsch, 1984).

As shown by X-ray crystallographic data, carbonic anhydrase contains one zinc atom per active site coordinated to three nitrogen atoms from imidazole side-chains of histidines residues 94, 96 and 119, and to one oxygen atom from a water molecule in an approximately tetrahedral conformation. The reaction mechanism has been the subject of much discussion over the years (reviewed in Silverman and Lindskog, 1988).

The zinc atom based in a hydrophobic pocket acts to lower the pK_a of the water molecule and binds a coordinated hydroxide ion although the pH of the surrounding medium may be neutral. Another enzyme residue, probably His 64, accepts the proton before transferring it to buffer. The hydroxide ion behaves as a nucleophile, by attacking carbon dioxide to form bicarbonate ion. A sigmoid pH profile is obtained for both hydration and dehydration reactions but in opposing senses, which may be a consequence of the requirement for a base such as the hydroxide ion or His 64 in the active site with a pK_a in the region of 7 as an intermediate for both reactions (Silverman and Lindskog, 1988). A 5-coordinated zinc atom (with a second water molecule attached via its oxygen atom to the zinc) has been shown to have theoretical advantages for the catalysis in that the second water molecule is ready to slip into the place of the first water molecule as soon as reaction has occurred, but it is at present difficult to be sure how many water molecules are at the active site (Cook and Allen, 1984).

The detailed interaction of the sulphonamides, used originally as diuretics, with carbonic anhydrase has been a subject of considerable interest. The sulphonamide group has to be unsubstituted for activity, with the amide nitrogen atom binding to the zinc atom at the fourth coordination site displacing the water molecule. One of the two sulphonamide oxygen atoms links with the fifth coordination site of the zinc (see Vedani and Meyer, 1984). Kinetic studies have shown that the sulphonamides are competitive inhibitors of the dehydration of carbonic acid as well as the hydration of carbon dioxide, as one would expect from the principle of microscopic reversibility. In this situation, competitive kinetics have shown that the inhibitor binds at the substrate binding site (Cook and Allen, 1984).

Carbonic anhydrase is normally present in tissue in great excess, as indeed are a number of other enzymes, and over 99 per cent of the enzyme has to be inhibited before physiological effects become apparent. The importance of this high concentration relates to bicarbonate ion transport across epithelial cell membranes where sodium is the counter-ion, in that a large excess of enzyme is required to act on the substrate.

One example of this occurs in the proximal tubule of the kidney where the majority of bicarbonate ion that has previously been filtered out in the glomerulus is reabsorbed (Kokko, 1984) (Fig. 8.1). In a coupled transport process, sodium ion is reabsorbed while hydrogen ion is secreted into the filtrate passing through the lumen of the proximal tubule (Fig. 8.2a). Hydrogen and bicarbonate ions form carbonic acid which disproportionates to carbon dioxide and water, catalyzed by carbonic anhydrase in the luminal membrane (Lucci *et al.*, 1983). Carbon dioxide diffuses across the luminal membrane of the proximal tubule cell and the cytoplasmic enzyme reconverts in into bicarbonate ion which is transported across the basement membrane of the cell into the peritubular fluid that drains into the blood. The proton that is also formed in the reaction is recirculated across the luminal membrane in exchange for sodium (DuBose *et al.*, 1981; Lucci *et al.*, 1983). The overall result is the net reabsorption of sodium and bicarbonate ion.

If insufficient hydrogen ion is secreted to react with all the tubular bicarbonate, then the urine will contain larger amounts of bicarbonate than usual and consequently will have a higher pH. The content of sodium ion will also increase (Kokko, 1984) and in order to maintain the osmotic pressure water will be secreted together with the ions (Fig. 8.2a). This situation occurs as the consequence of carbonic anhydrase inhibition because the supply of protons is greatly reduced. It appears that at least two-thirds of the carbon dioxide reabsorption is mediated by carbonic anhydrase because, when the enzyme is completely inhibited, the proportion of bicarbonate reabsorbed falls to about 37 per cent of its control value (Maren, 1984). Carbonic anhydrase inhibitors were originally developed for use as diuretics but their action is only weak because their major effect is in the proximal tubule (Kokko, 1984). As a consequence the greater volume of fluid reaching the distal tubule produces a compensatory reabsorptive effect there which markedly reduces the effect of the drug.

Another family of diuretics, the widely-used thiazides, were originally developed out of studies on the sulphonamides. Although the thiazides are weak inhibitors of carbonic anhydrase, they act primarily at the early portion of the distal tubule and, since carbonic anhydrase is mainly located in the proximal tubule, inhibition of this enzyme is unlikely, therefore, to play much

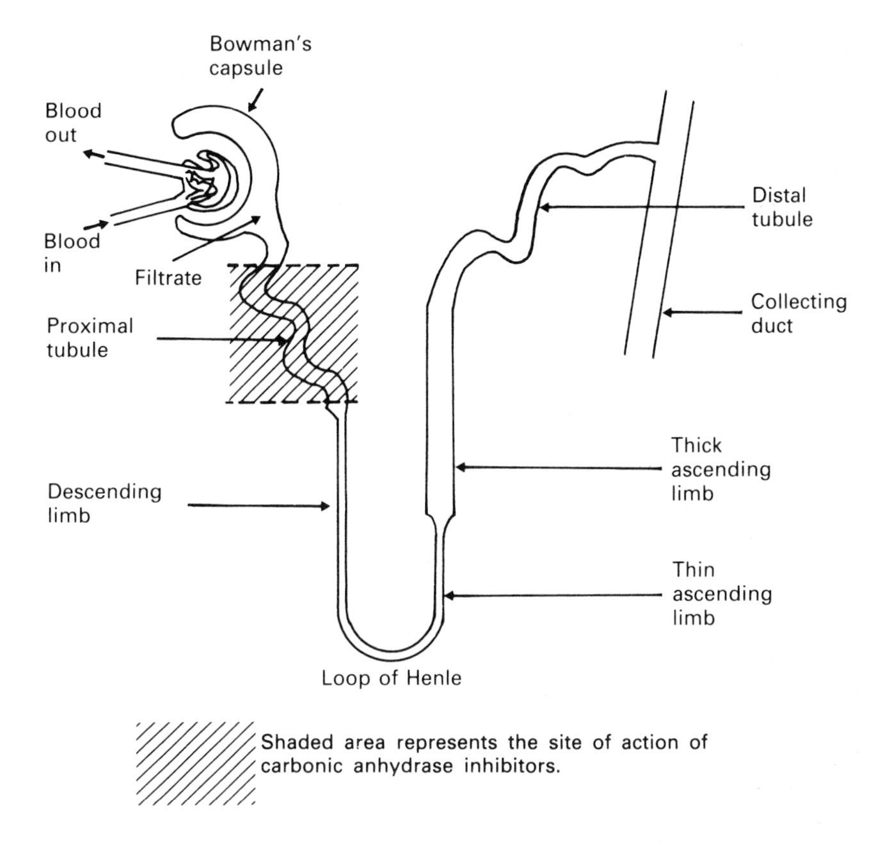

Low molecular weight substances are ultrafiltered from the blood through a capillary system in the glomerulus. The filtrate passes into Bowman's capsule and then through the proximal tubule, the loop of Henle and the distal tubule. It finally reaches a collecting duct that drains several nephrons.

Figure 8.1 Site of action of carbonic anhydrase inhibitors.

of a part in the mode of action of the thiazide diuretics. Inhibition of sodium chloride cotransport may be the key (Gesek and Friedman, 1992).

Although carbonic anhydrase inhibitors are now very rarely used as diuretics, they are sometimes used to increase the pH of the urine. This becomes necessary when, for example, drugs that cause tumours to diminish in size (oncolytic) release large amounts of the breakdown product of purines, namely uric acid. At normal pH this could begin to precipitate and so a carbonic anhydrase inhibitor is used, often in conjunction with sodium bicarbonate, to keep the acid in the form of the more soluble sodium salt.

Carbonic anhydrase inhibitors are also used for the treatment of glaucoma. The object is to reduce the volume of liquid in the aqueous humor of the eye where this has become excessive and is putting a great deal of pressure on the eyeball. The situation in the eye differs from that in the kidney in that the

(a) Proximal tubule of the kidney.

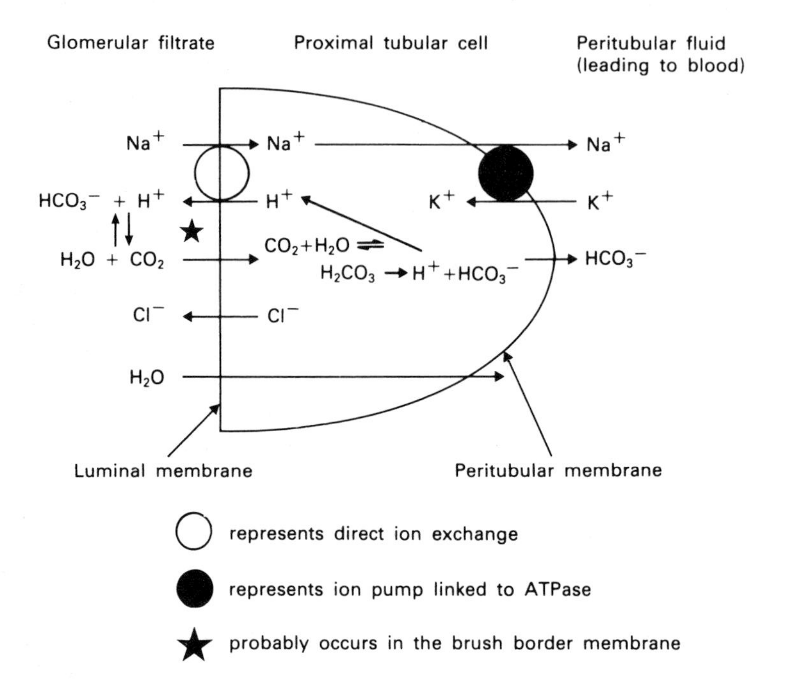

(b) Ciliary epithelium in the eye.

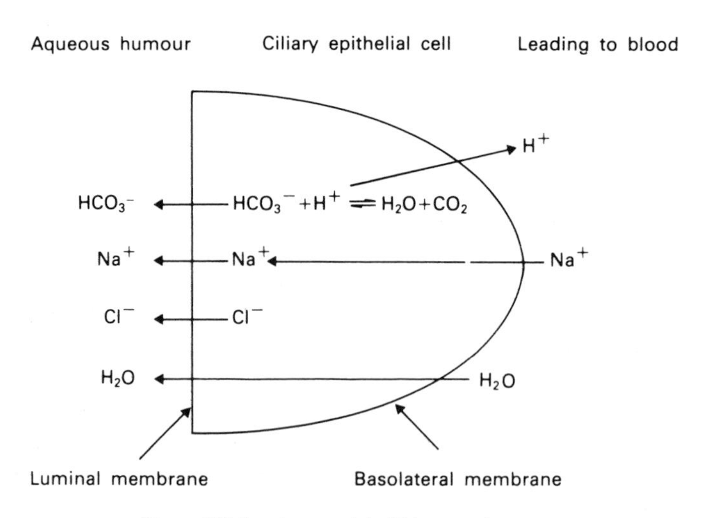

Figure 8.2 Ion transport in kidney and eye.

protons are secreted across the basolateral membrane into the blood and not into the lumen or aqueous humor. Bicarbonate ion is secreted into the aqueous humor, however, and inhibition of carbonic anhydrase reduces the level of the ion in the humor by about 50 per cent thus reducing the concomitant fluid outflow (Maren, 1984). The details of ion transport in the eye are less well understood, however, than in the proximal tubule cell of the kidney (Fig. 8.2b).

Carbonic anhydrase inhibitors are particularly useful for the treatment of glaucoma since they reduce the pressure behind the iris, and assist in opening the angle between the iris and the cornea thus allowing the aqueous humor to leak into the anterior chamber and be reabsorbed. Drugs are used in the acute condition in order to manage the attack while the patient is prepared for surgery - normally a necessity because a physical obstruction has usually induced the condition in the first place. The main intention of acute treatment is to reduce liquid volume, and thus intra-ocular pressure, so preventing damage to the optic nerve. The drugs are less useful in treating chronic primary open angle glaucoma, which slowly progresses as a consequence of a gradual closing of the channel through which the aqueous humor circulates until it may result in irreversible loss of vision.

Acetazolamide was originally used to treat glaucoma but its use has largely been superseded by the more lipophilic methazolamide. The major problem with carbonic anhydrase inhibitors given orally has been the occurrence of side-effects. Topically administered agents are now under development for glaucoma which should have less potential for causing adverse reactions (Hurvitz *et al.*, 1991).

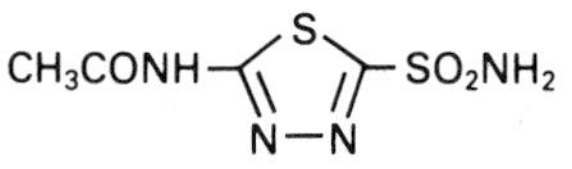

Acetazolamide

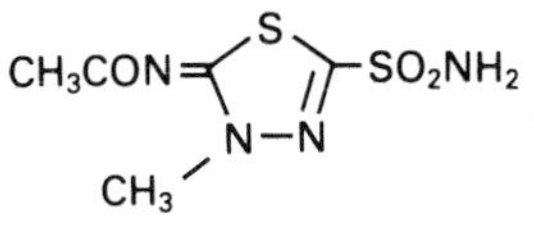

Methazolamide

8.3 Angiotensin-converting enzyme

The second zinc metalloenzyme to be discussed in this chapter is the metallopeptidase, angiotensin-converting enzyme (ACE), the search for inhibitors of which has turned out to be both a very interesting intellectual study and also of great importance medically. The search was initiated because ACE converts the biologically inactive decapeptide, angiotensin I, to the highly active pressor (blood pressure raising) octapeptide, angiotensin II, by removing the C-terminal dipeptide His-Leu (Fig. 8.3). The enzyme will also cleave the nonapeptide, bradykinin, which is a depressor agent, to inactive metabolites. ACE is still sometimes known as kininase 2, and it plays a pivotal role in the control of blood pressure.

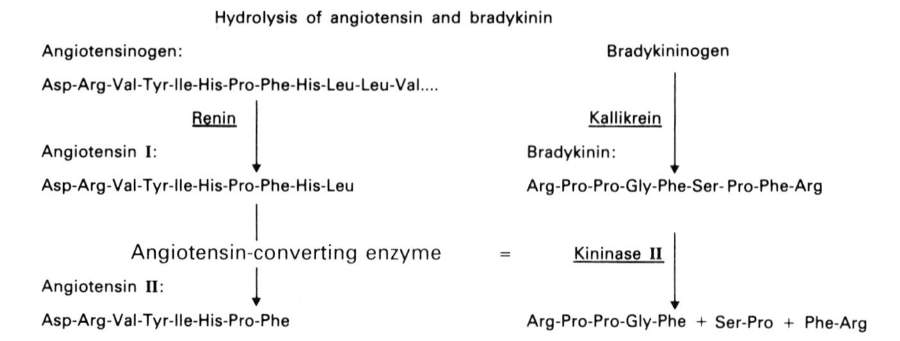

Angiotensinogen, the precursor protein, is cleaved at one specific site by renin, secreted by the kidney, to yield biologically inactive angiotensin I. This is converted by angiotensin-converting enzyme to the active octapeptide angiotensin II.

Precursor bradykininogen is converted to the active nonapeptide bradykinin by kallikrein and thence to the inactive pentapeptide by angiotensin-converting enzyme.

Figure 8.3 Hydrolysis of angiotensin and bradykinin.

In general, the enzyme prefers substrates with hydrophobic amino acids in the ante-penultimate position. It will cleave a variety of substrates, both peptide and ester, provided that the C-terminal amino acid has a free carboxyl group, and proline is not present in the penultimate position. Thus angiotensin II is not hydrolyzed further by ACE, whereas bradykinin is attacked in three places (Fig. 8.3), and so the enzyme may be regarded as not particularly specific (Ondetti and Cushman, 1984).

Angiotensin II is one of the most active biological agents known. It raises blood pressure by constricting blood vessels and, in addition, it causes aldosterone to be released from the adrenal cortex. The latter acts on the kidney to retain sodium ion, and therefore also water, thereby raising blood volume and pressure. The complex system for maintaining blood volume and pressure initially depends on the action of the endopeptidase renin. When blood volume or sodium ion concentration in the plasma falls below a certain critical value, renin is released from the kidney and acts on an α-globulin known as angiotensinogen - a glycoprotein of molecular weight 60 kDa - to yield angiotensin I. Fig. 8.4 outlines the interactions of the renin–angiotensin–aldosterone system.

It was clear in the 1970s from this knowledge that an inhibitor of ACE might be expected to have anti-hypertensive properties, and a study began to design agents that would bind to the active site. Although ACE had not been isolated at all at that time, advantage was taken of the much greater understanding of the active site of another zinc metallopeptidase, namely carboxypeptidase, an enzyme which is elaborated by the pancreas and cleaves the C-terminal amino acid from peptides and proteins.

The structures of enzymes where zinc has a catalytic role show three ligands from the protein attached to the zinc: in carboxypeptidase A they are His-69,

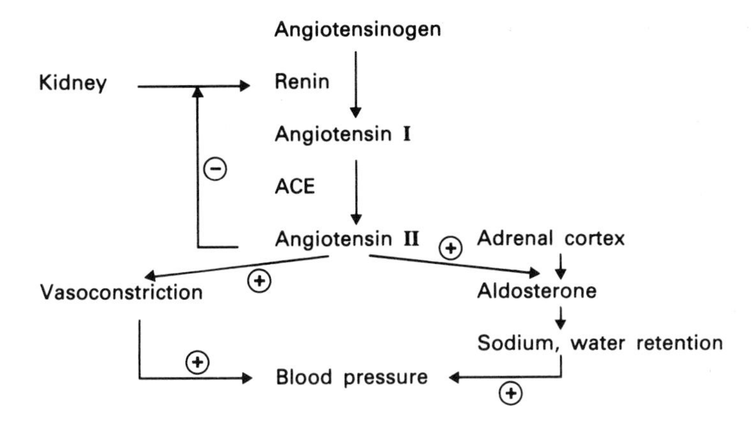

Figure 8.4 Interactions of the renin-angiotensin-aldosterone axis.

Glu-72 and His-196. The fourth ligand is the oxygen of a water molecule (reviewed in Vallee and Auld, 1990). This is replaced in the enzyme–substrate complex by the oxygen of the carbonyl group of the peptide link to be broken, and the latter is thereby polarized by the action of the zinc atom acting as a Lewis acid. Glu-270 then functions as a base or nucleophile to attack the carbon of the carbonyl group to yield a transient acyl-enzyme intermediate. Tyr-248 donates a proton to the departing amino group. The free carboxylate of the substrate is linked by attraction between a negative and positive ion (salt bridge) to Arg-145.

The nature of the group that, in effect, breaks up the enzyme–substrate complex by attacking the carbonyl carbon is likely to be a hydroxide ion, which may be activated by Glu-248 or possibly the zinc (Lipscomb, 1980). It would be intellectually satisfying if the zinc atom played a role in carboxypeptidase mechanism similar to that in carbonic anhydrase, in lowering the pK_a of the water molecule and allowing a hydroxide ion to attack a carbon of a carbonyl group. The presence of three histidine ligands (positions 94, 96 and 118) around the zinc suggests that this is the case. For a diagrammatic representation of a possible enzyme mechanism see Fig. 8.5.

Although the information regarding the active site of angiotensin-converting enzyme is still fairly sparse, it is clear that similar groups are involved in the catalysis, as shown by group-specific inactivating agents backed up by the use of ligands to protect against inactivation (Brunning, 1983; reviewed in Ondetti and Cushman, 1984). One difference between the two enzymes is the requirement for chloride ion demonstrated by ACE. Chloride ion has no effect on the

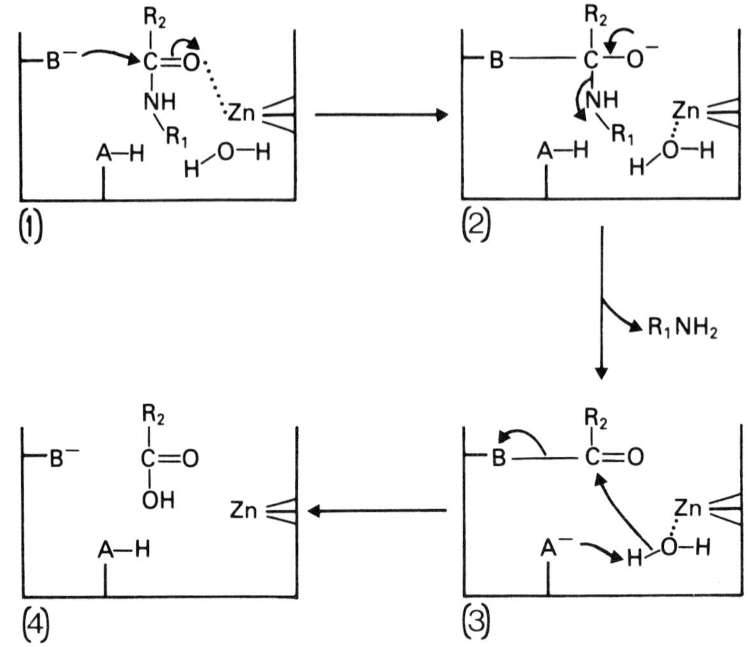

1. The base group B (Glu-270 in carboxypeptidase) attacks the carbon of the carbonyl group already distorted by activation in the zinc.
2. An acyl-enzyme intermediate is formed, and then rearranges to release the amine R_1NH_2 taking the proton from the acid group A (Tyr-248).
3. The hydroxyl of the water molecule activated by the zinc attacks the carbonyl group to yield the acid R_2COOH, while acid group A receives the proton.
4. The acid R_2COOH is released and the active site resumes its original format.

Figure 8.5 Putative Zn-metallopeptidase mechanism.

rate of modification of arginyl, carboxyl and tyrosyl residues by these reagents, but it does affect the rate of reaction of pyridoxal phosphate with the enzyme, which suggests that lysine, which forms a Schiff's base with the aldehyde group of pyridoxal, may bind the halide ion. Activation by chloride ion is seen with most substrates and at all pH values, and is characterized by a lowering of K_m but no change in K_{cat} (Ondetti and Cushman, 1984). Recent studies have defined the active site glutamic acid residue in bovine lung ACE in the sequence: H_2N-Phe-Thr-Glu-Leu-Ala-Ser-Glu. This sequence shows considerable homology with bovine carboxypeptidase A, but little with the bacterial zinc endopeptidase, thermolysin (Harris and Wilson, 1985).

8.3.1 Angiotensin-converting enzyme inhibitors - captopril and enalapril for hypertension

The story of the development of the ACE inhibitors that are marketed today is a triumph of logical and imaginative planning, combining features of both substrates and inhibitors. As mentioned above, the information about the active site of ACE was strictly limited. Nevertheless, the potential similarities between ACE and carboxypeptidase A were exploited to the full. Early studies with inhibitory peptides showed that having proline or an aromatic amino acid in the C-terminal position gave the most effective inhibitors, but proline inhibitors gave the better results *in vivo*; Val-Trp is some 150 times more active *in vitro* than Ala-Pro but no more active *in vivo* (Petrillo and Ondetti, 1982).

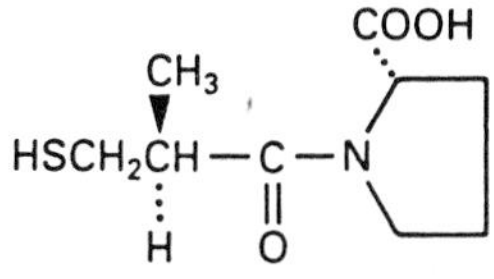

Captopril

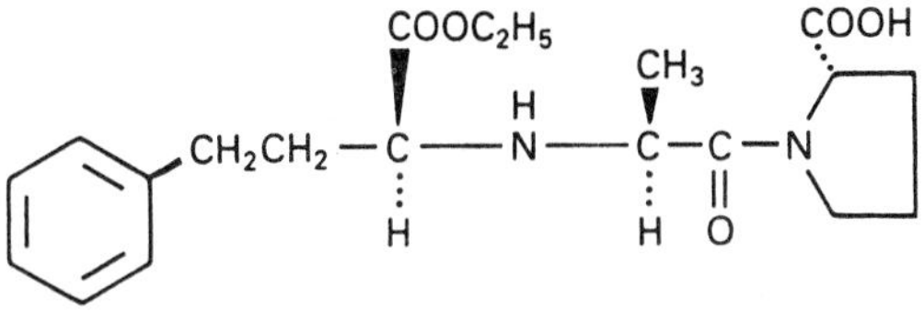

Enalapril

Succinylproline could be regarded as a good starting point for inhibitor development with four sites interacting with the protein; the carboxylate group of the proline, the proline ring itself, the carbonyl of the amide link and the carboxylate of the succinate all binding to the zinc atom (I_{50} of $3{\cdot}3 \times 10^{-4}$ M). A methyl group placed on the carbon atom next to the carbonyl group of the amide improved the binding markedly. Replacement of the carboxylate by a sulphydryl gave a further enormous improvement, to produce captopril with an I_{50} of $2{\cdot}3 \times 10^{-8}$ M. Enalaprilic acid was derived by three further changes; the addition of further binding affinity by a phenethyl group (possibly interacting with the binding site for the phenylalanine side-chain characteristic of angiotensin I), together with a restoration of the carboxyl group in place of sulphydryl, and the introduction of a secondary amino group. These changes lowered the I_{50} to 1×10^{-9} M (reviewed in Ondetti and Cushman, 1984; Mackaness, 1985) (Fig. 8.6). Captopril has two, and enalapril three, optically active carbon atoms, and interestingly all five centres are in the *S* configuration (Patchett *et al.*, 1980: Brittain and Kadin, 1990).

Detailed kinetic studies with different substrates and inhibitors, excluding captopril and enalaprilic acid, have revealed a complex pattern of interactions. It appears that an inhibitor can bind to more than one form of the enzyme and, in addition, there are parallel reaction pathways. Moreover, both captopril and enalaprilic acid are slow and tight binding inhibitors (Morrison, 1982). As a consequence, sufficient time must be allowed for complete binding to place, and, in addition, allowance must be made for the depletion of active enzyme by drug binding as both drug and enzyme concentrations are in the nanomolar

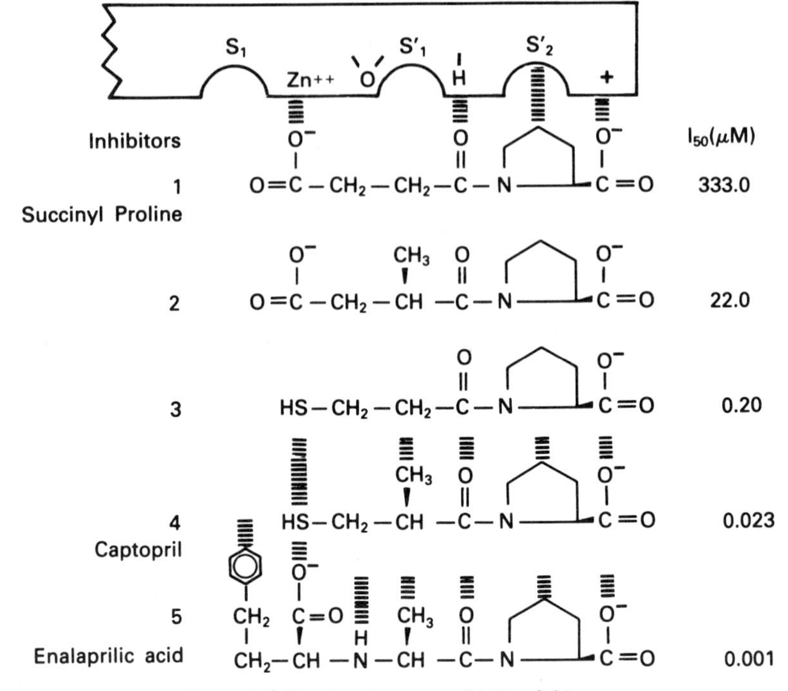

Figure 8.6 The development of ACE inhibitors.

range. K_i values of $3{\cdot}3 \times 10^{-10}$ M and $5{\cdot}0 \times 10^{-11}$ M have been calculated for captopril and enalaprilic acid respectively (Shapiro and Riordan, 1984). A two-step inhibition is identified characterized by a fast step followed by a slow second one which is probably an isomerization of the complex. The overall type of inhibition is judged to be competitive.

8.3.2 Pharmacology of ACE inhibitors

Captopril and enalapril have made a major contribution to our understanding of the mechanism for maintaining blood pressure as well as to the treatment of hypertension. These drugs lower angiotensin II levels in the peripheral network of blood vessels, causing them to dilate, and consequently lower blood pressure. There is no particular effect on the force or rate of contraction of the heart (Todd and Goa, 1992). Enalaprilic acid is given as the ethyl ester, enalapril, as the acid is poorly absorbed from the gastrointestinal tract. Subsequent hydrolysis by esterases yields the active agent.

In the clinic, captopril and analapril are extremely effective at lowering the blood pressure in hypertensive patients, particularly those in whom the renovascular system is the originator of the condition and the renin-angiotensin-aldosterone axis is involved. Elevated aldosterone and angiotensin II levels in the plasma lead to increased sodium and water retention and so to increased blood pressure. Treatment with ACE inhibitors lowers these levels to normal. The efficacy of the drugs extends to essential hypertension, i.e. of unknown origin, whether mild, moderate or severe, but in these cases a diuretic is usually prescribed, in addition, to lower sodium levels and blood volume and thereby assist the action of the ACE inhibitor (Brunner *et al.*, 1983; Ferguson *et al.*, 1984).

Since these drugs were designed to combat angiotensin II action, we might expect a correlation between initial plasma renin concentration and the blood pressure lowering effects, and this is indeed found to be the case for captopril (see Ferguson *et al.*, 1984) and enalapril (see Brunner *et al.*, 1983). Furthermore, the levels of angiotensin II prior to treatment correlate with falls in blood pressure as a consequence of treatment (MacGregor *et al.*, 1983). Plasma renin rises as a consequence of captopril treatment - not unexpectedly in view of the positive feedback effect of lowered sodium levels and blood volume on the secretion of renin. Plasma renin may not be the only target, however, because both renin and ACE have been found in the blood vessel walls, where renin levels are elevated as a consequence of captopril treatment (Unger *et al.*, 1983). In addition, the renin-angiotensin system has been found in brain and it is possible that ACE inhibitors that can cross the blood-brain barrier will have a further antihypertensive effect (Unger *et al.*, 1983).

In principle, we would expect that plasma renin would be an indicator of blood pressure, since elevated blood pressure usually acts to lower plasma renin levels. This does not always seem to occur, however, since a number of patients with high blood pressure have normal, instead of lowered, renin levels (Lindpainter *et al.*, 1988). ACE inhibitors are found to work even when renin levels are normal. As noted above, the renin-angiotensin system is found outside the kidney and a major contribution may be made by the heart. Angiotensin II produced in the heart may assist in contraction, but on the negative side produces vasoconstriction thus lowering the supply of oxygen to the heart with the risk of producing ischaemia and arrhythmias (Grinstead and Young, 1992).

The ACE inhibitors produce relatively few side-effects but dry cough and proinflammatory properties may result from inhibition of bradykinin hydrolysis, also carried out by ACE. Angiotensin II antagonists are being developed as more specific antihypertensives (Smith *et al.*, 1992).

There are two other effects of these drugs that have emerged as a consequence of their use. One is the interaction of angiotensin II with the transmission of adrenergic impulses, as the peptide stimulates the release of noradrenaline from presynaptic vesicles and inhibits the re-uptake of the neurotransmitter. Consequently, angiotensin II potentiates the vasoconstrictor response to sympathetic nerve stimulation as well as to exogenously applied noradrenaline. Clearly, the inhibition of angiotensin II production will limit this effect, and supplement the action of ACE inhibitors as vasodilators (reviewed in Unger *et al.*, 1983).

Another very interesting effect that seems to be associated with ACE inhibitor treatment is euphoria (Zubenko and Nixon, 1984). The reason for this is not entirely certain, but it may be due to the inhibition of the hydrolysis of the enkephalin precursor, [Met]-enkephalin-Arg^6-Phe^7 which is probably carried out by ACE in the brain (Normal *et al.*, 1985). It is already clear that enkephalinase, which cleaves [Met]-enkephalin at the Gly^3-Phe^4 bond, is not inhibited by captopril, and so it is intriguing to speculate that other opioid peptides are responsible for the euphoria; particularly as [Met]-enkephalin-Arg^6-Phe^7 is derived physiologically from the proenkephalin A precursor protein. Opioid peptides are discussed in section 9.7 in Chapter 9.

8.4 Endopeptidase 24.11 - thiorphan as analgesic

The treatment of pain in an effective but non-addictive manner remains one of the most important pharmacological goals of the future. The discovery of the opioid peptides as the endogenous ligands for the morphine receptor suggested that prolongation of their action could lead to a new treatment of pain, as an alternative to the use of morphine (Chapter 9). This approach has led to two structures under clinical trial: thiorphan for pain and acetorphan for diarrhoea. Acetorphan is the *S*-acetylphenyl ester of thiorphan and could be regarded as a pro-drug.

One enzyme that can hydrolyze enkephalins is endopeptidase 24.11 (it has been given this curious shorthand as an abbreviation of the Enzyme Commission number 3.4.24.11). The enzyme has been isolated from kidney microvilli and shown to be identical with an enzyme from pig brain. The enzyme should not, strictly speaking, be referred to as enkephalinase because it will hydrolyze various peptide substrates (Hersch, 1986). Nevertheless, in the central nervous system its main role does appear to be termination of the action of enkephalins by cleaving [Met]- and [Leu]-enkephalin to yield Tyr-Gly-Gly and either Phe-Leu or Phe-Met - acting, in fact, as a dipeptidylcarboxypeptidase (reviewed in Roques *et al.*, 1993).

Endopeptidase 24.11 is a zinc-containing enzyme and bears a considerable resemblance to the family of zinc metallopeptidases. Approaches to inhibition based on the strategy that originally led to the angiotensin-converting enzyme inhibitors has yielded a compound that is analogous to captopril but which shows considerable selectivity for endopeptidase 24.11 (Mumford *et al.*, 1982).

As in captopril binding to ACE, the sulphur atom of thiorphan binds to the zinc and the carboxylate to an arginine on the enzyme surface. Not all activity is lost, however, with these two sites blocked because acetorphan has a K_i of 5×10^{-6} M. The proposal is that the heteroatoms in the amide bond also make an important contribution to the binding by forming hydrogen bonds to the enzyme (Monteil *et al.*, 1992).

Enkephalinase also inactivates atrial natriuretic factor, a hormone produced by the heart which has sodium excreting (natriuretic) and diuretic activity amongst other attributes. Enkephalinase inhibitors may also be useful for potentiation of the effects of atrial natriuretic factor (Schwartz *et al.*, 1990; Roques and Beaumont, 1990).

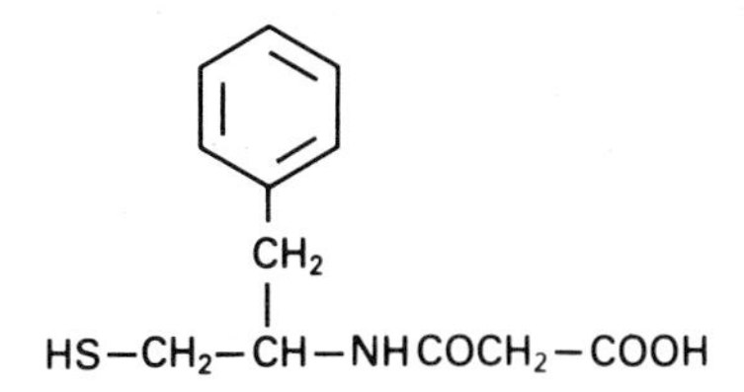

Thiorphan

Questions

1. What features of the zinc atom allow it to form covalent complexes?
2. Why do cells have carbonic anhydrase?
3. What is the present therapeutic use of carbonic anhydrase inhibitors?
4. What chiral centres does captopril have?
5. Why do ACE inhibitors lower blood pressure in patients whose renin levels are normal?
6. What biological effects does angiotensin show?
7. What therapeutic effects would an enkephalinase inhibitor be expected to show?

References

Brittain, H.G. and Kadin, H., 1990, *Pharm. Res.*, **7**, 1082-5.

Brunner, H.R., Turini, G.A., Waeber, B., Nussberger, J. and Biollaz, J., 1983, *Clin. Exper. Hypertens.*, **A5**, 1355-66.

Brunning, P., 1983, *Clin. Exper. Hypertens.*, **A5**, 1263-75.

Coleman, J.E., 1984, *Ann. N. Y. Acad. Sci.*, **429**, 26-48.

Cook, C.M. and Allen, L.C., 1984, *Ann. N. Y. Acad. Sci.*, **429**, 84-8.

Deutsch, H.F., 1984, *Ann. N. Y. Acad. Sci.*, **429**, 183-94.

DuBose, T.D., Pucacco, L.R. and Carter, N.W., 1981, *Am. J. Physiol.*, **240**, F138-46.

Ferguson, R.K., Vlasses, P.H. and Rotmensch, H.H., 1984, *Am. J. Med.*, **77**, 690-8.

Gesek, F.A. and Friedman, P.A., 1992, *J. Clin. Invest.*, **90**, 429-38.

Grinstead, W.C. and Young, J.B., 1992, *Amer. Heart J.*, **123**, 1039-45.

Harris, R.B. and Wilson, I.B., 1985, *J. Biol. Chem.*, **260**, 2208-11.

Hurvitz, I.M., Kaufman, P.L., Robin, A.L., Weinreb, R.N., Crawford, K. and Shaw, B., 1991, *Drugs*, **41**, 514-32.

Kokko, J.P., 1984, *Am. J. Med.*, **77**, (5A), 11-17.

Lindpainter, K., Jin, M., Wilhelm, M.J., Suzuki, F., Linz, W., Schoelkens, B.A. and Ganten, D., 1988, *Circulation*, **77**, Suppl. 1, 18-23.

Lipscomb, W.N., 1980, *Proc. Nat. Acad. Sci. (U.S.A.)*, **77**, 3875-8.

Lucci, M.S., Tinker, J.P., Weiner, I.M. and DuBose, T.D., 1983, *Am. J. Physiol.*, **237**, F443-9.

MacGregor, G.A., Markandu, N.D., Smith, S.J., Sagnella, G.A. and Morton, J.J., 1983, *Clin. Exper. Hypertens.,* **A5**. 1367-80.
Mackaness, G.B., 1985, *J. Cardiovasc. Pharmacol.,* **7**, Suppl. 1, S30-4.
Maren, H., 1984, *Ann. N. Y. Acad. Sci.,* **429**, 568-79.
Monteil, T., Kotera, M., Duhamel, I., Duhamel, P., Gros, C., Noel, N., Schwartz, J.C. and Lecomte, J.M., 1992, *Bioorg. Med. Chem. Lett.,* **2**, 949-54.
Morrison, J.F., 1982, *Trends in Biochem. Sci.,* **7**, 102-5.
Mumford, R.A., Zimmerman, M., ten Broeke, J., Joshua, H. and Bothrock, J.W., 1982, *Clin. Exper. Hypertens.,* **A5**, 1362-80.
Norman, J.A., Autry, W.L. and Barbaz, B.S., 1985, *Molec. Pharmacol.,* **28**, 521-6.
Ondetti, M.A. and Cushman, D.W., 1984, *CRC Crit. Rev. Biochem.,* **16**, 381-411.
Patchett, A.A., *et al.*, 1980, *Nature,* **288**, 280-3.
Petrillo, E.W. and Ondetti, M.A., 1982, *Medicinal Res. Rev.,* **2**, 1-41.
Roques, B.P. and Beaumont, A., 1990, *Trends in Pharmacol. Sci.,* **11**, 245-9.
Roques, B.P., Noble, F., Dange, V., Fournie-Zaluski, M-C. and Beaumont, A., 1993, *Pharmacol. Rev.,* **45**, 87-146.
Schwartz, J.C., Gros, C., Lecomte, J.M. and Bralet, J., 1990, *Life Sci.,* **47**, 1279-97.
Shapiro, R. and Riordan, J.F., 1984, *Biochemistry,* **23**, 5225-33.
Silverman, D.N. and Lindskog, 1988, *Acc. Chem. Res.,* **21**, 30-36.
Smith, R.D., Chiu, A.T., Wong, P.C., Herblin, W.F. and Timmermans, P.B.M.W.M., 1992, *Annu. Rev. Pharmacol. Toxicol.*, **32**, 135-65.
Todd, P.A. and Goa, K.L., 1992, *Drugs,* **43**, 346-81.
Unger, T., Ganten, D. and Lang, R.E., 1983, *Clin. Exper. Hypertens.,* **A5**, 1333-54.
Vallee, B.L. and Auld, D.S., 1990, *Proc. Nat. Acad. Sci. (U.S.A.),* **87**, 220-4.
Vallee, B.L. and Galdes, A., 1984, *Adv. Enzymol.,* **56**, 281-430.
Vedani, A. and Meyer, E.F., 1984, *J. Pharm. Sci.,* **73**, 352-8.
Zubenko, G.S. and Nixon, R.A., 1984, *Am. J. Psychiat.,* **141**, 110-11.

Chapter 9

Neurotransmitter action and metabolism

9.1 *Introduction*

In this chapter we consider the drugs that have been developed deliberately, or found in hindsight (serendipitously), to modulate neurotransmitter action or metabolism. In order to understand more about the way in which drugs can interfere with the transmission of nerve impulses we need to know how impulses are transmitted along a nerve.

The resting nerve cell or neurone maintains a potential difference between the cytoplasm and extracellular fluid of approximately 75 mV (cytoplasm negative). This is achieved largely by the functioning of a membrane-bound sodium pump acting to export positive sodium ions from the cell using the energy of hydrolysis of ATP. The membrane is practically impermeable to sodium ions at this potential, thus a tenfold concentration gradient is maintained between the nerve cell cytoplasm and the extracellular fluid.

An impulse may be generated by a variety of means such as heat, mechanical deformation, specific chemicals or neurotransmitters etc., depending on the type of neurone in question, and it produces a reduction in the membrane potential difference to approximately 60 mV. When this happens, 'gates' or channels in the membrane open at one end of the neurone, which are specific for the active transport of sodium ions, causing the sodium cation to rush in and thereby neutralize the negative charge inside the membrane (depolarization). The neurone remains in this state of depolarization, usually for 1 to 3 milliseconds in man, until sufficient potassium ion has left the cell through specific channels to repolarize the membrane. This condition is also known as refractory because no further impulses can pass until the resting potential of 75 mV is re-established. The intracellular concentrations of potassium and sodium ion are then restored by the action of the Na^{+}/K^{+}-adenosine triphosphatase which uses the energy of hydrolysis of ATP to drive an exchange of sodium inside for potassium outside the cell. This enzyme is discussed further in Chapter 10.

The impulse or action potential passes down the nerve cell until it reaches the end of the cell body opposite another nerve cell at a junction called a

Table 9.1 Drugs and their targets discussed in Chapter 9

Section	*Receptor/Enzyme*	*Type*	*Drug*	*Use in*
9.2	Adrenergic	α_2	Clonidine	Hypertension
		α_1	Prazosin	Hypertension
		β	Propranolol	Arrhythmia
		β_1	Metoprolol	Hypertension
		β_1	Atenolol	Hypertension
		β_2	Salbutamol	Asthma
			Salmeterol	
		α,β	Labetalol	Hypertension
9.2	Dopamine	D_1	Bromocryptine	Parkinsonism
		D_1	L-Dopa + carbidopa	Parkinsonism
		D_1	Fenoldapam	Hypertension
		D_2	Haloperidol	Schizophrenia
		D_2	Pimozide	Schizophrenia
9.4	Serotonin	5-HT_{1A}	Buspirone	Anti-anxiety
		5-HT_{1D}	Sumatriptan	Migraine
		5-HT_2	Ketanserin	Hypertension
		5-HT_2	Mianserin	Depression
		5-HT_3	Ondansetron	Migraine
			Granisetron	
9.5	Serotonin/ Noradrenaline uptake		Imipramine	Depression
			Iprindole	Depression
			Maprotiline	Depression
			Fluoxetin	Depression
9.6	Monoamine oxidase	MAOA	Clorgyline	Depression
		MAOA	Moclobemide	Depression
9.7	Cholinergic Muscarinic	M_1	Pilocarpine	Glaucoma
		M_1	Pirenzepin	Ulcers
		$M_{1,2}$	Atropine	Pre-anaesthetic medication
	Acetylcholinesterase		Pyridostigmine	Myasthenia gravis
			Neostigmine	Myasthenia gravis
9.8	4-Aminobutyrate	$GABA_A$	Chlordiazepoxide	Hypnotic/sedative/muscle relaxant
		$GABA_A$	Diazepam	Helminth infection
		$GABA_A$	Avermectin	Spasticity
		$GABA_B$	Baclofen	
9.9	Opiate	$\mu(\kappa,\sigma)$	Morphine	Pain
		$\mu(\kappa,\sigma)$	Pentazocine	Pain
		$\mu(\kappa,\sigma)$	Loperamide	Diarrhoea
9.10	Histamine	H_1	Mepyramine	Allergic reactions
		H_1	Chlorpheniramine	Hay fever
		H_2	Cimetidine	Ulcers
		H_2	Ranitidine	Ulcers

For some drugs there are additional receptors to which the drug may bind and which may contribute either to its mode of action and/or to its toxicity. Chlorpromazine, for example, is known to bind to several other types of receptor, possibly because of its affinity for membranes, and these effects are likely to contribute to its toxicity.

Furthermore, some of these drugs have other uses, as for example, neostigmine for the reversal of neuromuscular blockade in surgery.

synapse. The cell from which the action potential arrives at the synapse is known as the presynaptic cell. Communication with the second (post-synaptic) cell is performed by a chemical agent known as a neurotransmitter, specific for that synapse. The chemical is stored in vesicles immedately before the synapse and the release of these is triggered by the arrival of the action potential. The contents of the vesicles are extruded into the gap between the two nerve cells, known as the synaptic cleft, and the neurotransmitter passes across to bind to specific protein targets on the post-synaptic cell membrane, known as receptors. This complex formation may elicit a response of either an excitatory nature, i.e. a continuing impulse (accompanied by depolarization) or an inhibitory response, i.e. to prevent an impulse arising in the responding cell. The latter condition is accompanied by *hyper*-polarization up to -100 to -120 mV, usually by the inward transport of anions such as chloride ion, and this renders the cell insensitive to stimulation.

Clearly, the action of the neurotransmitter needs to be terminated as soon as it is completed, and hydrolytic enzymes accomplish this action; acetylcholine is hydrolysed by acetylcholinesterase in the synaptic cleft, while the transmitters that contain a primary amine group (aminergic) such as noradrenalin, dopamine and serotonin are taken up into the presynaptic cell and oxidatively deaminated to aldehydes by monoamine oxidase. Catecholamine transmitters may also be taken up outside the neurone where catechol *O*-methyltransferase renders the amine inactive by catalysing the transfer of a methyl group to the *meta*-hydroxyl on the catechol ring.

For a substance to be confirmed as a neurotransmitter, two major actions have to be seen: firstly the agent must be released from the appropriate nerve ending by an impulse, and secondly the action of the compound when applied to post-synaptic membranes must mimic the action that occurs when the nerve is stimulated in the normal way.

Furthermore, a number of other factors may be taken into account. The compound has to be present in the appropriate nerve terminals in sufficient quantity to act as a transmitter. This means that the enzymes for its biosynthetic pathway must occur in those terminals. Substances that interfere with the binding of the natural ligand to its receptor in subcellular preparations should have a similar action on the natural transmitter released in the normal way. In addition, there is usually a re-uptake mechanism and/or a metabolic enzyme for removal of the transmitter from the synaptic cleft. These conditions have been met for all of the transmitters covered in this chapter, with the possible exception of histamine. The similarities of drugs binding to histamine receptor compared with ligand binding to the other receptors discussed in this chapter suggest its inclusion here.

It should be noted that receptor binding requires the correct stereochemical form of an agonist. In the case of the catecholamines it is the naturally occurring L-form that is active while the D-isomers are at least ten times less active. Mere binding of an agent to a membrane preparation that contains a receptor does not guarantee that a pharmacological response will ensue. Parallel studies on isolated tissues or whole animals should also be carried out to confirm the receptor occupancy as a valid pharmacological result. The cloning of the requisite gene will confirm the existence of the receptor.

Membrane-bound receptors are discussed in this chapter (for intracellular receptors see Chapters 6 and 7). There are various receptor families now known: one is linked directly to an ion channel. Another very widespread theme is the receptor linked via a guanine nucleotide binding protein (G-protein) to an intracellular signalling system.

9.1.1 G-Protein linked receptor structure

Bacteriorhodopsin is the model protein for the G-protein-linked receptors because it is the only structure subjected to experiment. Bacteriorhodopsin is not a receptor at all but a visual pigment, acting as a proton pump, found in the membranes of the bacterium *Halobacterium halobium*. A characteristic feature of bacteriorhodopsin is seven α-helices arranged in a bundle perpendicular to the plane of the lipid bilayer (see Savarese and Fraser, 1992).

The structure of G-protein-linked receptors has been analysed by analogy with bacteriorhodopsin through the hydrophobicity of the amino acids in a given region of the protein. The hydrophobic regions would be expected to lie in the hydrophobic environment of the membrane, while the more hydrophilic or polar amino acids would be more favourable sited in the aqueous environment either inside or outside the cell. The receptors have seven domains, comprised mainly of hydrophobic amino acids which are modelled to span the membrane with intervening cytoplasmic and extracellular loops.

The receptor structure can be divided into three domains: extracellular which includes the amino terminus where there is a major site of *N*-glycosylation and the loops between transmembrane regions 2 and 3, 4 and 5, 6 and 7; the membrane domain which consists of the seven transmembrane regions and the intracellular area with the carboxyl terminus and the loops between transmembrane regions 1 and 2, 3 and 4, 5 and 6. They are often referred to as 7-transmembrane (TM) receptors and are common to a wide variety of ligands discussed in this chapter including noradrenaline, dopamine, serotonin and histamine.

From mutagenesis studies, the ligand-binding site is believed to lie in the transmembrane region. The structure of the aminergic hormones, noradrenaline, dopamine and serotonin all have a cationic amino function and an aromatic ring, while acetylcholine also has a positive charge based on the quaternary nitrogen. The anionic carboxylate of an aspartate is probably the binding site for this charge (Asp-113 in TM3 in β-adrenergic receptors). A serine (Ser-204 in transmembrane-5 TM5) is likely to form a hydrogen bond with the catechol moiety in noradrenaline and dopamine and the 5-hydroxy group in serotonin. A phenylalanine in TM6 probably stacks hydrophobically with the aromatic ring. The different receptors are similar but sufficiently different to bind different agonists. One of the most interesting aspects of these studies is to define the molecular difference between agonists and antagonists, i.e. which amino acid is involved in linking binding to G-protein activation. It appears that Asp-79 in TM2 may be the crucial residue in the β_2-receptor because the other interactions noted above are essential for agonists and antagonists alike, while alteration of Asp-79 only reduces agonist binding, not antagonist binding. The third cytoplasmic loop also plays a part in the activation of the G-protein leading to adenylate cyclase or phospholipase activation (Shih *et al.*, 1991; Saverese and Fraser, 1992).

9.1.2 Signal transduction and receptor occupancy

The chain of events that connect receptor occupancy with cell response is not fully understood. Nevertheless it is clear that there are at least three principal methods of converting the extracellular signal to activate intracellular pathways. Two methods are enzymic. One involves the activation of the membrane-bound enzyme adenylate cyclase which catalyzes the conversion of ATP to cyclic adenosine 3′,5′-monophosphate (cAMP) - this in turn activates a specific protein kinase which is crucial in regulating the activity of many enzymes by controlling their phosphorylation state. The second pathway involves hydrolysis of phospholipids in the membrane to yield prostaglandins and leukotrienes, as well as release of calcium from intracellular stores, leading to a wide variety of physiological effects. The third method is direct stimulation of a cation channel, whether sodium, potassium or calcium, which allows the cation to pass into the cell (see Chapter 10 for a discussion on ion channels). Usually a particular type of receptor will only activate one or other of these pathways directly, although there may be interaction between receptors which can lead to a more complex picture of pathway stimulation.

G-proteins play a crucial role in signal transduction; they act as intermediaries between receptor occupancy and either enzymatic pathways or the activation of some ion channels. There are six types of G-protein - three of which are involved in drug action: G_s, the G_i proteins, and G_O. G_s activates adenylate cyclase and calcium channels, the G_i family inhibit adenylate cyclase and activate phospholipase A_2 and potassium channels while G_O regulate voltage sensitive calcium channels (Birnbaumer *et al.*, 1990). A recently discovered family of G_q stimulate phospholipase C (Sternweis and Smrcka, 1992).

The G-protein is a heterotrimer consisting of an α, β and γ subunit. At rest, G-protein is linked to the receptor and GDP is bound to the α-subunit; when an agonist binds to the receptor, GTP is exchanged for GDP and the α-subunit dissociates from the rest of the trimer and activates the enzyme. GTP is hydrolysed to GDP after one catalytic turnover of the enzyme, thus rendering the complex inactive until GTP again replaces GDP (Savarese and Francis, 1992). The complex of $\beta\gamma$ is also able to regulate some forms of adenylate cyclase (Taussig *et al.*, 1993) and phospholipase C (Sternweis and Smrcka, 1992).

When adenylate cyclase is activated, the rate of breakdown of adenosine triphosphate to adenosine 3′,5′-cyclic monophosphate and pyrophosphate is raised from a very low, almost undetectable level in some cases, and the cyclic nucleotide is able to carry out various reactions - the most important of which is the activation of protein kinases. These exist as inactive tetramers composed of two catalytic and two regulatory subunits. Cyclic AMP binds to the regulatory subunits causing them to dissociate from the catalytic dimer, leaving the latter free to express its activity in phosphorylating a number of crucial proteins. These include triglyceride lipase - that catalyses the initial and rate-limiting step in lipolysis - and the phosphorylase kinase which activates phosphorylase, again by phosphorylation, to hydrolyse glycogen to glucose (Gilman, 1984).

The process as a type of cascade allows a considerable amplification of the original hormone signal through the catalytic activity of adenylate cyclase and the protein kinases. The adenylate cyclase system is outlined in Fig. 9.1, while

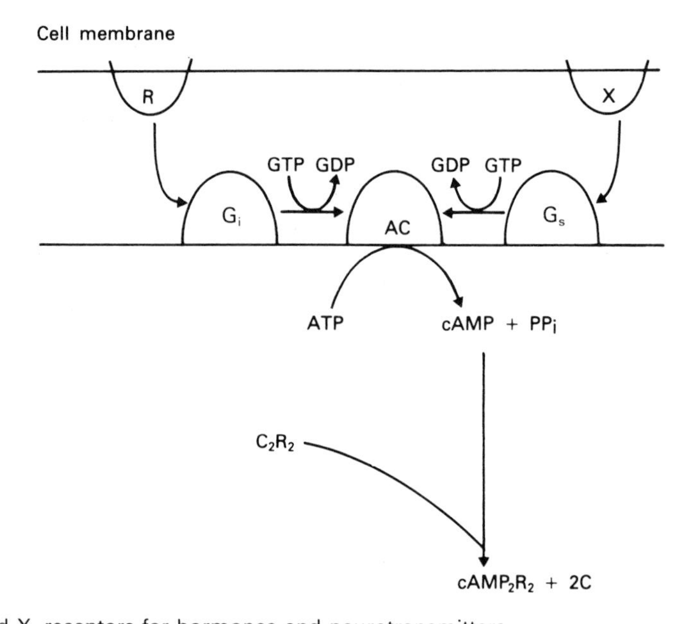

R and X, receptors for hormones and neurotransmitters
G_1, the inhibitory guanine nucleotide binding protein
G_s, the stimulatory guanine nucleotide binding protein
AC, adenylate cyclase catalytic subunit
cAMP, cyclic AMP; PP_i, pyrophosphate
C_2R_2, the tetramer of protein kinase A; C is the catalytic, R the regulatory, subunit

The binding of a hormone or neurotransmitter to receptors R or X causes release of G_i or G_s, thus allowing it to interact with adenylate cyclase. For every turnover of the enzyme, one molecule of GTP is hydrolysed to GDP. The enzyme catalyses the hydrolysis of ATP to cAMP and PP_i. cAMP binds to the regulatory subunit of protein kinase thus allowing the catalytic subunit to phosphorylate various proteins with crucial rate-limiting roles.

Figure 9.1 Adenylate cyclase modulation by receptor occupancy.

the receptors discussed in this chapter are listed by the signal transduction pathway that they activate in Table 9.2.

The second major method of signal amplification that the target cell uses is via hydrolysis of the phospholipid phosphatidylinositol that is found primarily in the plasma membrane of the cell. When membrane receptors are activated through ligand binding, a chain of events is set in motion whereby phospholipase C cleaves the phospholipid to release, amongst other products, arachidonic acid which is further metabolized to prostaglandins, thromboxanes and leukotrienes (Chapter 7).

The other products of phospholipid hydrolysis are diacylglycerol and either inositol 1-phosphate, the 1,4-diphosphate or the 1,4,5-triphosphate. Diacylglycerol activates protein kinase C in the presence of calcium ion and phospholipid to phosphorylate a number of proteins (Fig. 9.2). The triphosphate acts intracellularly to induce the release of calcium from the endoplasmic reticulum in order to produce a transient rise in the cytoplasmic concentration of the cation.

This transient rise in intracellular calcium ion level is sufficient, for example, to sustain the release of aldosterone by angiotensin II (reviewed in Rasmussen, 1986), but a more prolonged charge is required for smooth muscle contraction and other cellular processes (Berridge, 1993). This is obtained by a cycling

Table 9.2 Association of ligands with signal transduction pathways

	Receptor subtype	*GTP binding protein*
1. Adenylate cyclase		
Serotonin	5-HT_{1A}, 5-HT_{1B}, 5-HT_{1D}	G_i
Muscarinic	M_2, M_4	G_s
GABA	B	G_i
Opiate	μ, δ	G_i
Histamine	H_2	G_i
Dopamine	D_1	G_i
	D_2	G_s
Adrenergic	α_2	G_i
	$\beta_{1,2,3}$	G_i
		G_s
2. Inositol phospholipid		
Serotonin	5-HT_{1C}, 5-HT_2	G_q
Muscarinic	M_1, M_3, M_5	G_q
Histamine	H_1	G_q
Adrenergic	α_1	G_q

G_i represents the inhibitory guanine nucleotide regulatory protein leading to inhibition of adenylate cyclase.
G_s represents the stimulatory GTP binding protein.
G_q activates phospholipase C.
All the various receptors discussed in this chapter are listed, in order of the discussion, with respect to the signal transduction pathway that they activate. Exceptions are $GABA_A$ which does not appear to operate through either of these pathways and, moreover, unlike the receptors above, contains a chloride channel as part of the receptor; opiate (κ) and 5-HT_3 operates a certain channel.

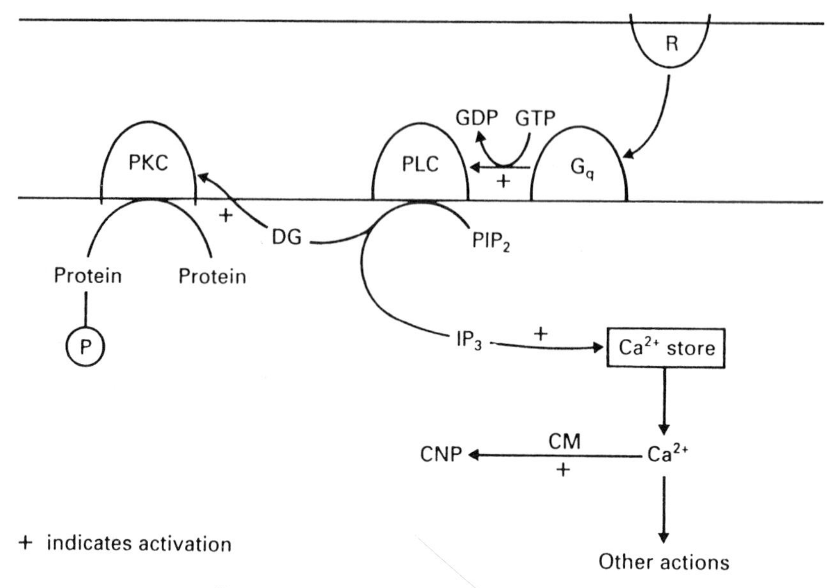

R, receptor
PIP_2, phosphatidylinositol 4,5-diphosphate
IP3, inositol 1,4,5-triphosphate
DG, diacylglycerol
PKC, protein kinase C
CNP, cyclic nucleotide phosphodiesterase
G_q, GTP binding protein
PLC, phospholipase C

Binding of a ligand to the receptor activates Gq which eliminates phospholipase C to hydrolyse PIP_2 to IP_3 and DG. IP_3 causes calcium to be released from intracellular stores such as mitochondrion and, in muscle, sarcoplasmic reticulum. DG activates PKC to phosphorylate various proteins and possibly to open a voltage-dependent channel to allow calcium influx into the cell. Calcium ion may activate certain processes directly or via its complex with calmodulin, e.g. CNP, one form of which is calcium-dependent, and often used to assay calmodulin.

Figure 9.2 Activation of inositol phospholipid pathway by receptor occupancy.

process, the details of which are not fully understood, but it is proposed that protein kinase C is involved in the latter case because polymyxin B, an inhibitor, also inhibits the release of catecholamine while phorbol diester, an activator, reverses this inhibition (Wakade *et al.*, 1986). Indeed, modulation by phorbol diester is usually a clear indication that protein kinase C and phosphatidylinositol hydrolysis are involved in a secretory process.

In the case of excitatory cells, such as those neurones from which acetylcholine is released, and also cardiac cells, the depolarization of the membrane is sufficient to open voltage-sensitive channels, so that as the voltage across the membrane falls calcium is drawn into the cell (reviewed in Rasmussen, 1986). Calcium ion activates a number of proteins either directly or as a complex with the specific binding protein, calmodulin. The latter complex frequently acts through protein kinases but may activate other proteins directly (such as cyclic nucleotide phosphodiesterase), thereby interacting with the cyclic nucleotide system.

9.1.3 Drug development

Drugs can interfere with the process of synaptic transmission in various ways. Occasionally, a false substrate can be provided for the biosynthetic pathway of the neurotransmitter, resulting in a false neurotransmitter with altered properties which competes with the natural agent. The mechanisms for removal of used neurotransmitter (both re-uptake and metabolism) have been found useful as targets for the rational development of drugs. Probably the major effort in drug development, however, has gone into the development of agents which act like the neurotransmitter (agonists) and those which antagonize its action.

It is a feature of pharmacology in particular, and possibly science in general, that initial concepts are simple but become progressively more complex as more information is obtained. Neuropharmacology is no exception - for every neurotransmitter discussed in this chapter, at least two kinds of receptor, with different effects, have been identified, although drugs for use as agonists or antagonists at some of the subtypes of receptor (see below) have not yet been devised!

For many of the receptors discussed in this chapter the endogenous ligand has been known for some time, as in the case of the receptors acted on by the catecholamines, but for others, e.g. the opiate receptors, drug binding studies suggested that there were tight binding sites before a natural ligand was discovered. The opiate peptides, such as enkephalin and endorphin, were subsequently found to fit the bill. At the present time, the diazepam binding site, although part of, and modulatory of, the 4-(or γ-)aminobutyric acid (GABA) receptor is clearly separate from the latter, and may also have an unknown endogenous ligand.

9.2 Adrenergic receptors

Adrenergic receptors occupy a privileged place in receptor pharmacology, in that the first separation of hormonal action into distinct activities was proposed on the basis of excitatory and inhibitory action of the catecholamines (named because of the 1,2-dihydroxybenzene or catechol moiety that they contain) on smooth muscle (Ahlquist, 1948). These activities were termed α and β, and from this arose the concept of a specific receptor for each type of physiological activity.

The three catecholamines [noradrenaline, adrenaline and isopropylnoradrenaline (isoprenaline); Fig. 9.3 shows the biosynthesis of the two former agents] differ markedly in their activities on the two receptor types. Noradrenaline and adrenaline are the most active on α receptors, while isoprenaline is the more potent agonist on β receptors. In cardiac muscle the predominant receptor is the β type, which in this tissue governs the increase in rate and force of contraction of the heart, known respectively as the positive chronotropic and inotropic effects.

Gene cloning has revolutionized our understanding of the adrenergic receptor – indeed the β receptor has often been the model. Until the middle of the 1980s there were only three known subtypes: α_1, α_2 and β. Recently discovered receptors have increased this number to nine by dividing each of the three subtypes into three, with a possible tenth in the α_{2D} (Bylund, 1992; Savarese and Fraser, 1992).

The receptor is one of the superfamily of receptors that couple to G-proteins and is composed of seven α-helical stretches of hydrophobic amino acids that cross the membrane separated by six loops, three extracellular and three cytoplasmic (for reviews see Kobilka, 1992; Savarese and Fraser, 1992). The C-terminus lies in the cytoplasm, and the N-terminal segment with glycosylation sites is outside the cell. The ligand binding domain lies in the transmembrane

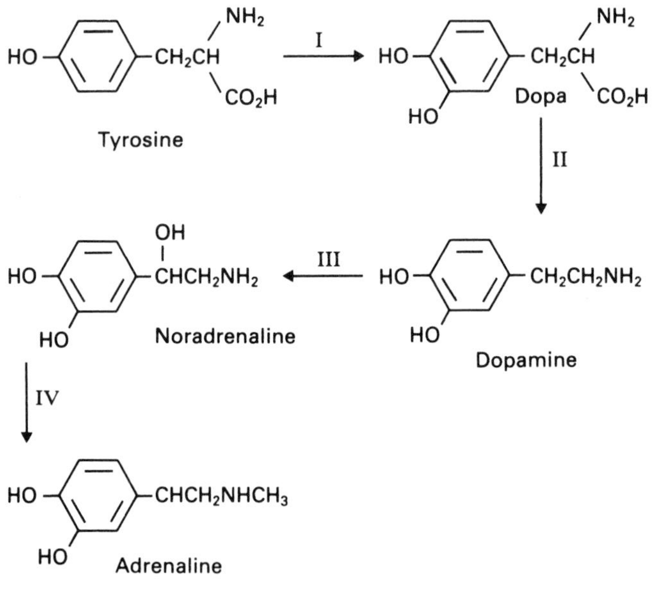

Tyrosine —I→ Dopa —II→ Dopamine —III→ Noradrenaline —IV→ Adrenaline

I: Tyrosine hydroxylase, tetrahydrobiopterin
II: Aromatic amino acid decarboxylase, pyridoxal phosphate
III: Dopamine β-hydroxylase, ascorbate
IV: Phenylethanolamine *N*-methyltransferase, *S*-adenosylmethionine
(Enzymes are noted together with their prosthetic groups)

Figure 9.3 Pathway of catecholamine biosynthesis.

region. The carboxylate group of an aspartate residue in the third transmembrane (TM3) domain binds the cationic amino group. Two serines in TM5 (Ser-204 and Ser-207 in the hamster β_2 receptor) form hydrogen bonds with the catechol hydroxyls; TM7 appears to differentiate between β_2 and α_2 antagonists.

The coupling of receptors to G-proteins and the subsequent chain of events leading to signal transmission is obviously of great interest. A second aspartate in TM2 (Asp-79 in the β_2 receptor) appears to be part of the agonist but not the antagonist binding site. Furthermore, mutation of this residue greatly reduces the ability of the receptor to activate (β_2) or inhibit (α_2) adenylate cyclase and therefore seems to be an integral part of this mechanism. The third cytoplasmic loop and C-terminus has been implicated in the coupling of the α_1 receptor to phospholipase C.

Repeated stimulation of a receptor can lead to a reduction in the signal sensitivity (a) by removal of the receptor from the membrane (sequestration), (b) by destruction of the receptor (down-regulation), or (c) by tachyphylaxis through phosphorylation. The β_2 receptor is phosphorylated at a site in the third cytoplasmic loop, thus impairing coupling to the G-protein.

At present we do not know the function of many of these recently discovered receptor subtypes - that is a challenge for the immediate future. We have, however, a range of relatively non-specific agents that are in therapeutic use for high blood pressure, asthma and other disorders.

9.2.1 α-Adrenergic receptors

The detailed study of the molecular pharmacology of the α receptor has lagged behind that of the β receptor, possibly because selective pharmacological modulators of the α receptors have only been discovered more recently. The α_1 receptor mediates most of the responses to sympathetic nervous system activation including smooth muscle contraction (Kendall, 1992). α_2 Receptors are also involved in sympathetic transmission; agonists reduce the passage of impulses to peripheral tissue. Autoreceptors at sympathetic nerve terminals regulate neurotransmitter release. Post-synaptic α_2 receptors modulate smooth muscle tone and control sodium ion excretion in the kidney (Kobilka, 1992). With regard to intracellular signal transduction, the α_{1A} receptor is unusual in that it activates calcium channels while the α_{1B} stimulates phospholipase C. All three α_2 subtypes inhibit adenylate cyclase (Bylund, 1992).

α_2-Agonists

The catecholamines by definition are agonists at the α-receptor and do not distinguish between the receptor subtypes. Clonidine, however, is specific for the α_2 receptor but again does not distinguish between the subtypes.

The binding affinity of clonidine at the α_2 receptor is high. Whether the receptor is in the intact membrane of the human platelet or solubilized in the

Structures of α-receptor; agonists and antagonists

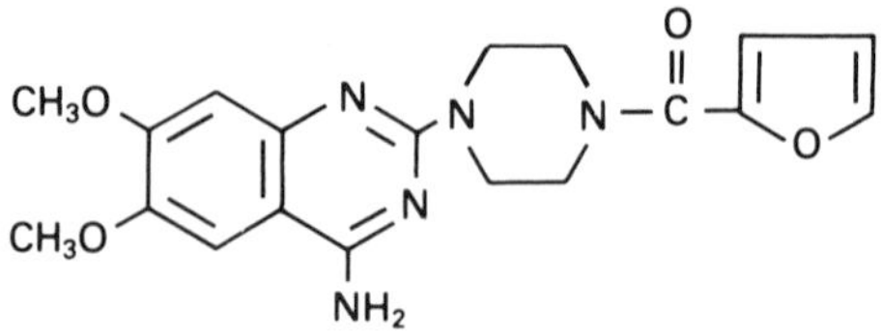

Prazosin

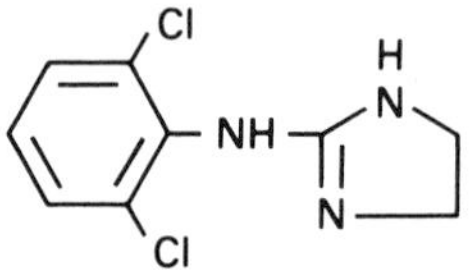

Clonidine

Prazosin antagonists. Clonidine is an agonist.

presence of digitonin, the binding constant lies between 4 and 9×10^{-9} M (Limbird *et al.*, 1982). A similar binding constant was obtained for the rat brain α_2 receptor in membrane-bound and detergent-solubilized form (Matsui *et al.*, 1984).

Clonidine is used to reduce blood pressure often in conjunction with another drug such as a thiazide diuretic (Rudd and Blaschke, 1980) or a β-adrenergic receptor blocker (Vauholder *et al.*, 1985). The advantage of the latter combination is that clonidine tends to induce a rapid fall in blood pressure while the reduction is much slower with β-blockers, so that an immediate but also longer lasting effect can be obtained with the drug combination.

α-Adrenergic antagonists

Blockade of the α_1 receptor antagonizes smooth muscle contraction; veins and other small blood vessels may dilate leading to a fall in vascular resistance and concomitant drop in blood pressure. Prazosin, an antagonist which does not distinguish between the α_1 receptor subtypes, is used for hypertension because of its action as an α_1 blocker particularly by post-synaptic receptors. Prazosin thereby inhibits the vasoconstriction produced by noradrenaline that is released from the sympathetic nerve endings, while the drug's lack of activity on α_2 receptors allows the neurohormone to exert negative feedback control of its own release, thus reducing the cardiac stimulation that follows non-selective α-adrenergic blockade (Rudd and Blaschke, 1985).

Prazosin binds very tightly to α_1 receptors with a binding constant of between 1 and 6×10^{-10} M to membrane-bound receptors, and appears to be some four orders of magnitude more selective for α_1 than α_2 receptors (see McPherson and Summers, 1982 and references therein).

9.2.2 β-Adrenergic receptors

β-Adrenergic receptors mediate many of the actions of the catecholamines; the action of β_1 receptors includes governing the force and rate of contraction of the heart beat, increasing secretion of renin from the kidney and antidiuretic hormone from the anterior pituitary. β_2 Receptors control relaxation in lung airways, but contraction in skeletal muscle. The β_3 receptor may mediate lipolysis in fat cells. All three receptor subtypes stimulate adenylate cyclase.

β-Adrenergic antagonists

The structure of isoprenaline underwent a number of modifications in order to devise a compound that would block the β activity of the catecholamines and be pharmacologically valuable as an anti-hypertensive. These modifications initially concentrated on extending the side-chain so that first three and then four atoms were interposed between the nitrogen atom and the benzene ring. Thus β-antagonists were derived from β-agonists. The β receptor is, therefore, the target for a number of drugs that are either agonists or antagonists. Propranolol, the first effective antagonist or β-blocker, is still used to lower blood pressure and in the management of cardiac arrhythmias (see Chapter 10 for a discussion on arrhythmia). The drug is, however, non-selective in that it blocks the β response in heart, lung and a variety of other tissues non-specifically (Lefkovitz, 1975), and this can lead to severe bronchoconstriction in the lung – highly undesirable in an asthma patient, where propranolol should not be used.

A search for more selective β-blockers has resulted in the development of, for example, metoprolol and atenolol which are more active on cardiac receptors (termed β_1) although they still inhibit lung β_2 receptors at higher doses. For example, metoprolol is 10 to 20 times mores selective for β_1 while atenolol is about three times more potent. Propranolol, on the other hand, shows no such selectivity. There is a problem in that the concentrations of drug required to have an effect on membrane preparations are much higher than those in intact tissue. This has been rationalized as being a consequence of some distortion of the normal receptor–enzyme coupling mechanism in the broken membrane fragments (Minneman *et al.*, 1979).

β-Adrenergic agonists

Isoprenaline is the prototype of β receptor agonists and found a use in earlier years as a treatment for asthma. It suffered, however, from a lack of specificity

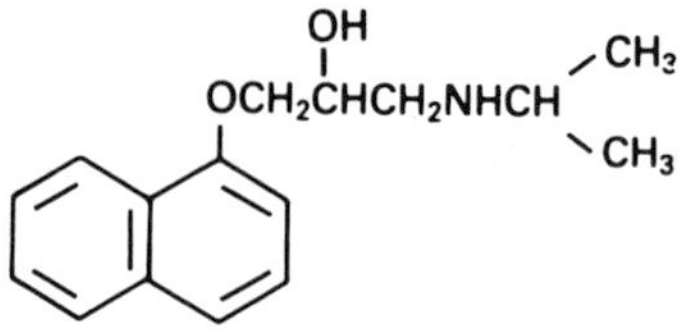

Propranolol

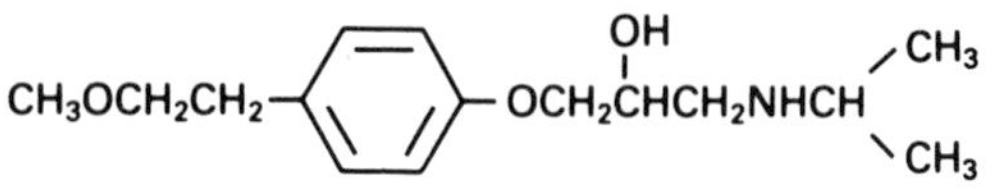

Metoprolol

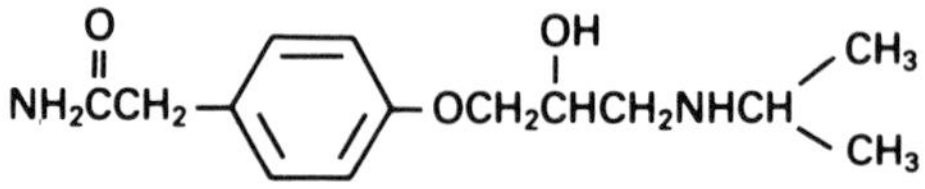

Atenolol

since it activated receptors in the cardiovascular system and gave rise to unpleasant side-effects, notably palpitations and occasionally cardiac arrhythmias (uneven beating of the heart) through its positive inotropic effect. In addition, the drug is rapidly inactivated by catechol *O*-methyltransferase and thus had a very short half-life. A longer-acting, more selective agent was required.

Accordingly, salbutamol was synthesized specifically for this purpose (Cullum *et al.*, 1969) and was shown to be much more specific for lung (β_2) than for cardiac (β_1) receptors. As a consequence, *in vivo* the drug showed lower activity in raising heart rate than in raising bronchial muscle tone in cats (Apperley *et al.*, 1976). *In vitro* the dose for half-maximal effect (ED_{50}) in contracting the guinea pig trachea was in the region of 3×10^{-9} M while ED_{50} for contraction on the right atrium was 10^{-7} M (Buckner and Abel, 1974).

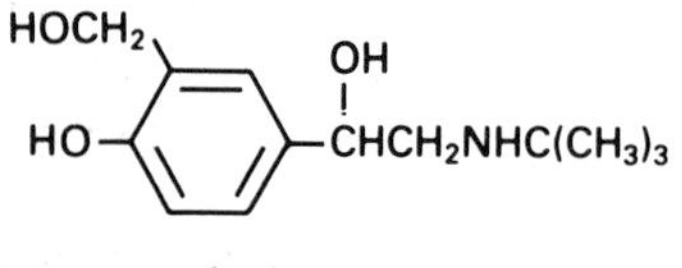

Salbutamol

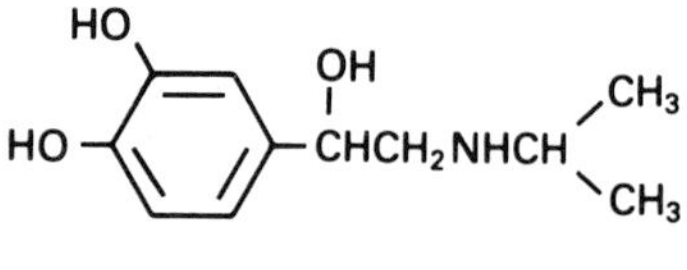

Isoprenaline

Salbutamol was only active for up to about six hours and a longer acting agent was needed for night-time asthma attacks. Eventually salmeterol, a much more lipophilic compound, was synthesized (see review by Jack, 1991) and was found to have a much greater duration of action in man; twice daily doses gave 24 hours coverage.

Salmeterol acts at β_2 receptors in the lung, activating adenylate cyclase and eventually producing bronchodilatation. Estimates of the ratio of efficacies of salbutamol and salmeterol vary from 0·1 to 3 according to the method of measurement (Dougall *et al.*, 1991; Jack, 1991). Salmeterol binds less tightly to β_1 receptors, however, than salbutamol and so is more selective. If (−)-

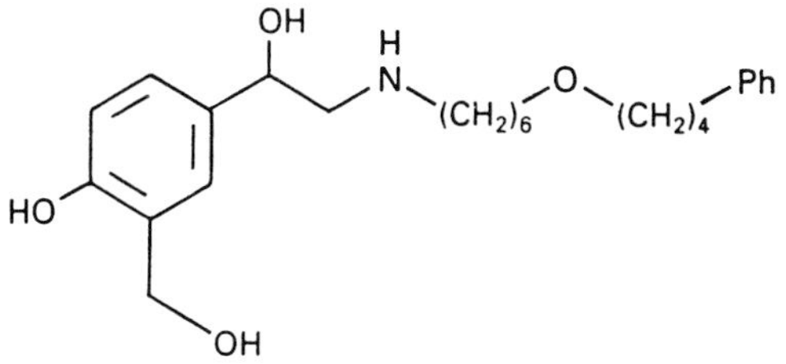

CN **Salmeterol**

isoprenaline is taken as being non-specific and has a ratio of 1 between agonism in airway smooth muscle (β_2) and cardiac atria (β_2), salbutamol has a ratio of 650, and salmeterol 50 000 (see Brogden and Faulds, 1991). With regard to binding to β_2 receptors, salbutamol is 50 times less tightly bound to rat lung membranes as shown by competition with ^{125}I-pindolol ($2{\cdot}5 \times 10^{-6}$ M versus $5{\cdot}3 \times 10^{-8}$ M; Coleman *et al.*, 1990).

The extremely potent binding of salmeterol to the β receptor is shown by attempts to wash the receptors free of the drug. ^{125}I-pindolol can remove isoprenaline and salbutamol from membranes but has little effect on salmeterol binding. Another unusual effect is shown by the β-blocker, sotalol. Relaxation of guinea-pig trachea by salbutamol and salmeterol is reversed by sotalol. When the sotalol is removed, salbutamol action does not reappear, whereas that of salmeterol does. This has been explained by suggesting that the polar headgroup of salmeterol remains bound to a site outside the ligand site, called an exo-site, while the β-blocker binds to the ligand site, and when the latter is washed off, salmeterol reasserts its binding at the ligand site (Brogden and Faulds, 1991; Jack, 1991).

If binding to the exo-site plays a large part in the overall binding and binding to the 'active' ligand site controls the efficacy, we can see how the membrane binding of salmeterol is so much greater than salbutamol, whereas they are much closer in efficacy (Dougall *et al.*, 1991).

In addition, the drug has an anti-inflammatory effect, like other β-agonists, through inhibiting the release of inflammatory mediators, including histamine and leukotrienes (C_4, D_4 and E_4) from rat mast cells and human lung fragments. This effect was antagonized by propranolol, suggesting that β receptors were involved (Butchers *et al.*, 1991). Whether salmeterol has any clinically important effect on the underlying inflammatory processes in asthma remains to be seen.

Asthma is characterized by early and late responses to challenge by an allergen, together with an increase in hyperresponsiveness of the bronchi to an allergen; corticosteroids and cromoglycate (Chapter 10) are both able to inhibit these effects, but this is not usually so for a short-acting β-agonist. Salmeterol is unusual among β-agonists in inhibiting these effects. Both salbutamol and salmeterol are optically active but the drugs are usually in a racemic mixture. They are normally prescribed in conjunction with steroids, as the latter are intended to stop the underlying inflammatory process, while the bronchodilatation produced by the β-agonists gives immediate relief and a feeling of well-being.

9.2.3 Mixed α- and β-antagonist

Labetalol has two asymmetric centres and thus has four stereoisomers; it is normally given as a mixture of all four isomers. They vary markedly in receptor activity blocking both α_1, β_1 and β_2 receptors. The *R,R* isomer is largely responsible for the β-blocking activity and is only weakly effective at α receptors. The *S,R* isomer is the most potent at α_1 receptors. The *S,S* isomer is relatively weak at all three subtypes of receptor (Brittain *et al.*, 1982). Furthermore, the *R,R* isomer has agonist activity at the β_2 receptor which can be blocked by propranolol (Prichard, 1992). This is not surprising, since the structure bears a resemblance to both noradrenaline and salbutamol, with the nitrogen atom only two carbon atoms away from the ring and a hydroxyl group on the side-chain α-carbon.

The advantage of this combination of multiple actions is that β blockade reduces blood pressure by slowing the heart rate while α_1 blockade results in relaxation of arterial smooth muscle. Furthermore, vasodilatation results from β_2 agonism (Prichard, 1992).

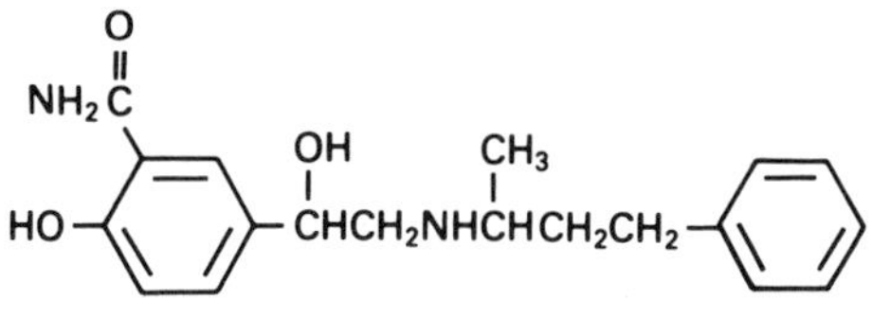

Labetalol

9.3 Dopamine receptors

In earlier years, radioligand binding studies found only two dopamine receptors: D_1 and D_2. D_1 was linked via the stimulatory G-protein, G_s, to adenylate cyclase whereas D_2 not only inhibited adenylate cyclase via G_i, but also activated K^+ (hence the hyperpolarization sometimes noted for this receptor) and inhibited calcium channels. As with most of the receptors noted in this chapter, however, molecular biological methods have since discovered three more receptors: D_3, D_4 and D_5. D_5 is similar to D_1 in stimulating, whereas D_3 and D_4 relate to D_2 in inhibiting, adenylate cyclase. The challenge the molecular biologists have set the pharmacologists is to find a role for these new receptors (Sibley and Monsma, 1992).

The D_2 receptor is one of the 7-transmembrane variety with the N-terminus lying outside the cell and the C-terminus in the cytoplasm. The N-terminus carries three potential glycosylation sites. The C-terminus is short and the third cytoplasmic loop between transmembrane regions (TM) 5 and 6 is large and carries a potential phosphorylation site for cyclic-AMP-dependent kinase - a characteristic of many receptors that inhibit adenylate cyclase. D_1 on the other hand, has a small third cytoplasmic loop and a long C-terminal section in the cytoplasm which is characteristic of receptors that activate adenylate cyclase. As with the β-adrenergic receptor, the 7-transmembrane domain carries the ligand binding site with a conserved aspartate residue in TM2 probably responsible for maintaining the active conformation, another aspartate in TM3 to bind the protonated amino group and two serines in TM5 to interact with the two catechol hydroxy groups.

The behavioural correlates of dopamine action are believed to be exemplified by the effects of dosage with apomorphine, a dopamine agonist (dopamine does not pass the blood–brain barrier), namely a stereotypic type of behaviour in rats that involves persistent biting, gnawing and licking, and hypothermia in mice etc. These activities can all be antagonized by ligands at both types of receptor. A further action of dopamine is to inhibit the production of prolactin which appears to be under the control of D_2 receptors since it is only antagonized by the neuroleptic drugs and not by SCH 23390 (Iorio *et al.*, 1983). Agonists and antagonists at D_1 and D_2 subtypes are used clinically for a number of conditions, including Parkinsonism, hypertension, and schizophrenia.

9.3.1 Dopamine agonists - L-dopa and bromocryptine for Parkinsonism, fenoldopam for hypertension

One of the most widely used dopamine agonists is bromocryptine, one of the ergot alkaloids derived from the growth of the fungus *Claviceps purpurea* on damp rye. The fungus forms a purple growth - the ergot - around the grain from which the ergot alkaloids are obtained. In the Middle Ages, ingestion of infected rye gave rise to a number of severe reactions including gangrene, abortion and, after high doses, death. A variety of pharmacologically active agents have been isolated and identified from the ergot, including the related structures, lysergic acid diethylamide and bromocryptine. The latter binds to D_2 receptors in the brain and pituitary, with affinities in the region of 10^{-9} M as measured by competition with ^{3}H-spiroperidol, a specific D_2 antagonist (Creese *et al.*, 1977).

Dopamine and acetylcholine exert balancing inhibitory and excitatory control over voluntary movements. When the dopamine neurones in a certain portion of the brain, known as the substantia nigra, serving the basal ganglia fail to inhibit these movements, Parkinsonism is the consequence. The brain can suffer the loss of many of these dopamine neurones without any apparent changes in behaviour, but when the loss reaches about 80 per cent, the symptoms of Parkinsonism appear, namely involuntary movements or tremors, tics, rigidity, a gradual loss of motor function and eventually, in the late stages of the disease, the inability to stand upright. Furthermore, the main psychic correlate is apathy (Bianchine, 1985). The substantia nigra governs the transmission of motor impulses that arise elsewhere in the brain, and contains a large number of pigmented neurones with several synaptic junctions with other neurones. Dopamine acts here as an inhibitory neurotransmitter.

The reason for this loss of neurones and the dopamine deficiency is not known but a toxin present as an impurity in some forms of heroin (1-methyl-4-phenyl-1,2,3,6-tetrahydropyridine: MPTP) can give rise to symptoms in animals indistinguishable from Parkinsonism in man. A similar destruction of dopaminergic neurones in the substantia nigra seems to occur (see section 9.6 where monoamine oxidase inhibitors are discussed). Some dopamine antagonists, specifically the drugs used to treat schizophrenia, can also give rise to symptoms of Parkinsonism by blocking the post-synaptic dopamine receptors in the basal ganglia. This condition is reversible, however, since it disappears when the drug is withdrawn (Borison *et al.*, 1973; Baldessarini, 1979).

The dopamine deficiency in Parkinsonism can be treated in a number of ways. Dopamine can be supplied to the remaining neurones or dopamine agonists can be used which bypass the damaged neurones and act on the post-synaptic receptors. Alternatively, the excitatory transmission governed by acetylcholine can be suppressed in order to achieve a more appropriate balance of inhibition and excitation (Pearce, 1984; Calne, 1984). The latter approach is discussed below (section 9.7).

Dopamine cannot be supplied directly to the brain since systemically administered catecholamines do not cross the blood–brain barrier. The precursor of dopamine, L-dopa (L-*d*ihydr*o*xy*p*henyl*a*lanine), can however be administered as a pro-drug (see Fig. 9.3 for the biosynthetic pathway). L-Dopa is decarboxylated by aromatic amino acid decarboxylase to yield dopamine. Pyridoxal phosphate is the cofactor. Unfortunately, much of the amino acid is decarboxylated in the liver and only about 5 per cent is available to cross the blood–brain barrier. Accordingly, a decarboxylase inhibitor, which itself cannot cross into the brain is usually co-administered with L-dopa. The most useful is carbidopa, which inhibits dopa decarboxylase by forming a Schiff's base between the terminal amino moiety of its phenylhydrazine group and the aldehyde of pyridoxal. Although daily dosage with L-dopa is often effective for a number of years, it gradually loses its efficacy and must be supplemented in various ways.

Bromocryptine can be used in conjunction with L-dopa in order that the dose of the latter may be reduced (see Lieberman and Goldstein, 1985, for a review on the use of bromocryptine in Parkinsonism). The use of an additional drug is also helpful in reducing the side-effects of L-dopa treatment, namely psychosis, hallucinations, nausea and respiratory distress.

Another condition where bromocryptine has found a medical role is in the treatment of infertility due to the hypersecretion of prolactin (Weinstein *et al.*, 1981). Prolactin secretion is suppressed by dopamine, and bromocryptine

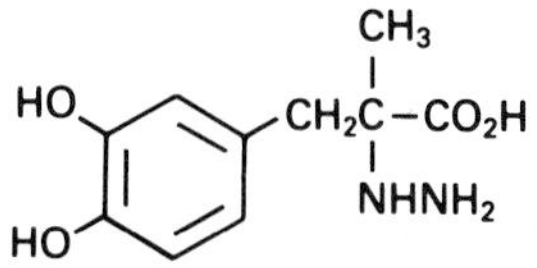

Carbidopa

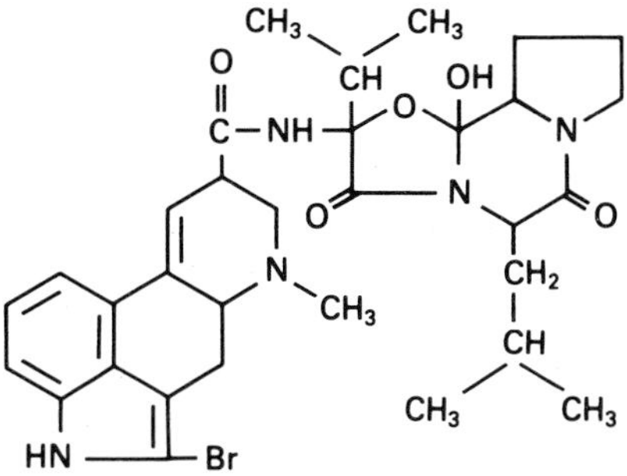

Bromocryptine

can also inhibit this secretion at the level of the pituitary; this is consistent with the view noted above that bromocryptine acts at D_2 receptors. High levels of prolactin will inhibit the release of gonadotrophins in humans and, consequently, will suppress ovulation. This effect is clearly of great biological value when nursing mothers are feeding their children because it blocks the possibility of another pregnancy, but in cases of inappropriate prolactin release infertility is the result. Bromocryptine can also be used to treat the symptoms of excessive secretion of prolactin by a pituitary tumour. In general, therefore, bromocryptine is widely used in situations where an agent that acts like dopamine is required.

Fenoldapam, an agonist at D_1 receptors in the periphery (often referred to as DA1) is under development for the treatment of malignant hypertension. Fenoldapam dilates blood vessels in the periphery, especially in the kidney, to induce secretion of water (diuresis) and sodium ion (natriuresis) (Hegde *et al.*, 1989), thus lowering the blood volume and reducing pressure. The drug is optically active and most of the activity resides in the *R* isomer (Holcslaw and Beck, 1990).

9.3.2 Dopamine antagonists - phenothiazines, butyrophenones and diphenylbutylpiperidines for schizophrenia

A large number of drugs that are used for the treatment of psychosis, notably schizophrenia, are antagonists of dopamine at the D_2 receptor. Schizophrenia is a complex disorder over the diagnosis of which there has been considerable confusion in the past, and so an outline is presented here. Recent attempts to categorize the diagnosis more accurately have apparently led to greater treatment success (Manschrek, 1981). The Diagnostic and Statistical Manual of the American Psychiatric Association (DMS-III) has laid down a number of criteria which must be satisfied before schizophrenia can be diagnosed. These include:

(a) a period of at least 6 months in which social withdrawal, impaired self-

Ligand structures

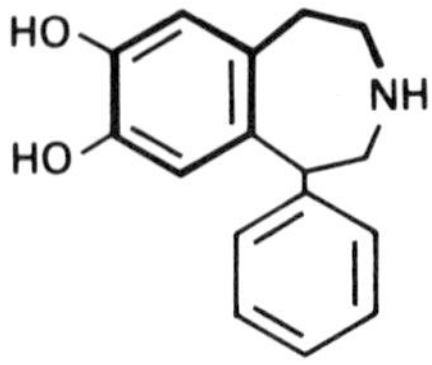

HO, HO, NH

SKF 38393

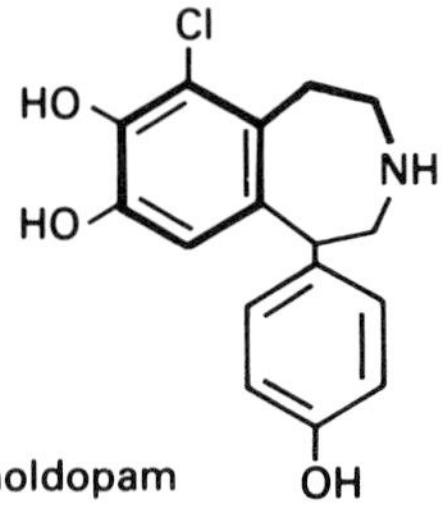

Cl, HO, HO, NH, OH

Fenoldopam

These structures contain the framework of the dopamine molecule drawn in heavy type.

awareness and markedly peculiar behaviour predominate - this is known as the 'prodromal phase';

(b) an active phase with delusions and hallucinations, possibly associated with incoherent and fragmented thinking, together with flatness of emotional expression;

(c) a phase similar to the prodromal phase in which the symptoms subside (residual phase);

(d) at least 6 months continuous signs of the condition in addition to (a), and

(e) the onset must occur under 45 years of age and not be due to any organic disorder.

The course of the disease varies widely but the active phase may be interspersed with residual phases over a number of years. The active phase can be treated and the signs greatly ameliorated by antipsychotic drugs, and in this respect the treatment has been greatly improved. The residual phase, social withdrawal and emotional aspects of the condition are much more difficult to treat. The anti-psychotic agents ameliorate the condition but do not effect a cure, although remission can occur (Manschrek, 1981).

A large number of anti-psychotic agents have been developed for the treatment of schizophrenia. The phenothiazine group of drugs, of which the first was chlorpromazine, interact with both types of dopamine receptor (and a number of other receptors as well). In fact, these drugs are extremely nonselective in their widespread biological actions. The more selective agents such as the butyrophenones (e.g. haloperidol) and diphenylbutylpiperidines (e.g.

pimozide) interact with D_2 receptors more specifically, although not totally selectively, and the relative binding affinities of the ligands correlate with their anti-psychotic potency in man (Snyder, 1981). The term neuroleptic is often used to describe these drugs, mainly to distinguish them from the other types of psychoactive drugs that depress the central nervous system such as general anaesthetics, sedatives, etc.

Phenothiazines

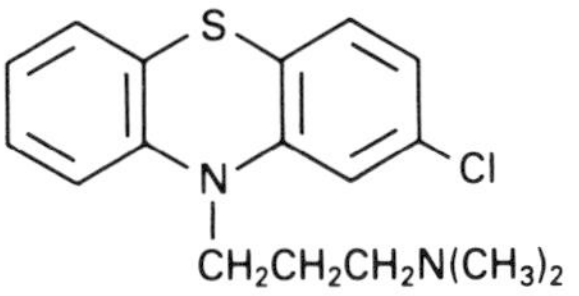

Chlorpromazine

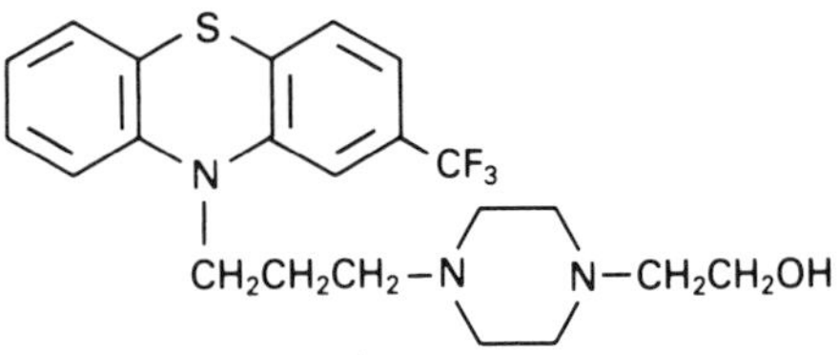

Fluphenazine

Diphenylbutylpiperidine

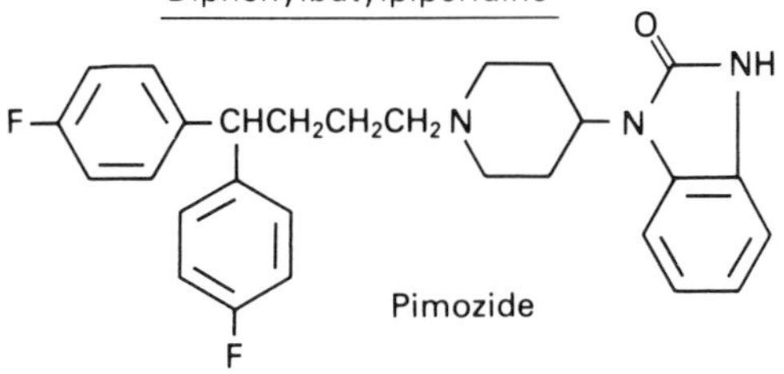

Pimozide

Butyrophenones

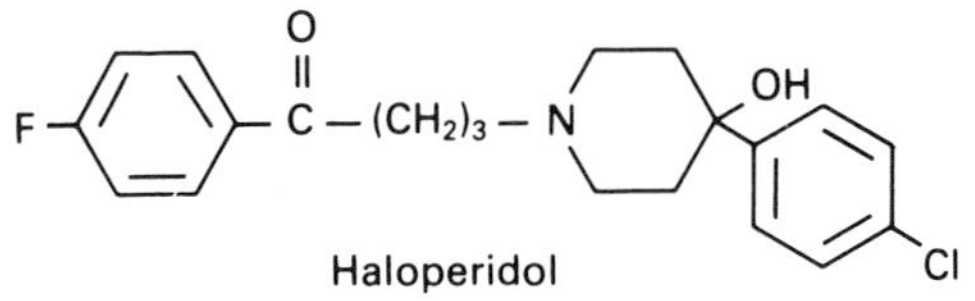

Haloperidol

Other characteristic behaviour of the neuroleptics stems from their binding to dopamine receptors in that they reduce (a) the firing of dopaminergic neurones (Bunney *et al.*, 1973), and (b) the rates of both biosynthesis of dopamine from tyrosine and of its metabolism to 3-methoxytyramine, homovanillic and dihydroxyphenylacetic acids (Fig. 9.4) (Anden and Stock, 1973).

It is probably incorrect, however, to assume that dopaminergic systems are overactive in schizophrenia - on the other hand a reduction in dopaminergic activity is certainly beneficial to the patient. The effect is reversed on withdrawal of the drug and so they are at best a palliative for the condition. An associated feature of the action of these drugs is the slow onset of clinical

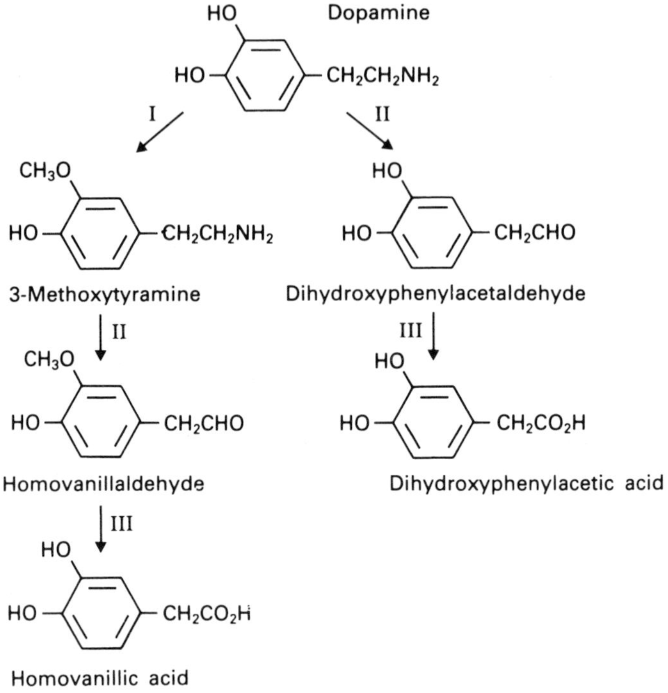

Dopamine

3-Methoxytyramine

Dihydroxyphenylacetaldehyde

Homovanillaldehyde

Dihydroxyphenylacetic acid

Homovanillic acid

I: Catechol *O*-methyltransferase, *S*-adenosylmethionine
II: Monoamine oxidase, FAD prosthetic group
III: Aldehyde dehydrogenase, NAD

Figure 9.4 Dopamine metabolism.

effect in man compared with the rapid onset of their pharmacological action *in vitro*. This suggests that, perhaps like the tricyclic antidepressants (section 9.5), secondary, indirect changes produced by the drug administration may make a major contribution to their clinical efficacy. In addition, the use of these drugs increases the number of D_2 receptors in the brain, as shown by measurement of dopamine binding to post-mortem material (Snyder, 1981) - an interesting result of the brain trying to nullify the effect of the drug.

Although in general the discussion of drug side-effects is not intended to be a major feature of this book, the neuroleptic drugs show at least three types of side-effect that are a consequence of dopamine receptor blockade in other parts of the brain. On the one hand, the therapeutic effects of the neuroleptic drugs result from their blockade of receptors in the projections to the limbic system that regulates emotion in the brain. Parkinsonism, however, can result as a consequence of post-synaptic blockade of the basal ganglia in the corpus striatum which are served by nerve fibres from the substantia nigra (Snyder, 1986). Neuroleptic treatment for schizophrenia can, therefore, cause some similar effects. Tardive dyskinesia, which is characteristic of Parkinsonism, and prolactin release may also result as side-effects of neuroleptic treatment. The former is characterized by a variety of movement disorders such as:

1. constant chewing movements with the tongue intermittently darting out of the mouth (fly-catcher tongue):
2. body rocking while sitting; and
3. marching on the spot.

These symptoms take time to develop, hence the term tardive. They are induced by receptor supersensitivity caused by prolonged drug treatment, in other words the receptors that are not blocked by the drug are supersensitive to dopamine (Borison *et al.*, 1983).

Prolactin release from the pituitary is inhibited by dopamine acting at D_2 receptors. Dopamine blockade has the effect, therefore, of increasing prolactin levels in the plasma leading to breast engorgement and a persistent or recurrent discharge of milk from the breast (galactorrhoea).

The hope that clarification of the mode of action of the neuroleptic drugs might lead to a greater understanding of the aetiology of schizophrenia has not yet been realized. Hyperactivity of the dopaminergic receptor is not a feature of the condition. Various attempts to demonstrate the formation *in vivo* of hallucinogenic by-products, e.g. by methylation of the neurotransmitter amines to form such compounds as bufotenin or mescaline, have failed. Nevertheless, we desperately need a much greater understanding of a disease which is very complex and difficult to treat.

Recent so-called 'atypical' neuroleptics have been developed of which one of the leading examples is clozapin - atypical because it does not produce the side-effects characteristic of D_2 blockade such as tardive dyskinesia (reviewed in Fitton and Heel, 1990). It binds only weakly to D_2 receptors compared with a variety of other receptors including D_1, histamine, serotonin ($5HT_2$) and α_2. Various suggestions have been made as to the basis of clozapin action; the D_1 (Ellenbroek *et al.*, 1991), the newly discovered D_4 receptor (see Sibley and Monsma, 1992), $5HT_2$ (Meltzer, 1992) and the α_2 adrenergic receptor (Pickar *et al.*, 1992), but at present the mode of action is not clear.

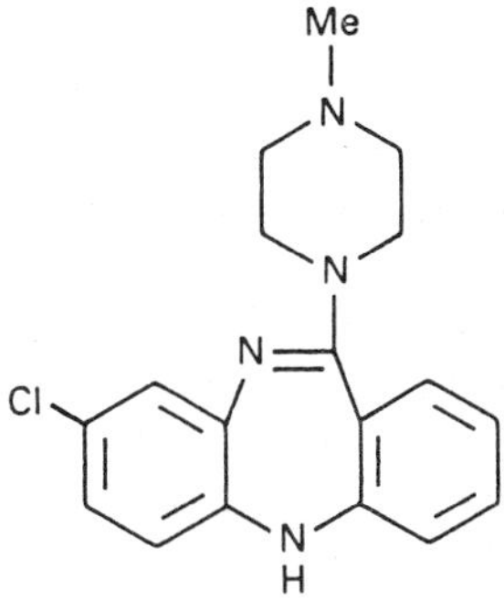

Clozapin

Although clozapin only shows extrapyramidal side-effects at a low level it does, however, have the serious side-effect of agranulocytosis (i.e. it can destroy the granulocytes - white blood cells such as neutrophils and basophils) and under treatment the white blood cell count has to be monitored weekly.

9.4 Serotonin receptors

Serotonin (5-hydroxytrytamine, 5HT) is an indolealkylamine which is mainly located in the chromaffin cells of the gut (so-called because they readily stain brownish-yellow with chromium salts); there are smaller concentrations in platelets and the brain. The physiological effects of serotonin are widespread including regulation of gastric motility and effects on platelet aggregation. The central effects include the perception of pain, sleep, control of behaviour and affective disorders and release of other hormones.

Recently, our understanding of serotonin receptors has greatly increased, partly by the discovery of specific ligands for the receptor subtypes and partly

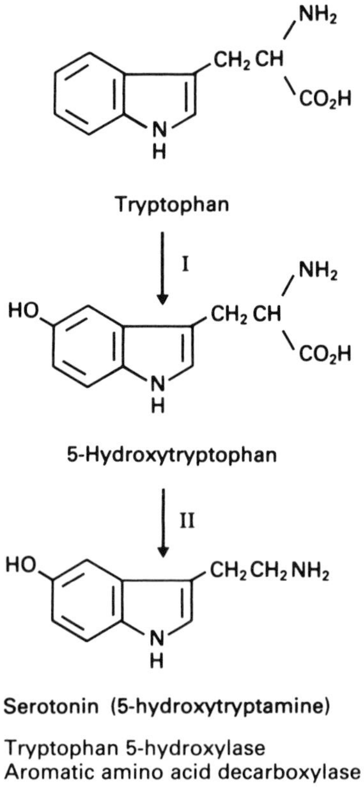

Tryptophan

5-Hydroxytryptophan

Serotonin (5-hydroxytryptamine)

I: Tryptophan 5-hydroxylase
II: Aromatic amino acid decarboxylase

Figure 9.5 Serotonin biosynthesis.

by the techniques of gene cloning (see Harrington *et al.*, 1992, for a review on the molecular biology of serotonin receptors). There are now known to be at least four major subtypes, $5HT_1$ to $5HT_4$. The $5HT_1$ subtype is further subdivided into six (A to F), $5HT_2$ into three (A to C) and whether $5HT_3$ and $5HT_4$ are homogeneous is still under study. The ligands which interact with these receptors are numerous, but only those that are in use as drugs will be considered in this chapter.

The 5HT receptors are coupled to G-protein-mediated pathways with the exception of $5HT_3$ which is linked to a 'fast' monovalent cation channel - fast because the effects occur more swiftly than with enzyme-linked pathways. Cloning of receptors has confirmed the existence of almost all the receptors defined by pharmacology. For example, the $5HY_3$ receptor has been cloned and expressed in *Xenopus* oocytes (large egg cells from the South African frog). The expected cellular response to 5HT was obtained, namely influx of cations, which could be blocked by specific antagonists (Maricq *et al.*, 1991). The relationship of receptor to signal pathway is listed in Table 9.3.

$5HT_{1c}$ and $5HT_2$ show 49 per cent homology overall - not surprising as the receptors couple to the same intracellular pathway - rising to over 80 per cent in the transmembrane regions. These receptors govern different functions, however, because, they are found in different sites and furthermore, serotonin has two orders of magnitude greater affinity for the $5HT_{1c}$ receptor than for the $5HT_2$ (Julius *et al.*, 1990).

9.4.1 $5HT_{1A}$ receptors, anxiety and depression

Buspirone - the most developed member of a group of compounds known as the azapirones - is a partial agonist at $5HT_{1A}$ receptors with an K_i of $2{\cdot}0 \times 10^{-8}$ M with some activity at D_2 receptors (see Hamik *et al.*, 1990). The drug is used for the treatment of anxiety (anxiolytic) and is under study for the treatment of depression (Lucki, 1991; Napoliello and Domantay, 1991). The azapirones apparently do not act by the same mechanism as the anxiolytic benzodiazepines, although both types of drug may reduce serotonin trans-

Table 9.3 Serotonin receptor classification

Receptor	*Signal Pathway*	*Cloned (AA)*	*7-Transmembrane*
$5HT_{1A}$	AC inhibited	Yes (421)	Yes
$5HT_{1B}$	AC inhibited	Yes (??)	Yes
$5HT_{1C}$	PC stimulated	Yes (460)	Yes
$5HT_{1D}$	AC inhibited	Yes (377)	Yes
$5HT_2$	PC stimulated	Yes (471)	Yes
$5HT_3$	Cation channel	Yes (487)	No
$5HT_4$	AC stimulated	No	?

Key: AC, adenylate cyclase; PC, phospholipase C; AA, number of amino acids in the cloned receptor.

mission. The azapirones also do not show addiction which has been recently recognized as a serious side-effect of the benzodiazepines.

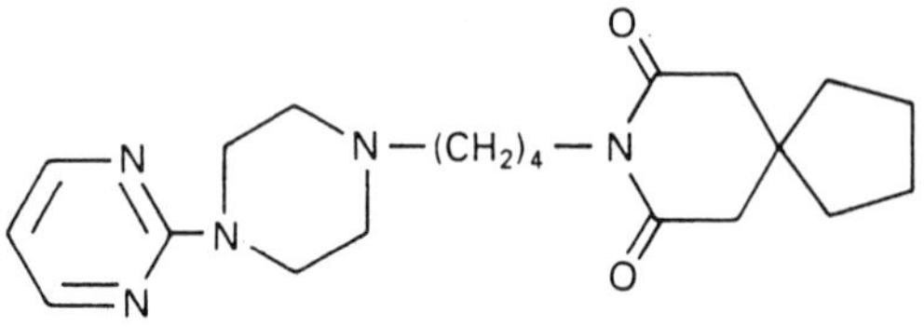

Buspirone

$5HT_{1A}$ receptors are found both pre- and post-synaptically in the brain. The presynaptic receptors appear to inhibit the firing of 5HT neurones, i.e. they are known as auto-receptors. It is an intriguing possibility that agonism at the presynaptic receptors is responsible for the anxiolytic effects while partial agonism at the post-synaptic receptors controls anti-depressive effects (Lucki, 1991). Section 9.5 discusses the relative importance of serotonin and noradrenaline in the aetiology of depression.

9.4.2 $5HT_{1D}$ receptors and migraine

Migraine is a condition of a unilateral and pulsating headache often accompanied by nausea and vomiting and occasionally by characteristic visual effects. There are two theories for the origin of migraine; nervous or vascular. The vascular theory (Lance *et al.*, 1989) proposes that blood vessels, inside the skull but outside the brain, become distended; protein is lost across the vessel membrane (extravasation) putting further pressure on the surrounding tissue. These sensory stimuli activate the trigeminal (fifth) cranial nerve which takes impulses to the pain centres in the cortex. This nerve also has connections with (a) the chemoreceptor trigger zone which is linked to the part of the brain that controls vomiting (see below) and (b) the hypothalamus which controls the visual stimuli (Saxena and Ferrari, 1989; Humphrey and Feniuk, 1991).

The involvement of serotonin in migraine attacks was suggested by the lowering of serotonin levels in the brain at the beginning of an attack. Furthermore, levels of the metabolite of serotonin, 5-hydroxyindoleacetic acid, increase in the urine after an attack, suggesting that the hormone has been released in the brain. Serotonin is able to constrict cranial arteries and shunts between arteries and veins. Vasoconstrictors, such as methysergide, known to be a (non-specific) antagonist at 5HT receptors, have been used to treat migraine. Although an infusion of serotonin does alleviate migraine it produces too many side-effects to be acceptable. The need was for a specific vasoconstrictor to operate only on the cranial blood vessels (Lance *et al.*, 1989).

Detailed studies with ligands showed that the $5HT_{1D}$ subtype is the most common subtype in human brain. Sumatriptan is a potent and specific agonist

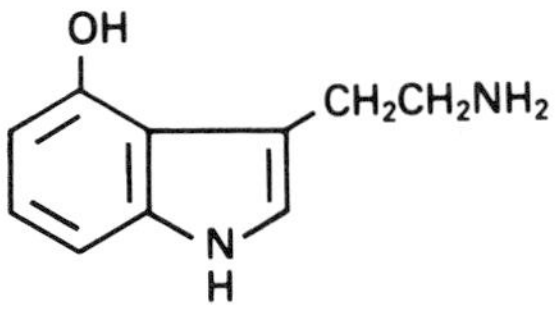

Serotonin

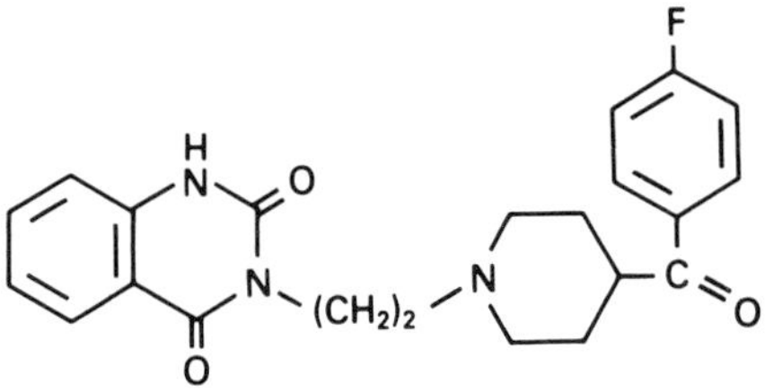

Ketanserin

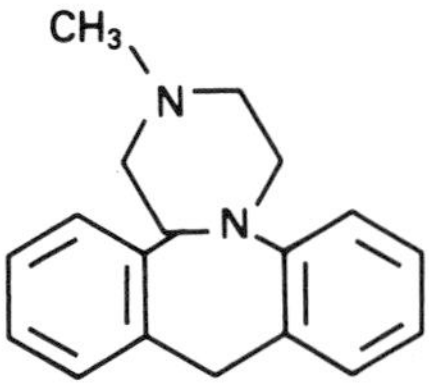

Mianserin

at the $5HT_{1D}$ receptor subtype with a pK_d of 7·54 - only five times less active than serotonin itself. Sumatriptan can also inhibit forskolin-activated adenylate cyclase activity in membranes from the bovine substantia nigra (Schoeffter and Hoyer, 1989), which is to be expected as $5HT_{1D}$ receptors interact with the adenylate cyclase pathway. Furthermore, sumatriptan does not cross the blood-brain barrier, which supports the peripheral aetiology of migraine (Humphrey and Feniuk, 1991), and is effective even after an attack has started.

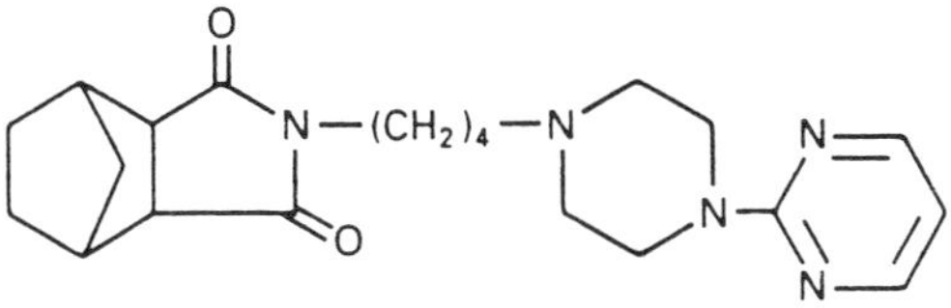

Tandospirone

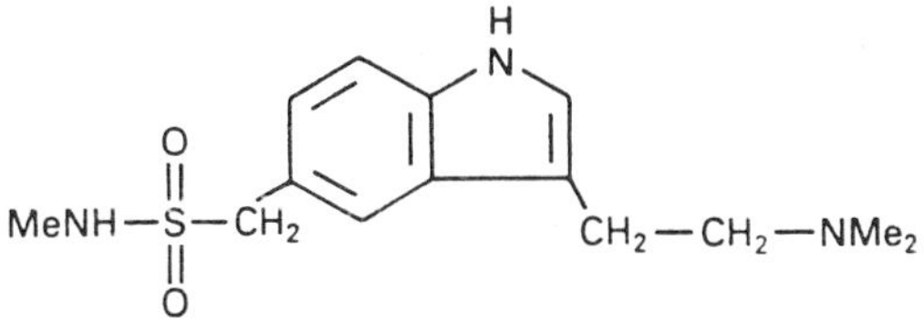

Sumatriptan

Sumatriptan is believed to act by activating the $5HT_{1D}$ receptors in blood vessels in the dura mater (the membrane surrounding the brain). The vessels become constricted, thereby easing the pressure on the vessel wall, preventing extravasation and stimulation of the sensory fibres of the fifth cranial nerve (Humphrey and Feniuk, 1991). This may inhibit release of neuropeptides, which play a role unknown as yet, blocking neurogenic inflammation.

9.4.3 $5HT_2$ receptors

$5HT_2$ receptors mediate the contraction of gastrointestinal smooth muscle, as well as platelet aggregation and other inflammatory responses (Leysen *et al.*, 1984) by raising intracellular levels of calcium through the phospholipase C pathway. Ketanserin is a powerful antagonist of serotonin binding at this subtype (K_d 3×10^{-10} M), and stimulates the phospholipase C pathway. Ketanserin is also a weak antagonist of α_1 adrenergic receptors.

Serotonin causes platelets to change shape and aggregate to produce clots, and also augments the effects of other agents such as ADP and collagen. In addition, serotonin causes contraction of smooth muscle cells and amplifies the contraction produced by other agents. Ketanserin can prevent both actions (see Brogden and Sorkin, 1990).

Ketanserin is used to treat hypertensive conditions in which vasoconstriction and platelet aggregation are due to serotonin. The relative importance of adrenergic blockade to the action of ketanserin is still under investigation. A close analogue of ketanserin, ritanserin, which is devoid of α_1-blocking activity is not effective in lowering blood pressure, which suggests that binding at both receptors may be required.

Central effects of $5HT_2$ receptor antagonists are shown by mianserin and ritanserin. Mianserin, a clinically effective anti-depressant, is also a potent antagonist at $5HT_2$ receptors with the +-isomer being the more effective (Alexander and Wood, 1987). Ritanserin has been developed for depression and as an anxiolytic.

9.4.4 $5HT_3$ receptors

The nausea and vomiting associated with a number of anti-cancer drugs, principally cisplatin, causes a great deal of distress to the patient and places much strain on their compliance with the treatment. For some years attempts were made to prevent emesis by co-administration of dopamine antagonists such as metoclopramide because some forms of vomiting responded to this treatment. This drug is only effective at high doses against cisplatin-induced emesis, however, suggesting that another receptor was involved. Eventually this was identified as the $5HT_3$ receptor.

The nausea was believed to stem from the damage done to the epithelial lining of the gut by the anti-cancer drug, resulting in release of serotonin from the chromaffin cells. Serotonin activates the afferent fibres of the vagal nerve and allows the influx of sodium and potassium, thus depolarizing the nerve by neutralizing the normal negative charge. The vagal nerve leads to a sector of the brain known as the nucleus tractus solitarius where the complex motor activity required for vomiting appears to be controlled. Alternatively, the signal could be relayed to this sector via a section of the area postrema, called the chemoreceptor trigger zone, which lies outside the blood–brain barrier and responds to toxic materials in blood or cerebrospinal fluid (Barnes *et al.*, 1991; Hesketh and Gandara, 1991).

The involvement of serotonin in the emetic response was suggested by the following data:

1. Administration of 5-hydroxytryptophan, the precursor of serotonin (Fig. 9.5), to man resulted in nausea and vomiting.
2. Depletion of the serotonin stores in the gut, either by the action of reserpine or inhibition of synthesis, prevented the emetic action of cisplatin.
3. Levels of 5-hydroxyindoleacetic acid in the urine of cisplatin-treated subjects rose sharply during emesis (see Barnes *et al.*, 1991).

A number of chemical synthetic programmes were undertaken to find a serotonin antagonist which would block the action of the transmitter. Several compounds of different structural types were eventually identified, including granisetron, ondansetron, tropisetron and zacopride, which were specific antagonists at $5HT_3$ receptors and inhibited cisplatin-induced emesis (Hesketh and Gandara, 1991; Aapro, 1991). These antagonists also control radiation-induced emesis.

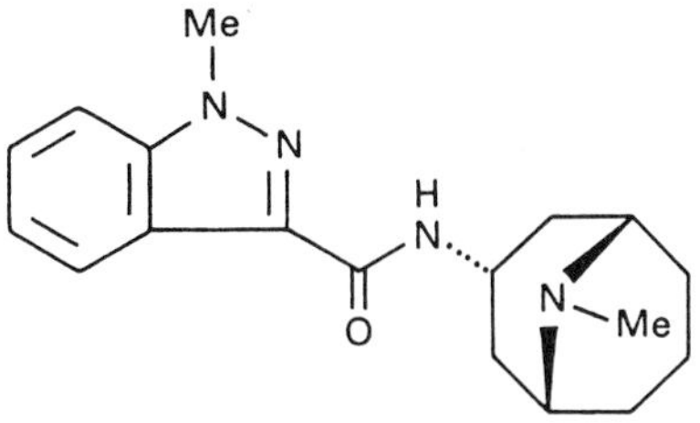

Granisetron

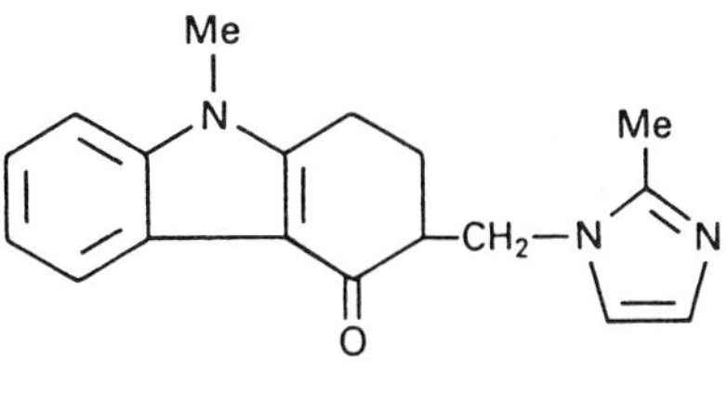

Ondansetron

Radiolabelled versions of these agents showed the high density of $5HT_3$ receptors in the area postrema and nucleus tractus solitarius (Barnes *et al.*, 1991). These regions receive the highest density of vagal nerves which originate from the gut and so it is possible to reconcile the peripheral and central theories of emetic action.

As shown by molecular modelling techniques, the structural requirements for antagonism included an aromatic ring, a carbonyl group and a basic nitrogen at the correct distances from each other (Evans *et al.*, 1991). An indole ring is not necessary.

$5HT_3$ receptor antagonists are also being studied for potential use as antipsychotic agents, as it appears that $5HT_3$ receptors modulate dopamine D_2 receptors by reducing activity without causing sedation - unlike the neuroleptic agents that are D_2 receptor antagonists (Barnes *et al.*, 1992). In addition, $5HT_3$ antagonists show anxiolytic activity that differs from the benzodiazepines in not inducing withdrawal symptoms when the drug is withdrawn.

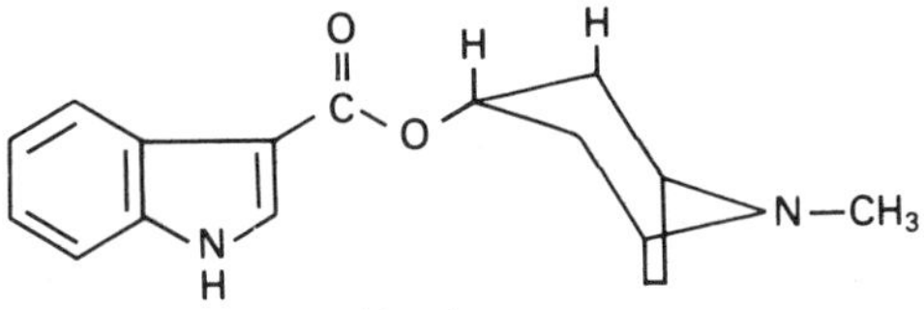

Tropisetron

9.5 Serotonin and noradrenaline re-uptake mechanisms - tricyclic anti-depressants

As noted in the introduction, there are uptake systems that efficiently remove the amine neurotransmitters from the synaptic cleft in order to terminate the signal. These mechanisms are believed to be related in the case of serotonin to re-uptake mechanisms in the blood platelet and the latter are often used as models of the neuronal system, even though platelets and synapses differ markedly in other ways.

Depression is a condition that can be treated by different types of drug, notably the monoamine oxidase inhibitors and the tricyclics in addition to electroconvulsive therapy. The aetiology of the condition, in spite of intensive research over the last 20 years, still defies a full understanding. A reduction in activity in either serotonin or noradrenaline transmission has been variously canvassed and the question is still open at the present time.

The inhibition of monoamine oxidase A, the isoenzyme that primarily catalyses the oxidation of noradrenaline and serotonin, in principle raises the brain levels of both monoamines. It does not, however, answer the question as to which, if either, monoamine is functionally inadequate in depression. Indeed, there is a further question which needs to be answered; although depression can be treated by drugs that promote monoaminergic transmission, is this a necessary and sufficient explanation of their anti-depressive activity, particularly in view of the long time interval between potentiation of monoamine activity and the onset of clinical relief of symptoms? This interval may be of the order of days or even weeks, and has led to a shift in research emphasis from acute effects to a study of slower adaptive changes induced by chronic anti-depressive therapy.

The discussion was initiated in 1967 by Coppen who suggested that noradrenaline levels were associated with drive and energy, and serotonin with mood. This view may owe a lot to the age-old connection between the catecholamines and 'fight or flight' that is a classic feature of school biology textbooks. There are difficulties here, in that the effect of monoamine oxidase inhibitors in animal tests that measure activity (e.g. tetrabenazine-induced sedation) have been shown to correlate with raised levels of serotonin rather than with noradrenaline (Christmas *et al.*, 1972). Nevertheless, there are always problems in extrapolating from the results of animal tests to effects in man in any aspect of drug development, particularly in the area of depression because of the difficulty in designing tests that measure mood.

The debate has been thrown into sharp relief by studies on a group of drugs known as the tricyclic anti-depressants, of which the best known is probably imipramine. These compounds bear a structural resemblance to the phenothiazines, since they have three aromatic rings fused in a linear fashion as well as a side-chain containing an amine group, but they do not affect dopamine receptors. They do, however, interfere with the re-uptake of serotonin and noradrenaline into the presynaptic nerve terminal and in some cases have a variety of effects on receptors for other neurotransmitters. One view is that tertiary amines tend to favour the inhibition of serotonin uptake while secondary amines affect noradrenaline re-uptake (Carlsson, 1984). A newer group of agents that specifically inhibit serotonin uptake but are not tricyclics are now available, of which one of the best known is fluoxetine. Although optically active the + and − forms are almost equiactive as inhibitors of serotonin uptake at $2{\cdot}1 \times 10^{-8}$ M and $3{\cdot}3 \times 10^{-8}$ M respectively (Schmidt *et al.*, 1988). These drugs are less toxic than the tricyclics but are just as effective (Leonard, 1992).

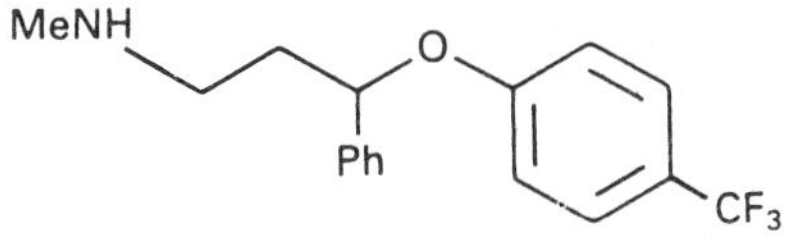

MeNH O Ph CF_3

Fluoxetine

On the other hand, there are other agents which are effective anti-depressants and inhibit the re-uptake of noradrenaline selectively. One such compound is maprotiline which has a secondary amine side-chain attached to a tetracyclic structure, not greatly different from the classical tricyclics (Pinder *et al.*, 1977).

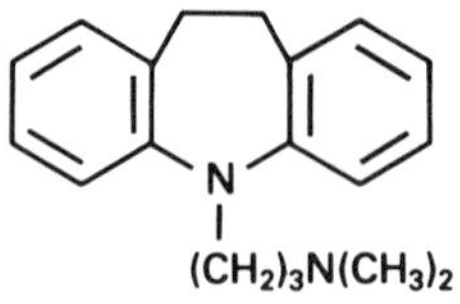

N

$(CH_2)_3N(CH_3)_2$

Imipramine

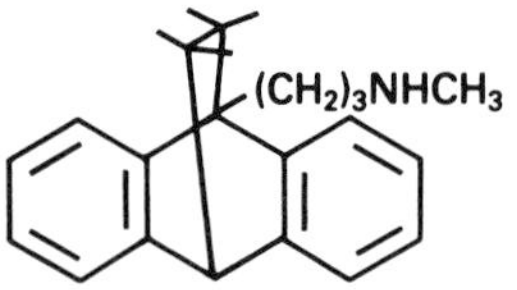

$(CH_2)_3NHCH_3$

Maproteline

$CH_2CH_2CH_2N(CH_3)_2$

N

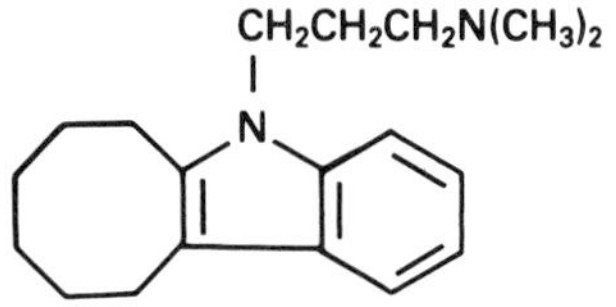

Iprindole

Another approach has been to investigate the brain levels of serotonin, and its metabolite 5-hydroxyindoleacetic acid (to indicate turnover), in the brains of patients who suffered from depression, suicide etc. (reviewed in Goodwin and Post, 1983). Clear indications have been found of a lowered level of serotonin *activity* at least, and in some cases a lowered *level* of amine. The concept of neuronal activity has been extended by Blier *et al.* (1990) who measured the effect of all the anti-depressive treatments on the firing of brain serotonin neurones. Both tricyclic drugs and electroconvulsive therapy sensitized post-synaptic neurones to serotonin (the post-synaptic receptor participates in the neurotransmission of the nerve impulse, whereas the presynaptic receptor or autoreceptor regulates neurotransmitter release and re-uptake). Furthermore, serotonin re-uptake blockers allowed serotonin to desensitize the autoreceptors more efficiently, while $5HT_{1A}$ agonists activated the post-synaptic receptors. In all these cases the therapy increased the neuronal activity of serotonin.

In sharp contrast Lipinski *et al.* (1987), have proposed that the post-synaptic α_1 receptor is the common pathway for anti-depressive action as noradrenaline-containing neurones are able to reach most parts of the brain. Chronic stimulation of this pathway, whether directly or by interaction with other neurotransmitter receptors, eventually leads to supersensitivity, while the other adrenoceptors appear to be desensitized by anti-depressive treatments.

The net effect of these changes is enhanced neurotransmission, and a correction of a subnormally functioning receptor–transmitter relationship. Whether this involves serotonin or noradrenaline depends on the drug involved, but it is possible that there are interactions between the two systems which may lead to a similar result after long-term administration. To discover how these neurotransmitters interact is clearly of paramount importance in understanding the aetiology of the condition and in the development of new and more effective drugs.

9.6 Monoamine oxidase - tranylcyromine and moclobemide for depression, deprenyl for Parkinsonism

The importance of monoamine oxidase (MAO) lies in its function of destroying excess neurotransmitter after action. Clearly, any interference with MAO activity will prolong the action of the transmitter and such interference has been found of therapeutic value in the treatment of depression.

The initial discovery came after two drugs, isoniazid and iproniazid, were put on the market in 1951 for the treatment of tuberculosis (iproniazid is the isopropyl analogue of isoniazid; the latter is still used for the therapy of tuberculosis). Eventually, it was found that the tubercular patients became euphoric (i.e. highly elated) on iproniazid. Depressed patients subsequently were also found to respond. Iproniazid was eventually withdrawn from the market in the USA because of liver toxicity, although it is still available in the UK, and a number of other compounds were developed in its wake. One characteristic they all have in common is the irreversible inhibition of MAO.

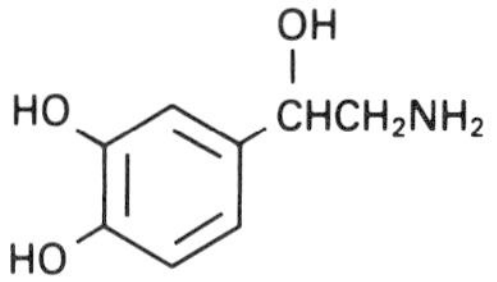

Noradrenaline

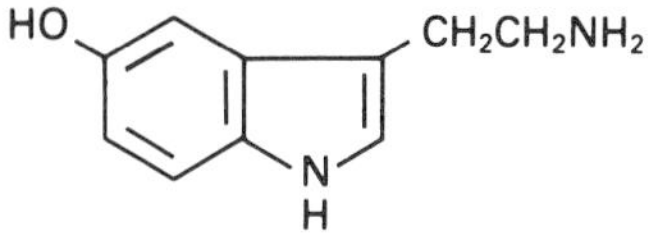

Serotonin

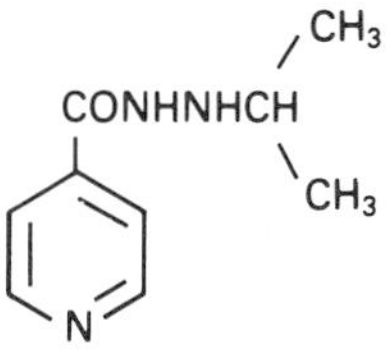

Iproniazid

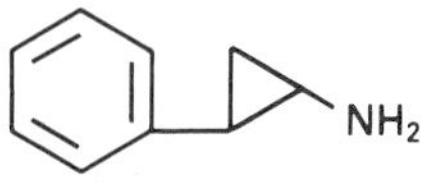

Tranylcypromine

MAO is a flavoprotein, present in the outer mitochondrial membrane, which catalyses the oxidation of the primary amine group of neurotransmitters (e.g. noradrenaline, dopamine and serotonin) to an aldehyde; hydrogen peroxide is formed in the process (Figs 9.6 and 9.7). Serotonin is thereby oxidized to 5-hydroxyindoleacetaldehyde and noradrenaline to 3,4-dihydroxymandelaldehyde. The mechanism requires a flavin to become reduced by abstracting

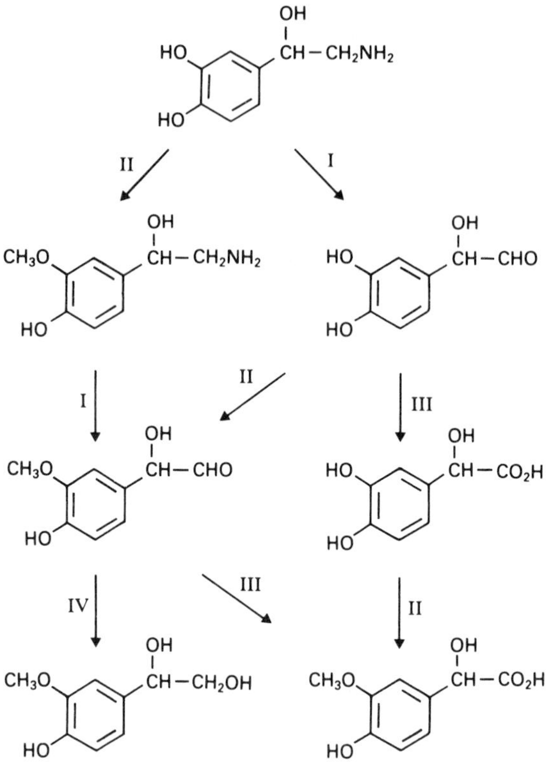

I: Monoamine oxidase
II: Catechol *O*-methyltransferase
III: Aldehyde dehydrogenase
IV: Aldehyde reductase
VMA: Vanillylmandelic acid
MOPEG: 3-Methoxy-4-hydroxyphenylethyleneglycol

Figure 9.6 Noradrenaline metabolism.

an electron from the amine to give a radical cation; subsequent loss of a hydrogen from the carbon produces a carbon radical and rearrangement of the carbon produces an imine which hydrolyses to the aldehyde (Silverman, 1991). Subsequent conversions yield either acid or alcohol, and in the case of the catecholamines, methylation of one of the catechol hydroxyl groups takes place primarily in the liver by catechol *O*-methyltransferase (COMT). COMT will also catalyse the methylation of the original monoamines (the methyl group is obtained from *S*-adenosylmethionine).

A detailed study of MAO was carried out using a number of acetylenic inhibitors. Two isoenzymes were identified (called A and B by Johnston, 1968). MAO-A was primarily responsible for the oxidation of serotonin and noradrenaline and was inhibited specifically by clorgyline, whereas MAO-B was respon-

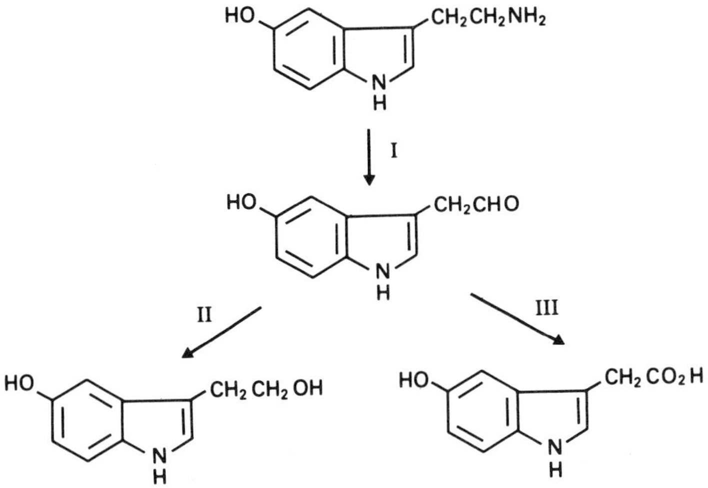

I

II III

5-HTP 5-HIAA

I: Monoamine oxidase
II: Aldehyde reductase
III: Aldehyde dehydrogenase
5-HTP: 5-Hydroxytryptophol
5-HIAA: 5-Hydroxyindoleacetic acid

Figure 9.7 Serotonin metabolism.

sible for the oxidation of phenylethylamine (benzylamine, although not occurring physiologically, is often used as the substrate *in vitro*). Deprenyl was subsequently found to be specific for MAO-B (Knoll and Magyar, 1972). Other naturally occurring monoamines, such as dopamine, tryptamine and tyramine are substrates for both isoenzymes. Most tissues contain both isoenzymes but there are cases where only one isoenzyme is present that is able to catalyse, slowly, the oxidation of substrates of the other isoenzyme.

The mechanism underlying this specificity has been addressed by Fowler *et al.* (1982) who showed that, since the structures formed with the selective acetylenic inhibitors are similar, the selectivity must derive from a kinetic difference. The inhibitors are K_{cat} or suicide inhibitors (Chapter 1) and the reaction can be represented as follows:

$$E + I \underset{k_{-1}}{\overset{k_{+1}}{\rightarrow}} EI \overset{k_2}{\rightarrow} EI^* \quad (K_i = k_{+1}/k_{-1})$$

where E and I represent the free enzyme and inhibitor respectively, EI represents the non-covalently bound inhibitor/enzyme complex and EI^* is the covalent complex. In the case of clorgyline the K_i for MAO-A form is $5{\cdot}4 \times 10^{-8}$ M and $5{\cdot}8 \times 10^{-5}$ M for MAO-B. This difference in the rate of formation of the reversible complex is sufficiently large to account for almost all the specificity, although k_2 for the inactivation of MAO-A is greater by one order of magnitude than k_2 for MAO-B. Deprenyl, on the other hand, owes much of its selectivity for MAO-B to a much higher k_2 for that form, i.e. the rate of formation of the irreversible adduct is much faster.

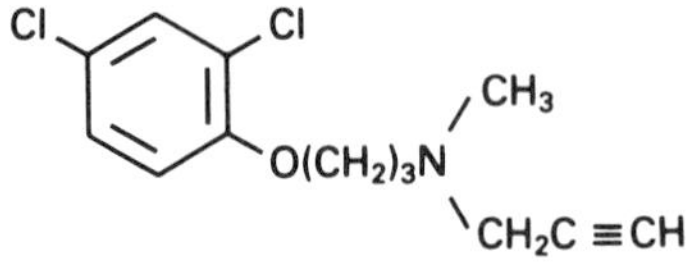

Clorgyline

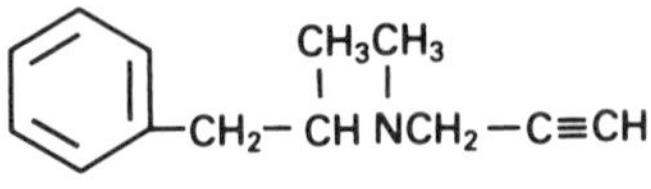

Deprenyl

A detailed study of the interaction of acetylenic irreversible inhibitors with MAO has shown that the flavin portion is the target for the inhibitor. The inhibitor initially may be oxidized to an allenic grouping that subsequently reacts irreversibly with the N^5 position of the isoalloxazine ring (Maycock *et al.*, 1976) to give a structure of the type:

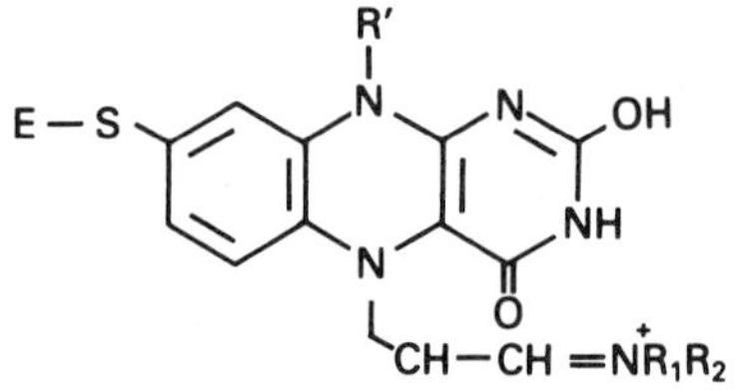

R_1, R_2 represent alkyl groups
R′ the ribityl side-chain and
E the enzyme

(the flavin is linked covalently to the enzyme through a sulphur atom). In contrast, the cyclopropylamine type of inhibitor, e.g. tranylcypromine, although also a suicide substrate, yields a product that can be attacked by a nucleophilic sulphhydryl group and no binding at all occurs to the flavin (Silverman, 1983):

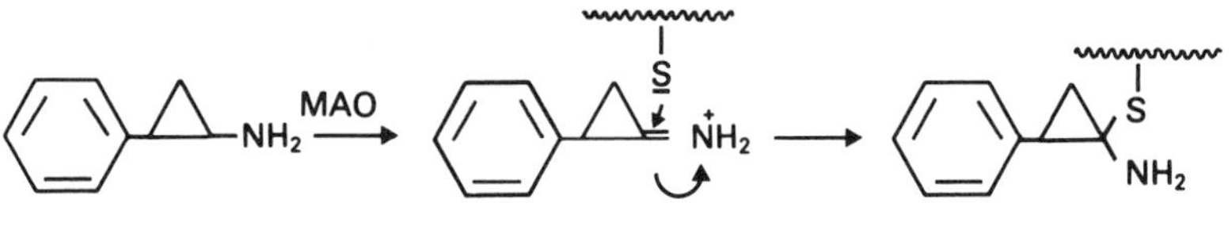

This adduct can be broken down more easily at neutral pH after the protein has been denatured.

The two human isoenzymes show 70 per cent homology and derive from two distinct genes (see Shih, 1991). Each polypeptide contains a FAD (flavin adenine dinucleotide) residue but there is some doubt as to whether the enzyme is a monomer or homodimer. A stretch of about 20 amino acids near the cysteine at the C-terminus that binds the flavin is conserved, while other conserved residues near the N-terminus may bind the adenine of FAD. The flavin is linked to the cysteine via a thioester bond. Regions of the protein that anchor it in the membrane have also been proposed (Powell, 1991).

Monoamine oxidase is also found in many other tissues and one of its principal tasks in the gut is to detoxify monoamines absorbed in the diet as, for example, tyramine from cheese, pickled herrings, chianti etc., or dopamine in broad beans. These amines will otherwise reach the circulation and are likely to produce a hypertensive crisis by releasing noradrenaline from nerve endings (section 9.2). Clearly, any drug that irreversibly inhibits MAO non-specifically, and is given on a daily basis, will result in a greatly reduced level of the enzyme, since it takes almost three weeks to regain its original level by re-synthesis after a single dose. This was found to be the case with the earlier non-specific irreversible inhibitors such as tranylcypromine, and use of these agents was eventually restricted to a hospital environment where close supervision of the patient could be maintained. Even the use of selective but irreversible inhibitors could cause hypertensive crises in patients brought on by eating cheese. Recent re-evaluation suggests that concerns over the 'cheese effect' may have been too great. Irreversible MAO inhibitors may be of considerable value in 'atypical', as opposed to 'endogenous', depression where electroconvulsive therapy and the tricyclics may not be effective (Bass and Kerwin, 1989).

More recently, a new class of reversible inhibitor has been described that shows specificity for MAO-A, but does not apparently potentiate tyramine's ability to raise blood pressure. Moclobemide, for example, shows an IC_{50} for MAO-A of $6{\cdot}1 \times 10^{-6}$ M and $>10^{-3}$ M for MAO-B (reviewed in Haefely *et al.*, 1992). The inhibition for MAO-A is initially competitive but gradually changes over time to a more tightly bound complex. Moclobemide is more effective *in vivo* than these figures would predict, possibly because of the increase in inhibition or alternatively through MAO metabolism to a more active agent. Moclobemide is equally effective in endogenous and atypical depression but shows only a very slight 'cheese' effect (Fitton *et al.*, 1992).

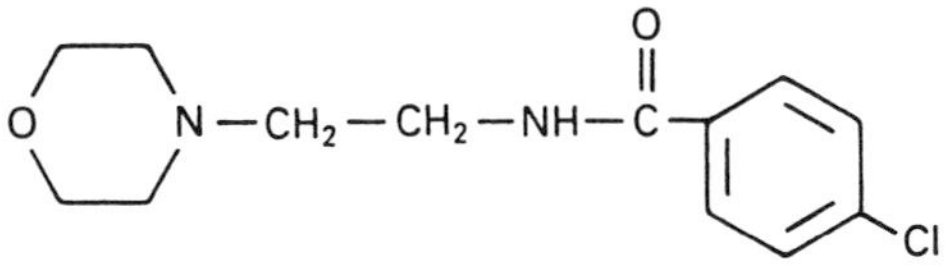

O N—CH_2—CH_2—NH—C(=O)— Cl

Moclobemide

9.6.1 Monoamine oxidase and Parkinsonism

An interesting recent development involving monoamine oxidase is the discovery that it may be involved in the generation of some types of Parkinsonism. Various heroin addicts injected synthetic heroin contaminated with 1-methyl-4-phenyltetrahydropyridine (MPTP). They subsequently developed irreversible symptoms characteristic of Parkinsonism (slowness of movement, tremor and rigidity) normally confined to those over 55 years of age. Parkinsonism develops as a consequence of the destruction of the dopamine neurones in the nigro-striatal region of the brain and at least 80 per cent of these must be inactivated before the disease manifests – hence the connection with old age. There is now considerable concern that other chemicals in the environment may also be responsible for the development of Parkinsonism.

MPTP itself does not produce the effect. The compound is oxidized by MAO-B, surprisingly fast for a tertiary amine, via a two electron oxidation to the dihydropyridine (MPDP) which subsequently disproportionates into MPTP and the aromatic *N*-methyl-4-phenylpyridinium ion (MPP^+; Fig. 9.8). This MPP^+ is taken up by the catecholamine uptake system into the nigro-striatal region and probably reacts with the surrounding tissue. This transformation also takes place in other parts of the brain and it is not clear why the nigro-striatal area should either be particularly sensitive to the toxin, or should concentrate it so effectively.

L-(−)-Deprenyl is an irreversible inhibitor of MAO-B in a similar fashion to clorgyline for MAO-A (p. 185). L-Deprenyl first forms a non-covalent complex with the enzyme; the flavin is then reduced while the drug is oxidized and the product reacts covalently with the N^5 position of the isoalloxazine ring (Gerlach *et al.*, 1992).

For much of its life, L-deprenyl had been a drug in search of a condition. The discovery that L-deprenyl inhibited the conversion of MPTP to MPP^+ triggered clinical trials in Parkinsonism, notably the DATATOP study (Deprenyl and Tocopherol Antioxidative Therapy of Parkinsonism). The drug greatly prolonged the initial phase of the disease before L-dopa had to be given, presumably by protecting the dopamine neurones against damage (LeWitt, 1991; Tetrud and Langston, 1989).

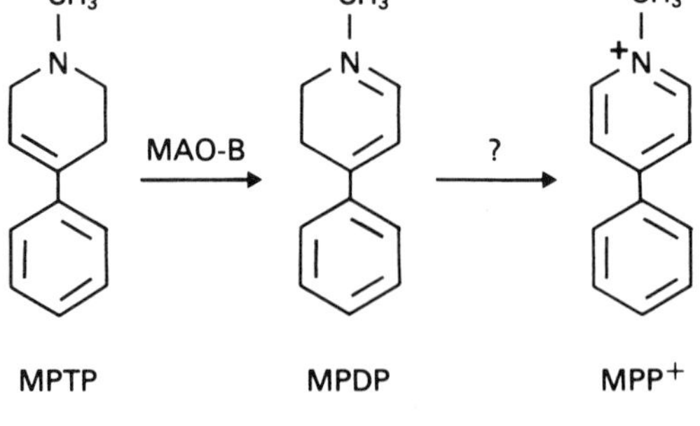

MPTP MAO-B MPDP ? MPP^+

MPTP: 1-Methyl-4-phenyl-1,2,3,6-tetrahydropyridine
MPDP: 1-Methyl-4-phenyl-2,3-dihydropyridine
MPP^+: 1-Methyl-4-phenylpyridinium ion

Figure 9.8 The oxidation of MPTP by MAO-B.

9.7 Acetylcholine action

Acetylcholine is probably the most widely found neurotransmitter in the human body as it serves large areas of the central and peripheral nervous systems, and mediates a variety of actions both inhibitory and excitatory. In the autonomic nervous system, i.e. the system that governs involuntary functions such as heart rate, digestion etc., acetylcholine is the predominant neurotransmitter. Acetylcholine is, therefore, responsible for carrying messages to a variety of organs including heart, blood vessels, glands and smooth muscles.

$$CH_3CO_2CH_2CH_{2}\overset{+}{N}(CH_3)_3$$

Acetylcholine

Two major subdivisions of acetylcholine receptors have been defined on the basis of agonist and antagonist action, namely nicotinic and muscarinic. The former are characterized by the action of nicotine as an agonist and of tubocurarine as an antagonist, and are located in autonomic ganglia (i.e. where two neurones or sets of neurones interact), skeletal muscle and some of the synapses in the central nervous system. The nicotinic receptors on ganglia and skeletal neuromuscular junctions differ, however, in some respects; tubocurarine blocks both but its action is much more marked on the latter, and there are other more selective antagonists.

Muscarinic receptors, on the other hand, are found in smooth muscle, cardiac muscle and glands, and are the predominant form of the acetylcholine receptor in the central nervous system. Muscarine and pilocarpine are agonists; atropine and hyoscine antagonists.

9.7.1 Muscarinic receptor - pirenzepin for ulcers, atropine for pre-anaesthetic medication, pilocarpine for glaucoma

Gene cloning has emphasized the heterogeneity of muscarinic receptors, as with other receptors discussed in this chapter. Five genes have been found to encode muscarinic receptors where two had originally been defined by ligand affinities. As in other cases, not all the proteins expressed have pharmacological functions yet assigned to them. The muscarinic receptors mediate a wide variety of functions including phosphoinositol hydrolysis, adenylate cyclase inhibition, mobilization of calcium from intracellular stores and modulation of potassium channels (reviewed in Hulme *et al.*, 1990; Lambert *et al.*, 1992).

Muscarinic receptors are one of the many groups of receptors that are coupled to G-proteins and structurally have seven transmembrane (TM) regions. As with the other receptors, the N-terminus is extracellular and the C-terminus cytoplasmic thus forming six loops, three inside and three outside the cell. Agonists bind primarily to TM3, TM6 and TM7, although there is a suggestion that other transmembrane domains may help to form a hydrophilic ligand binding pocket (Hulme *et al.*, 1990). The third intracellular loop and possibly the C-terminus are involved in G-protein binding.

Agonists bind to an aspartate in TM3 and a tyrosine in TM7 which are believed to hydrogen bond in the ground state. When the amino group of the agonists binds to the aspartate carboxyl, the hydrogen bond is broken and a conformational change may result, allowing the G-protein to bind (Hulme *et al.*, 1990). Another aspartate residue is conserved in TM2 and is involved in signal transduction, as mutation greatly reduces the ability of the M1 receptor to activate phospholipase C. Two cysteines in the first and second extracellular loops form a disulphide bridge which may act to stabilize the receptor (Saverese and Fraser, 1992).

G-Protein G_i couples muscarinic receptors M_2 and M_4 to adenylate cyclase in an inhibitory fashion, while (probably) G_q links M_1, M_3 and M_5 to phospholipid hydrolysis. The results for agonists binding to receptor have given a wide range of results, probably due to varying levels of guanine nucleotides because the presence of guanine nucleotides can reduce the affinity of the agonist for the receptor markedly. The G-protein–receptor complex has a much higher affinity for the agonist than the receptor alone. This may reflect the energy of another hydrogen bond between, possibly, the ester function of acetylcholine and an amino acid residue. Antagonists do not show this range of effect (Hulme *et al.*, 1990).

The M_1, M_3, M_4 and M_5 receptors are normally found in neural tissue and the M_1 and M_3 in exocrine glands as well. M_2 is a cardiac and smooth muscle receptor. Receptors are found both pre- and post-synaptically, the presynaptic receptors or autoreceptors (usually M_2) inhibit the release of acetylcholine.

Muscarinic antagonists

Pirenzepin as a relatively selective ligand compared with atropine (see below) has been used for the treatment of peptic ulcers (Carmine and Brogden, 1985).

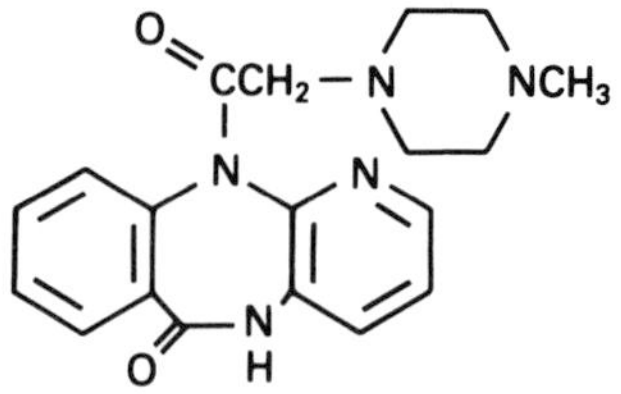

Pirenzepin

Some side-effects were noted, characteristic of anti-muscarinic drugs and due to blockade of the parasympathetic nervous system, notably dry mouth and blurred vision, but they were of low incidence and severity. The lack of central nervous system effects suggests that although the drug has a great affinity for neuronal receptors, it does not cross the blood–brain barrier because it is too hydrophilic.

As well as stimulating peristalsis in the gut, acetylcholine releases gastric juices from the glands by activating M_1 receptors, both on postganglionic neurones and the parietal cells of the gastric mucosa. Since the latter have a low affinity for pirenzepin, the drug target is believed to be the postganglionic neurone receptor (Goyal, 1988). Pirenzepin shows some selectivity for the M_1 receptor with a binding constant of $4{\cdot}6 \times 10^{-8}$ M to M_1 and $\approx 3 \times 10^{-7}$ M for M_3 (Lambert *et al.*, 1992). It is interesting that efficacy and binding studies give results that are very close unlike the β_2 agonists noted earlier in this chapter.

DL-Atropine, an alkaloid derived from *Atropa belladonna* or deadly nightshade, is non-selective but a very tight binder nevertheless, having a binding constant in the region of 10^{-10} M; the levorotatory form is the more active. Atropine is used to counteract the effects that stem from the stimulation of the vagus nerve during anaesthesia, and so is usually part of operation premedication.

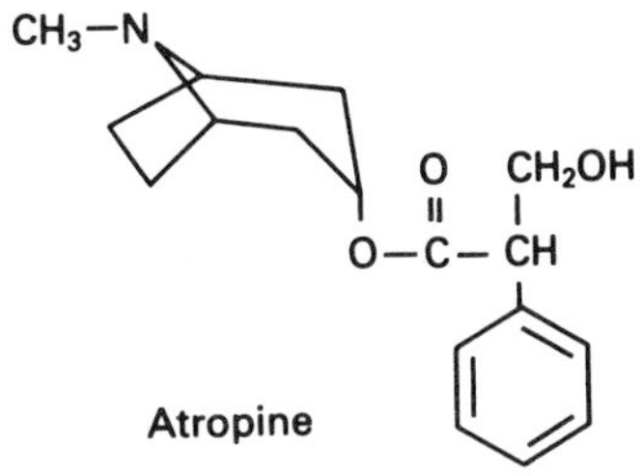

CH_3—N

O CH$_2$OH

O—C—CH

Atropine

Muscarinic agonists

One type of glaucoma is characterized by a build-up of water in the aqueous humor of the eye, a condition which is particularly prevalent in the elderly. The iris and the cornea are normally separated by a channel through which the aqueous humor can filter and be dispersed by absorption into the nearby blood vessels. In later life the iris can be pushed forwards, possibly because of inflammation or increasing size of the lens, thus closing the channel. The aqueous humor can no longer filter out and pressure on the eyeball increases, with particular danger of destruction of the optic nerve fibre leading to blindness.

Pilocarpine, an alkaloid derived from the leaves of the South American shrub *Pilocarpus*, is one of the drugs recommended for this condition (Shapiro and Enz, 1992). The drug acts as a miotic, i.e. it narrows the pupils by causing them to contract, thus drawing the iris away from the cornea and allowing the fluid to drain away. The active form of pilocarpine is the isomer where the side-chains at C-2 and C-3 of the imidazole ring are in the *cis* position. Pilocarpine does not distinguish between the species of muscarinic receptor, with binding and efficacy measured on ganglion depolarization (M_1) and guinea-pig ileum (M_3) in the region of 10^{-6} M (Shapiro and Enz, 1992).

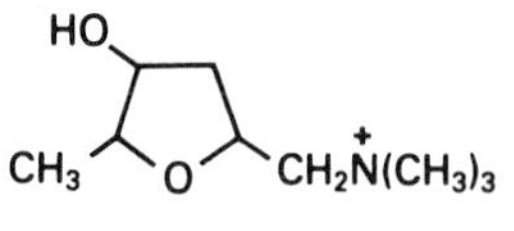

Muscarine

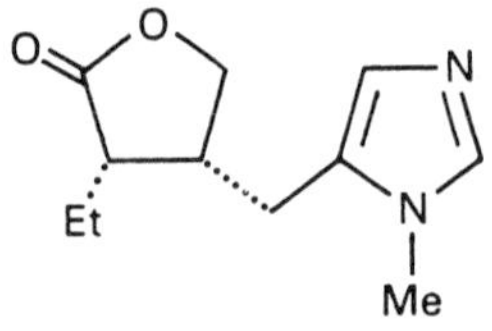

Pilocarpine

9.7.2 Acetylcholinesterase - pyridostigmine for myasthenia gravis

Closely associated with cholinergic synapses is the enzyme acetylcholinesterase, which rapidly hydrolyzes the neurotransmitter and is essential to ensure that the signal is terminated as soon as it has passed. The enzyme is attached to the lipid environment of the membrane via a link with inositol phosphate and diacylglycerol because phospholipase C will liberate the enzyme from its normal membrane-bound state (reviewed in Low *et al.*, 1986).

Acetylcholinesterase hydrolyses the ester link of the neurotransmitter to yield acetate and choline with a very high turnover number. The enzyme has two well-defined parts in its binding site for substrates; one anionic which binds the cationic head of the substrate (and of inhibitors), and an esteratic site at which hydrolysis takes place with the formation of an acyl-enzyme intermediate. There is apparently another anionic binding site which can be occupied by bis-quaternary ligands (see the review by Soreq *et al.*, 1992). A charge-relay system similar to that defined for trypsin, chymotrypsin and elastase with glutamate carboxyl, histidine imidazole and serine hydroxyl is proposed at the active site of acetylcholinesterase.

Glutamate Histidine Serine

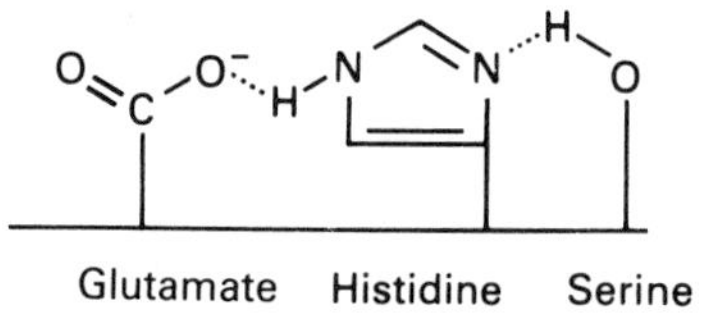

As with the proteases, the serine hydroxyl is activated by the charge relay system sufficiently to be a strong nucleophile, thus forming a transient acyl-enzyme intermediate, which is easier to detect with acetylcholinesterase as the complex dissociates more slowly.

A family of compounds with structures related to acetylcholine and including a cationic head group and carbamate ester, inhibit acetylcholinesterase by acting as false substrates, but with very low turnover numbers, so that the true substrate is unable to be hydrolysed. One of the best known is pyridostigmine which inhibits acetylcholinesterase by binding to the anionic site with its cationic pyridine nitrogen, while the carbamate ester is available to be hydrolysed by the esteratic site. The tetrahedral complex formed on hydrolysis, a dimethylcarbamoyl-enzyme, resembles the transition state of the normal reaction but is extremely slow to hydrolyse and prevents the hydrolysis of the normal substrate. The drug is therefore a type of suicide inhibitor. The IC_{50} for the reaction is $1{\cdot}6 \times 10^{-6}$ M (Wilson *et al.*, 1961) which is effectively k_1k_2 in the following scheme:

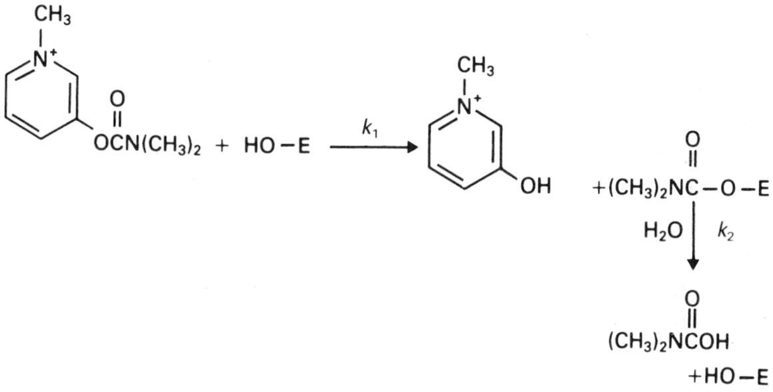

$$\text{(N-methylpyridinium-3-yl)}\text{-}OC(=O)N(CH_3)_2 + HO{-}E \xrightarrow{k_1} \text{(N-methylpyridinium-3-yl)}\text{-}OH + (CH_3)_2NC(=O){-}O{-}E$$

$$(CH_3)_2NC(=O){-}O{-}E \xrightarrow[H_2O]{k_2} (CH_3)_2NC(=O)OH + HO{-}E$$

Inhibitors of acetylcholinesterase would be expected to potentiate the action of acetylcholine and this has been made use of in the development of the organophosphate insecticides. In medicine, less toxic inhibitors have been used in conditions where the receptors are damaged or diminished in number, as in myasthenia gravis. This condition is a genuine auto-immune disease. The damage is caused by auto-antibodies to the nicotinic acetylcholine receptors at the motor end-plate where the nerve connects with striated muscle. The receptor has been cloned and consists of five subunits, α_2, β, ϵ and δ. The auto-antibodies bind to the α subunit as a type of antagonist preventing the binding of acetylcholine; the transmission of stimulatory signal from nerve to muscle fails resulting in muscle weakness (reviewed in Vincent, 1991). In addition, a degree of membrane lysis occurs, and a considerable amount of debris is found in the synaptic cleft. The net result is a widening of the cleft and elongation of the synaptic membrane, further reducing neuromuscular transmission (Scadding and Havard, 1981).

Myasthenia gravis is most obviously recognized by muscle weakness which characteristically starts around the eyes and face and then works its way down via the limb girdle, outer limbs and trunk in that order. There is a marked fatigability of skeletal muscle, in that if a motor nerve of a patient is stimulated, synaptic transmission rapidly falls away, unlike the results obtained with a normal subject. Pyridostigmine and neostigmine are currently widely used for this condition (Havard and Fonseca, 1990). They compensate for the deficit of acetylcholine receptors by slowing down the rate of degradation of acetylcholine and so increase the duration of its activity.

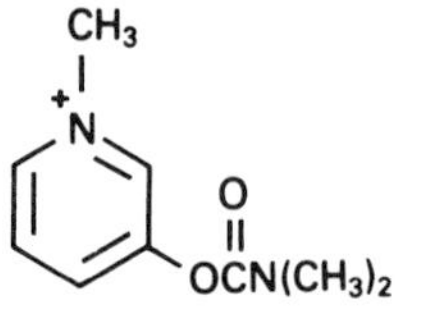

Pyridostigmine

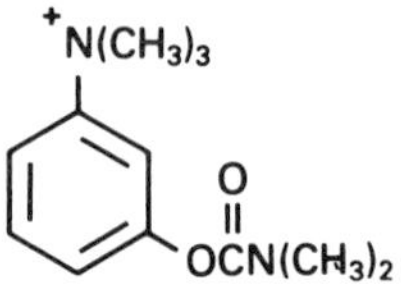

Neostigmine

9.8 4-Aminobutyric acid receptor - benzodiazepines as hypnotics, avermectin as anthelminthic, baclofen for spasticity

4-Aminobutyrate (γ-aminobutyrate; GABA) is a neurotransmitter that acts at inhibitory synapses, often presynaptically, i.e. the inhibitory nerve terminal secretes GABA on to a neighbouring excitatory nerve terminal and thereby reduces the output of the excitatory transmitter (often acetylcholine). By this means, specific neuronal pathways converging on another neurone can be eliminated without influencing the efficacy of others. The relevant biosynthetic pathway is found in GABA neurones, as found with other neurotransmitters; GABA neurones can synthesize GABA from glutamate with the help of glutamate decarboxylase (Fig. 9.9).

$H_2NCH_2CH_2CH_2CO_2H$

4-Aminobutyrate

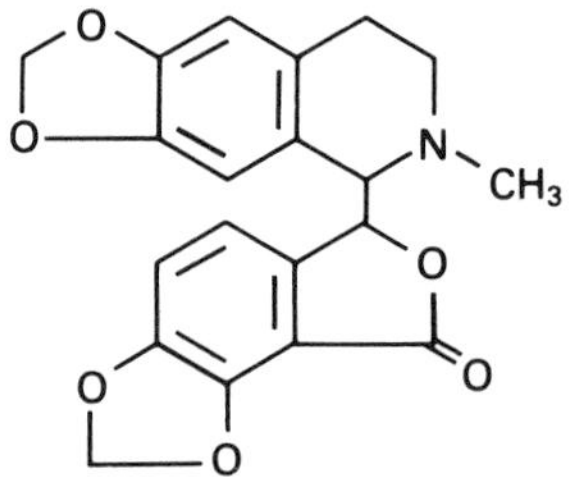

Bicuculline

GABA receptors are divided into two major types. $GABA_A$ receptors are unusual in that they contain a chloride ion channel in the receptor itself. The binding of agonists thus allows negative ions to pass into the cell and increases the negative membrane potential (hyperpolarization) so rendering it unable to transmit the nerve impulse or action potential. The consequence of this action is to induce muscle relaxation, to reduce anxiety and to block convulsions.

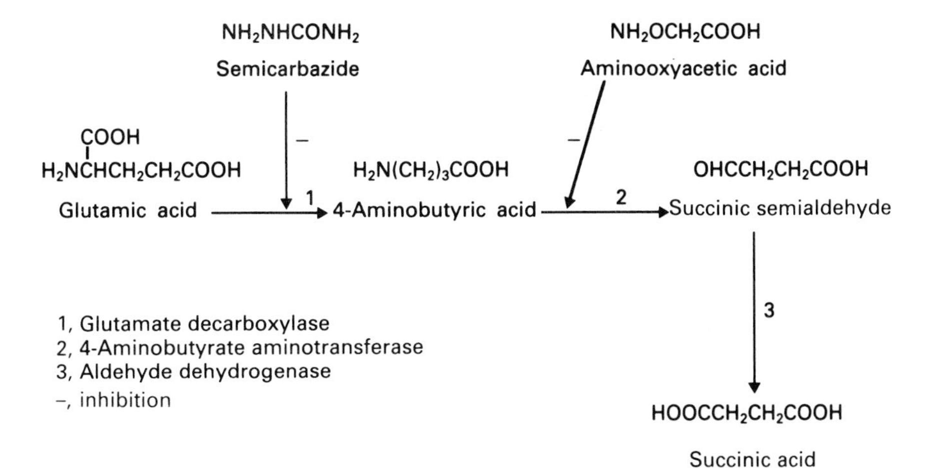

Figure 9.9 The pathways of GABA synthesis and breakdown.

Sedation may also be a consequence (Sanger, 1985). $GABA_B$ receptors are discussed under baclofen (see section 9.8.3).

H_2NCH_2CH — Cl; CH_2CO_2H

Baclofen

H, N, O, O; H_2NCH_2

Muscimol

9.8.1 Benzodiazepines

A well known family of psychoactive drugs used as hypnotics, sedatives and tranquillizers, the benzodiazepines, that were originally marketed in the late 1960s mainly for the treatment of anxiety, were subsequently discovered to bind very tightly to specific sites in the brain (Speth *et al.*, 1978). A considerable body of evidence linked this site with the $GABA_A$ receptor and, not for the first time, the experimental use of a drug advanced the state of scientific knowledge, having initially been of therapeutic value (Sanger, 1985). The best known benzodiazepines are probably the anxiolytic (tranquillizer) drugs diazepam and chlordiazepoxide (these are generic names; the drugs may be more familiar under the trade names of Valium and Librium).

Benzodiazepines act in such a way as to simulate the effects of exogenously administered GABA, but if GABA levels are depleted there is no such effect. This occurs, for example, when semicarbazide is co-administered; glutamate decarboxylase, an enzyme with pyridoxal phosphate as a prosthetic group, is inhibited through the formation of a Schiff's base and GABA levels are lowered.

Conversely, benzodiazepine effects are potentiated by aminooxyacetic acid, an inhibitor of GABA oxidation to succinic semialdehyde by GABA aminotransferase.

The best characterized benzodiazepine ligand is flunitrazepam which binds to homogenates of human brain with a binding constant in the range 1–3×10^{-9} M (Speth *et al.*, 1978). This was measured both by kinetic analysis of the rates of association and dissociation, and also by Scatchard analysis. Inhibition of flunitrazepam binding by a number of benzodiazepines correlated well with a number of pharmacological tests (notably inhibition of convulsions induced by pentylenetetrazole in mice and muscle relaxation in cats) and also with the dose used to treat anxiety in man. Braestrup *et al.* (1977) obtained similar results using ^{3}H-diazepam binding to a membrane preparation from human brain. They also showed that GABA receptors varied in density between different parts of the brain and this density paralleled the distribution of benzodiazepine receptors.

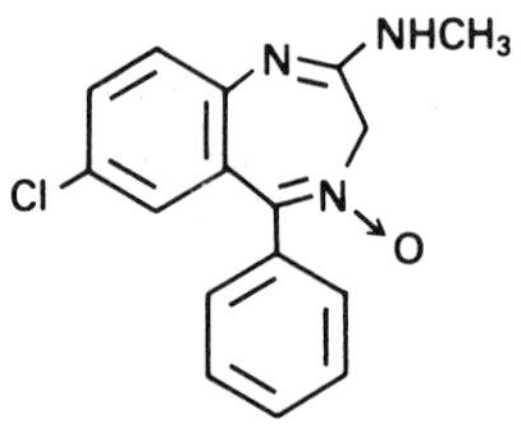

Chlordiazepoxide

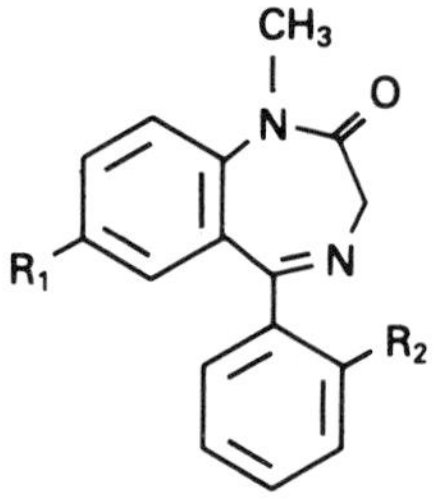

	R_1	R_2
Flunitrazepam	$-NO_2$	F
Diazepam	$-Cl$	H

Conversely, the binding of GABA and GABA agonists, such as muscimol, potentiates the binding of benzodiazepines to their receptors, an action which can be reversed by the $GABA_A$ antagonist bicuculline (Blaestrup *et al.*, 1980). The receptor complex has been cloned and expressed in *Xenopus* oocytes to give receptors that control chloride channels; there are three families of subunits (α, β, γ). There are six versions of an α subunit, two of the β subunit and two γ subunits, and possibly others. Benzodiazepine binding occurs on a trimer (α_1, α_2, α_3 or α_5) with β_2 and γ_2. The α subunit carries the drug site, although the γ subunit is also required for maximum effect. The GABA site resides mainly on the β subunit, although there is some overlap between them (Sieghart, 1989; Vicini, 1991). The variation in the α subunits gives rise to $GABA_A$ receptors with differing GABA sensitivity which may relate to previously defined benzodiazepine receptors, although the receptors all control GABA-gated chloride channels (Sieghart, 1989).

A model has been proposed for the receptor that contains four transmembrane regions and bears a resemblance to another gated ion channel, the nicotinic receptor (Vicini, 1991). A histamine residue appears to be crucial for the binding of benzodiazepines (Wieland *et al.*, 1992).

Other studies, however, have cast doubt on this behavioural correlation (Sanger, 1985). If rats are trained to respond to food and drink but responding, in addition, is linked to an aversive stimulus such as an electric shock, benzodiazepines greatly increase what would otherwise be a low rate of response. Furthermore, the magnitude of the drug effect correlates highly with the clinical potency in treating anxiety. Not all agonists increase the rate of response when given systemically, but do if injected into the appropriate part of the brain. These differences may arise as a consequence of non-uniform distribution of agents in the brain, or it may be that some of the anxiolytic actions of benzodiazepines are not involved with GABA - unlike the muscle relaxant and anti-convulsant actions (Sanger, 1985).

9.8.2 Avermectin

GABA acts as a neurotransmitter in a number of other organisms besides mammals, including arthropods and nematodes, which are themselves pathogenic for man. Conspicuous among these infections is one caused by a nematode worm, *Onchocerca volvulus*, a native of West and Central Africa, which infects man to produce the disease known as river blindness or onchocerciasis. The worm in the larval form (microfilaria) is ingested from humans by a bite from the blackfly of the genus *Simulium*, which breeds in fast-flowing streams. After about a two-week incubation, the larvae can be injected by bite back into man in an infective form. The microfilaria migrate in the surface tissue where they are found in nodules, particularly around the joints, and to the eye where they usually cause blindness by attacking the optic nerve. The larvae mature over 1 to 3 years into adult worms (macrofilaria), the female of which can release thousands of microfilariae daily. These migrate to the surface tissues and the eye and either degenerate or are ingested by the blackfly bite. The blackfly is the vector without which the worm could not attain such high numbers, as neither the larvae or adult worms multiply in humans (Goa *et al.*, 1991). The only drugs that have been used are suramin, which is extremely toxic, and diethylcarbamazine, which is less toxic but is liable to kill the adult worms swiftly and precipitate an extreme allergic reaction by virtue of the large amount of nematode protein rapidly entering the circulation. This reaction can cause more damage than the nematode itself, particularly in the eye (Mazzotti reaction).

Recently, however, a major improvement has been effected by the use of ivermectin. This is a member of the avermectin family of compounds, first obtained from the fermentation of *Streptomyces avermitilis* (Egerton *et al.*, 1979). The avermectins which possess a 16-membered lactone ring with a spiroacetal system involving C-17 to C-25 and a disaccharide substituent at position C-13 (see structures for details). The naturally occurring avermectins fall into four major (A_{1a}, A_{2a}, B_{1a} and B_{2a}) and four minor (A_{1b}, A_{2b}, B_{1b} and B_{2b}) series (Campbell *et al.*, 1983; Fisher and Mrozik, 1992). Ivermectin is

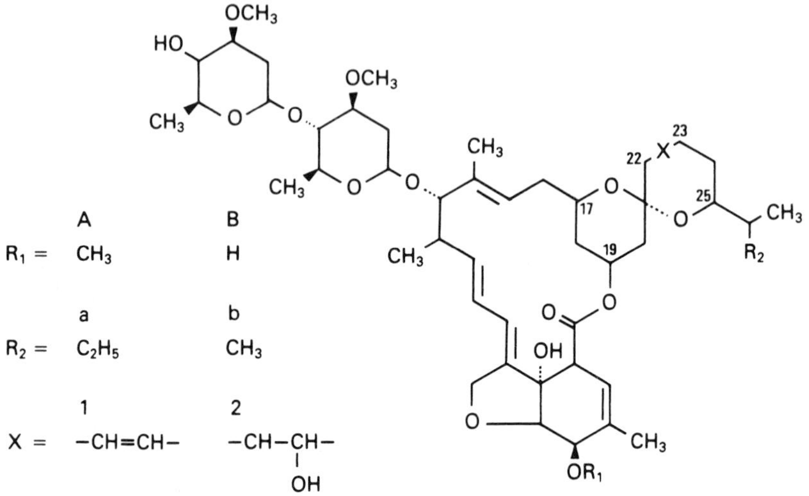

Avermectins

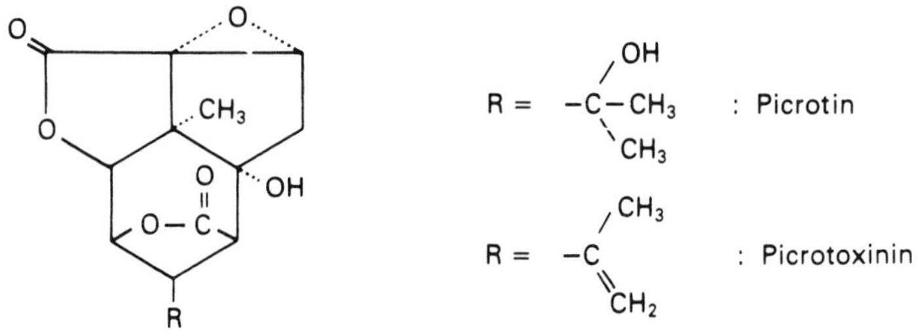

Picrotoxin (1 mol picrotin:1 mol picrotoxinin)

synthesized from avermectin B_1 by reduction of the 22,23 double bond (see structure). Ivermectin is now regarded as the front-line treatment for river blindness. Annual doses of 150 μg/kg are sufficient to reduce the microfilarial load by 80–99 per cent in the surface tissues, but more slowly than diethylcarbamazine, so the Mazzotti reaction is much less marked. The drug does not kill the adult worms, but it prevents the female from releasing the larvae and they gradually degenerate in the womb. As a result, the blackfly transmits the parasite in far smaller numbers. Another advantage is that the infrequent dose is much easier for health workers to organize in the African bush (Goa *et al.*, 1991).

Ivermectin was originally introduced to the veterinary market for the treatment of internal parasites (endoparasitic nematodes) of mammals of economic importance such as sheep, cattle and horses and of external parasites (ectoparasites), namely arthropods such as ticks and mites. Lobsters are also sensitive to the drug! The common link between these organisms is their possession of neuronal synapses at which GABA acts as a transmitter (see Campbell *et al.*, 1983; Wright, 1987 for reviews on avermectin and its biological action).

Most of the experimental studies have been carried out with avermectin B_{1a}, referred to as avermectin below. In *Ascaris suum*, a relatively large nematode that infects pigs, there are at least two sites of drug action: one is a block between a nerve cell (known as an interneurone) that forms a synapse with an excitatory motor neurone on the dorsal side of the nematode and leads to depolarization of the muscle membrane. The second site is at the neuromuscular junction of an inhibitory motor neurone on the ventral side and muscle, the activation of which leads to hyperpolarization. At the former site the action of 6×10^{-6} M avermectin B_{1a} in reducing the depolarization can be mimicked by the agonists muscimol and piperazine, and reversed by the antagonist picrotoxin, which indicates that the interaction is with the $GABA_A$ receptor (Kass *et al.*, 1984). On the other hand, avermectin's action in reducing hyperpolarization at the second site does not respond to picrotoxin reversal and is presumably nothing to do with $GABA_A$ receptors.

In sharp contrast, at 1×10^{-7} M concentration, the drug appears to antagonize the action of GABA in reducing the length of time that ion channels are open, and the probability of their opening - measured by using microelectrodes implanted into the nematode (Martin, 1987). The apparent contradiction between these findings remains to be resolved, although the use of a low physiological level of the drug may be crucial.

The arthropod target for avermectin is also a neuromuscular junction as shown by experiments in lobsters where the action of the drug was synergistic with GABA. The neuromuscular junction is known to be controlled by one excitatory axon, with glutamate as the transmitter, and one inhibitory axon, with GABA as the transmitter. Avermectin reduced both excitatory and inhibitory post-synaptic potentials, probably by opening chloride channels and hyperpolarizing the membrane. The response was restored by picrotoxin. It appears likely that avermectin acts as a GABA agonist either by potentiating the binding of GABA to its receptor (Campbell *et al.*, 1983) and/or by releasing GABA presynaptically. The net effect of drug on both nematode and arthropod is muscular paralysis.

In vitro studies indicate that avermectin B_{1a} can stimulate the high affinity binding of GABA to rat brain membranes by increasing the number of available binding sites, rather than the binding constant, at an IC_{50} of 7×10^{-6} M. This effect is chloride-ion-dependent and is antagonized by picrotoxin and bicuculline (Pong and Wang, 1982). Avermectin cannot reach these sites *in vivo* as it does not cross the blood–brain barrier. Avermectin also enhances the binding of diazepam to rat brain membranes at micromolar concentrations, and potentiates the pharmacological activity of the tranquillizer, notably a decrease in motor activity and muscle power (Williams and Yarbrough, 1979).

More recent studies in locust leg muscle (the exterior tibiae) have shown that there are at least two other chloride channels which can be modulated by avermectin besides that connected to the GABA receptor: one is probably linked to glutamate receptors and the other is not linked to any known receptor (Duce and Scott, 1985). Other avermectin binding sites have also been found on chloride channels in non-pathogenic arthropods. In crayfish stomach muscle there are sites that are extremely sensitive to the drug at concentrations below 10^{-12} M. The channels can be opened directly and reversibly at these levels and irreversibly above 10^{-11} M. This effect can be blocked by picrotoxin but is not sensitive to GABA. Avermectin is thus acting as an inhibitory neurotransmitter without the need for the mediation of a second messenger system (see Fisher and Mrozik, 1992).

In conclusion, the relatively simple picture of avermectin acting to potentiate GABA requires some qualification. In some cases this view is correct, but in other situations avermectin may act as an antagonist, and in yet others may have actions which do not relate to GABA action but may still be connected with chloride channel opening.

9.8.3 Baclofen

$GABA_B$ receptors differ from $GABA_A$ in that (a) no chloride channel is involved, (b) they are insensitive to bicuculline and muscimol (Allan and Harris, 1986) and (c) the benzodiazepines do not bind. The best known ligand at the $GABA_B$ receptor is the anti-spastic agent baclofen. Baclofen is 3-*p*-chlorophenyl-GABA and was synthesized as a more lipophilic analogue to pass the blood–brain barrier and mimic the actions of GABA. L-Baclofen is more potent as an agonist at $GABA_B$ receptors than the D-isomer by a factor of about 1000 (reviewed in Wojcik and Holopainen, 1992).

Like $GABA_A$ receptors, $GABA_B$ are found presynaptically and can thus inhibit the release of several neurotransmitters. The anti-spastic action of baclofen probably derives from inhibition of many of the sensory fibres bringing impulses into the dorsal side of the spinal cord at vertebrae I to IV (Allerton *et al.*, 1989). Normally the nerve impulse originating from the sensory neurone is passed either directly or via interneurones to the motoneurone which then activates the muscle reflex. Spastic neurones are hyperactive. The drug reduces the hyperactivity by lowering the excitatory post-synaptic potential on the motoneurone through two possible mechanisms: acting presynaptically to reduce the release of the excitatory transmitter from the sensory nerve fibres, or post-synaptically by hyperpolarizing the neurone.

The detailed mechanism of baclofen is not yet fully understood, but in some cells the drug causes hyperpolarization (and thus prevents transmission) by stimulation of an (outward) K^+ current; in other cells baclofen inhibits an inward calcium channel (Bowery, 1989). These actions are both mediated by G-proteins and could explain the inhibition of nerve transmission. Inhibition of adenylate cyclase and activation of inositol phospholipid hydrolysis, however, have also been described (Wojcik and Holopainen, 1992) but the relevance of these to the anti-spastic action of baclofen is not clear.

9.9 Opiate receptors - morphine for pain

The opiates are a class of naturally occurring opium alkaloids of which the best known is probably morphine. Opioids are chemically synthesized agents such as meperidine, methadone and naloxone which mimic or antagonize the pharmacological actions of morphine. Both opiates and opioids bind to receptors designated as opioid.

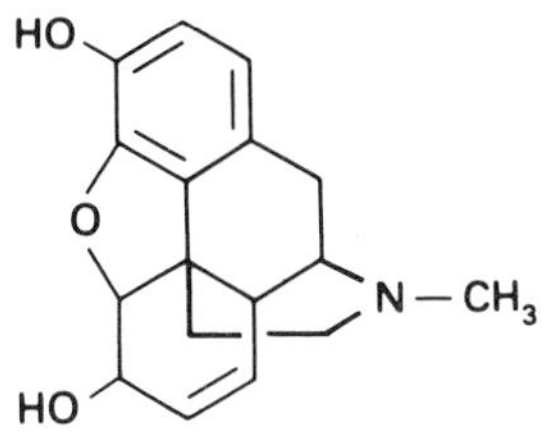

Morphine

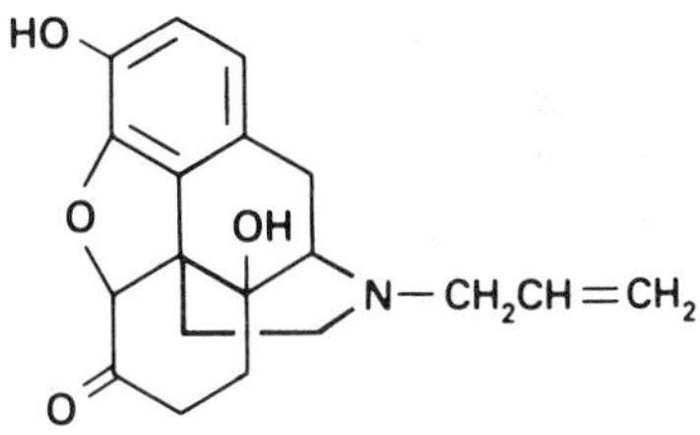

Naloxone

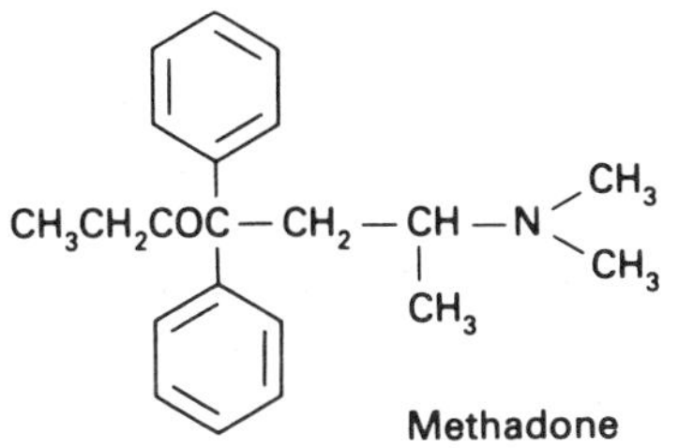

Methadone

Drugs of this type are powerful pain killers (analgesics); morphine, in particular, derived from the opium poppy, has been in use for this purpose for a very long time. The sort of pain that these agents are used to treat is the chronic nerve pain that often accompanies terminal cancer and recovery from operations. Less severe pain is countered by the use of non-steroidal anti-inflammatory drugs such as aspirin, which inhibit the biosynthesis of prostaglandins (Chapter 7).

A large number of analogues have been synthesized in order to improve on the properties of morphine by reducing undesirable side-effects such as drowsiness, nausea, vomiting, constipation and, most serious of all, respiratory depression that can be lethal if a sufficiently high dose is given. In addition, there is also the grave problem of physical dependence on the opiate family of drugs.

There are at least three and possibly four major types of opioid receptor: μ, κ, δ and possibly σ. Stimulation at all four receptors appears to relate to relief of pain. Morphine and its analogues are primarily μ agonists although they often have appreciable activity for κ and δ receptors. The μ receptor controls respiratory depression (slowing of the rate of breathing), reduced intestinal motility, nausea and vomiting and also drug dependence. κ receptors are primarily concerned with diuresis by inhibiting the release of anti-diuretic hormone and sedation. The δ receptor, which responds more markedly to opioid peptides such as enkephalin than opioid alkaloids, appears to exist in the guinea-pig ileum but not in the mouse vas deferens.

Some opioids also bind to other receptors known as σ, which appear to mediate disorientated and depersonalized feelings; the σ receptor may be regarded as non-opioid, as morphine has a very low affinity and naloxone, the classical opioid antagonist, does not block morphine's action. There appears to be some cross-over between σ and dopamine D_2, however, because a number of atypical neuroleptics, including haloperidol, bind to both receptors in the region of 10^{-8} M (Bowen *et al.*, 1990).

Opioid receptors couple to G-proteins (mainly the G_i family - see Childers, 1991), although μ may also couple through G_o. μ, κ and δ receptors inhibit adenylate cyclase. In common with other G-protein coupled receptors GTP reduces agonist, but not antagonist, binding.

μ and δ agonists stimulate the opening of inward potassium currents, thus hyperpolarizing the membrane and reducing the rate at which signals are transmitted, while κ agonists close calcium channels. These actions seem to be mediated directly by G-proteins and not through adenylate cyclase inhibition. The opioids are transmitters at inhibitory synapses, usually by binding to receptors located presynaptically and suppressing the release of excitatory transmitters. μ appears to inhibit noradrenaline release, κ that of dopamine and both μ and δ that of acetycholine (Schoffelmeer *et al.*, 1992).

The finding that there were saturable, stereospecific binding sites for opioid drugs in the central nervous system prompted a search for endogenous ligands to bind at these receptors. Very soon two naturally occurring peptides (the enkephalins) were discovered in pig brain. Structurally they are very alike and are known as Met-enkephalin (Tyr-Gly-Gly-Phe-Met) and Leu-enkephalin (Tyr-Gly-Gly-Phe-Leu), differing only at the carboxyl terminus (Hughes *et al.*, 1975). Subsequently other, larger opiate peptides were isolated from the pituitary gland and were called endorphins to distinguish them from the enkephalins (Li *et al.*, 1976; Chretien *et al.*, 1976). Figure 9.10 shows opioid structural relationships. It is interesting that Met-enkephalin forms the amino-terminus of the endorphins, which in turn represent residues 61–91 at the carboxyl terminus of the pituitary hormone, β-lipotrophin, although β-lipotrophin does not possess any opiate properties.

A further 17-amino-acid, peptide was isolated by Goldstein *et al.* (1977) and named dynorphin. Dynorphin is Leu-enkephalin with an extension of 12 amino acid residues at the carboxyl terminus. There are at least four peptides with

Aminoacid sequence of the opioid peptides.

Precursor	Peptides	Sequence
Pro-opiomelanocortin	α-Endorphin	Tyr-Gly-Gly-Phe-Met-Thr-Ser-Glu-Lys-Ser-Gln-Thr-Pro-Leu-Val-Thr-Leu: 17 residues
	४-Endorphin	Tyr-Gly-Gly-Phe-Met-Thr-Ser-Glu-Lys-Ser-Gln-Thr-Pro-Leu-Val-Thr-Leu: 17 residues
	β-Endorphin	Tyr-Gly-Gly-Phe-Met-Thr-Ser-Glu-Lys-Ser-Gln-thr-Pro-Leu-Val-Thr-Leu-Phe-Lys-Asp-Ala-Ile-Ile-Lys-Asn-Ala-His-Lys-Lys-Gly-Gln: 31 residues
Proenkephalin	Met-enkephalin	Tyr-Gly-Gly-Phe-Met: 5 residues
	Leu-enkephalin	Tyr-Gly-Gly-Phe-Leu: 5 residues
	Met-enkephalin	Tyr-Gly-Gly-Phe-Met-Arg-Phe:. 7 residues
	Met-enkephalin	Tyr-Gly-Gly-Phe-Met-Arg-Gly-Leu: 8 residues
Prodynorphin	Dynorphin A	Tyr-Gly-Gly-Phe-Leu-Arg-Arg-Ile-Arg-Pro-Lys-Leu-Lys-Trp-Asp-Asn-Gln: 17 residues
	Dynorphin B	Tyr-Gly-Phe-Leu-Arg-Arg-Ile: 8 residues
	α-Neoendorphin	Tyr-Gly-Gly-Phe-Leu-Arg-Lys-Tyr-Pro-Lys: 10 residues
	β-Neoendorphin	Tyr-Gly-Gly-Phe-Leu-Arg-Lys-Tyr-Pro: 9 residues

Tyrosine is always at the amino-terminal end of the sequence. The classic sequence of the first five amino acids (Tyr-Gly-Gly-Phe-Leu-) can be seen running through the entire group of peptides. Further congruence can be seen within each of the three groups as we move towards the carboxyl terminus.

Melanocyte stimulating hormones (MSH), both α, β and γ, are also derived from pro-opiomelanocortin, and so is adrenocorticotrophic hormone (ACTH), α-MSH, indeed, is residues 1 to 13 of ACTH.

Figure 9.10 Amino acid sequence of the opioid peptides.

opioid activity coded for by the cDNA for the dynorphin precursor (Kakidani *et al.*, 1982). That these peptides are endogenous ligands for the opiate receptor was shown early on since they closely mimic the action of morphine and could be antagonized by the morphine antagonist naloxone (Frederickson, 1977; Fig. 9.10).

The endogenous ligands are not absolutely specific for the different receptors, for example β-endorphin appears to bind equally well to μ and κ receptors but has no affinity for δ. Leu-enkephalin, on the other hand, binds preferentially to the δ-site but binds less well to the μ-site. Dynorphin A favours the κ site, while still giving measurable binding at μ. Morphine is also not entirely selective in that it favours the μ site but still has measurable affinity for both κ and δ. Naloxone, likewise, is more potent at the μ site but can also bind at the δ and κ sites. This lack of selectivity allows the opioid peptides to appear similar to morphine in pharmacological terms, although they do not necessarily favour the same receptor. Other agonists are being synthesized that are more specific (Kosterlitz, 1985).

9.9.1 Opioid dependence

Narcotic dependence is thought to relate in some way to the chronic inhibition of adenylate cyclase activity, since levels of this enzyme are lowered in membranes from morphine-dependent rats and, as noted above, GTPase levels from these animals are lowered. Naloxone increases these levels towards the normal values (Barchfeld and Medzihradsky, 1984). It is interesting to note that methadone, although addictive itself, is often used to treat withdrawal symptoms of heroin users (Jaffe and Martin, 1985).

A possible mechanism for the derivation of withdrawal symptoms is suggested by the interaction of opioids with gonadotrophins. A long-acting analogue of Met-enkephalin was shown to inhibit the release of follicle-stimulating hormone and luteinizing hormone whilst naloxone produced a significant rise in levels of gonadotrophins in both male and female subjects (Grossman *et al.*, 1981). The symptoms of withdrawal and naloxone treatment are very similar to the symptoms of premenstrual syndrome (Reid and Yen, 1983), and opioid agonists and antagonists have their greatest effect on serum luteinizing hormone levels in the premenstruum. These symptoms may be a consequence of high gonadotrophin levels, resulting partly from inadequate feedback inhibition by ovarian steroids and also from continuous stimulation from the hypothalamus by release of luteinizing-hormone-releasing hormone. The sudden withdrawal of opiate narcotics, therefore, may sharply raise gonadotrophin levels that can give rise to the withdrawal symptoms (Coulson, 1986).

9.9.2 Pentazocine as analgesic

In order to develop pain-killing drugs that do not cause dependence and painful withdrawal symptoms when their use is discontinued, considerable synthetic effort has been expended. This eventually led to the development of pentazocine among other drugs - structurally a derivative of benzomorphan. When given orally this drug does not cause dependence, but when extracted from the tablets and injected intravenously with the anti-histamine tripelennamine, it may do so. Nevertheless, pentazocine is still one of the least addictive morphine-like drugs (reviewed in Brogden *et al.*, 1973).

Pentazocine shows a mixed type of action at opiate receptors, being a moderately weak antagonist at μ and a relatively powerful agonist at κ and σ. The antagonism of μ receptors helps to explain why the compound is less likely to cause dependence, while the agonism at κ receptors is probably responsible for its analgesic effects. This type of analgesia differs from that induced by morphine, since it only affects the spinal cord whereas morphine is able to act at supra-spinal loci, presumably as a consequence of agonism at μ receptors. At high doses (60–90 mg in man), actions characteristic of stimulation of the σ receptors occur, i.e. hallucinations and dysphoria.

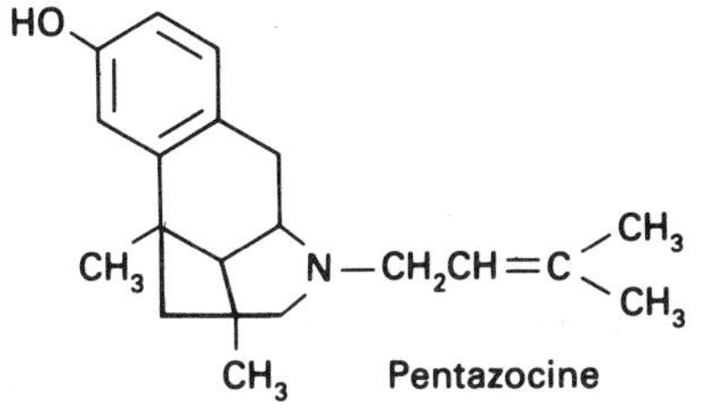

Pentazocine

9.9.3 Loperamide as anti-diarrhoeal

One of the most marked side-effects of opiate agonist usage, constipation, probably arises through action at the opiate receptor in the gastrointestinal tract. This has been put to good use in the development of drugs for the treatment of diarrhoea. Morphine has the ability to increase muscle tone, diminish the amplitude of contractions and markedly reduce the propulsive activity in the intestine. Consequently, over-the-counter preparations of morphine adsorbed to kaolin are often effective in treating diarrhoea. Efforts have been made, however, to find an opiate that does not suffer from the central nervous system side-effects of morphine such as drowsiness etc.

Loperamide was developed as an answer to this problem and has proven to be a safe and effective remedy for diarrhoea as well as being relatively free of side-effects (Ericcson, 1990). Loperamide does not cross the blood–brain barrier and thus does not give rise to sedation, although it will bind to opioid receptors in brain homogenates. The drug slows gastrointestinal motility (peristalsis) and is of particular value in cases where the bowel action is excessive, e.g. overactive bowel syndrome (Hughes *et al.*, 1982).

Diarrhoea may, however, arise not only through excessive peristalsis but also through reduced absorption or increased secretion of water and ions at the mucosal surface. Loperamide antagonizes the secretion induced by prostaglandin E_1, theophylline and cholera toxin (Awouters *et al.*, 1983) but apparently not that induced by vasoactive intestinal polypeptide (Schiller *et al.*, 1984). Both cholera toxin and prostaglandin E_1 induce the activation of adenylate cyclase, thus raising cyclic AMP levels, and the evidence suggests that cyclic AMP is the mediator in the release of chloride ion, followed by water, to maintain osmotic balance.

The mechanism of action of loperamide is probably via the opioid receptor since naloxone antagonizes its action. On the other hand, a direct inhibitory effect of the opioid receptor on adenylate cyclase is unlikely because loperam-

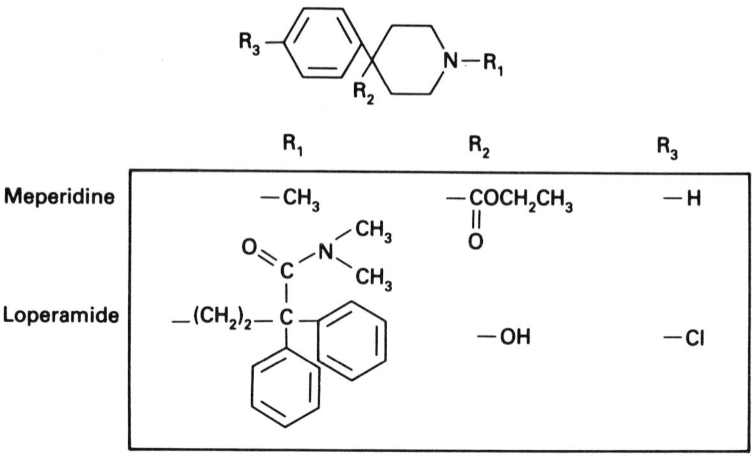

	R_1	R_2	R_3
Meperidine	$-CH_3$	$-COCH_2CH_3$ (C=O)	$-H$
Loperamide	$-(CH_2)_2-C-$ (with $C(=O)N(CH_3)_2$ and two phenyl rings)	$-OH$	$-Cl$

ide does not antagonize the rise in cyclic AMP induced by both prostaglandin E_1 and cholera toxin. The suggestion has been made that calmodulin may be involved (Awouters *et al.*, 1983) because:

(a) The secretory process is calcium-dependent and calmodulin mediates many of the intracellular actions of calcium in secretion.
(b) Loperamide binds to other moderately high affinity sites in the gut wall, where calmodulin is highly concentrated.
(c) Loperamide binds to calmodulin in the presence of calcium *in vitro* with a binding constant of $1{\cdot}2 \times 10^{-5}$ M (Zavecz *et al.*, 1982).

This situation remains unresolved at the present time. What is clear, however, is that loperamide suppresses the secretion of chloride ion from the mucosal cell into the gut lumen (Hughes *et al.*, 1982).

9.10 Histamine receptors - mepyramine as anti-allergic, cimetidine as anti-ulcer

The response of mammals to histamine has been studied extensively ever since the work of Sir Henry Dale, one of the great pioneers of pharmacology. There are three subtypes of histamine receptor: H_1, H_2 and H_3. The H_1 receptors govern the contraction of smooth muscle in lung bronchi and gut, the increase in capillary permeability and inflammatory response. H_2 receptors control the action of histamine in inducing the secretion of gastric acid and positive chronotropic and inotropic effects in the heart. H_3 receptors are located, amongst other places, presynaptically (autoreceptors) in the brain and govern the synthesis and release of histamine and other neurotransmitters. Both H_1 and H_2 play a part in the dilation of finer blood vessels and are also found in the brain where histamine functions as a neurotransmitter (Garrison, 1990).

All three receptors are linked to G-proteins which activate different cell-signalling systems. Agonism at H_1 receptors activates the phospholipase C pathway, while H_2 receptors, through G_s, are linked to adenylate cyclase in a positive fashion. H_3 may be linked negatively to adenylate cyclase (Arrang *et al.*, 1987) or possibly to an ion channel (Hill, 1990).

The H_1 and H_2 receptors have been cloned and expressed in various tissues. Biogenic amines require an anionic ligand to bind the cationic amino group which, in every case investigated so far, turns out to be an aspartate residue in the third transmembrane segment. The histamine receptors are no exception. The fifth segment in the catecholamine receptors is important for catechol binding. Histamine also requires an aspartate to bind the monocationic nitrogen in the imidazole ring. For the H_1, that appears to be sufficient to bind the imidazole. The H_2 receptor, however, recognizes both the nitrogen atoms – the (neutral) second forms a hydrogen bond with the threonine residue only present in TM5 of the H_2 receptor. Another aspartate appears necessary for antagonist, but not agonist, binding in the H_2 receptor (Gantz *et al.*, 1992). Both receptor subtypes are glycosylated with the carbohydrate residues being sited on asparagine residues at the N-terminal segment. The large third intracellular loop and the cytoplasmic C-terminus probably carries the G-protein recognition site. The homology between the two receptor subtypes is, however, only 40·7 per cent which is less than between H_1 and the muscarinic M_1 (44·3 per cent) (Gantz *et al.*, 1992; Yamashita *et al.*, 1991).

9.10.1 H_1 receptor antagonists

H_1 antagonists, such as mepyramine and chlorpheniramine, have been available for several decades and are used to block the action of histamine released endogenously in allergic reactions, particularly in the upper respiratory tract, as in hay fever. Skin allergies also respond favourably in some cases. Histamine is one of the mediators (autacoids) of the allergic response generating itching and some aspects of the inflammation that accompany this reaction. The lower part of the respiratory tract is involved in the genesis of asthma which does not respond to H_1 antagonists in man, since leukotrienes are the major cause of allergic bronchoconstriction (see section 10.6 for a discussion on asthma). In guinea-pigs, however, histamine is the major effector and H_1 antagonists are protective (Garrison, 1990).

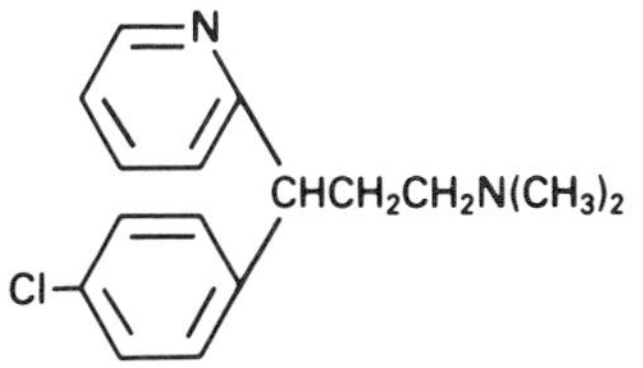

N
$CHCH_2CH_2N(CH_3)_2$
Cl

Chlorpheniramine

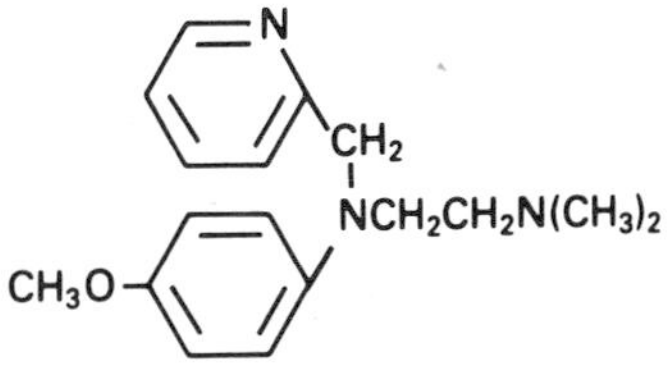

N
CH_2
$NCH_2CH_2N(CH_3)_2$
CH_3O

Mepyramine

Mepyramine has a low binding constant with H_1 receptors as measured by binding to homogenates of, for example, guinea-pig cortex ($1{\cdot}5 \times 10^{-9}$ M; Daum *et al.*, 1982) and rat cerebral cortex ($2{\cdot}1 \times 10^{-9}$ M; Hall and Ogren, 1984). Chlorpheniramine has a chiral centre: the (+)-isomer is more tightly bound to the isolated bovine receptor by about 100-fold more effectively than the (−)-isomer ($4{\cdot}3 \times 10^{-9}$ M and $4{\cdot}6 \times 10^{-7}$ M, respectively; Yamashita *et al.*, 1991). (+)-Chlorpheniramine was the most potent isomer, by three orders of magnitude, in antagonizing contractions produced by histamine in the hamster smooth muscle from the vas deferens (White *et al.*, 1993). The drug was both effective and bound at similar concentrations. The H_1 receptor activates turnover of phosphatidylinositol, as measured in rabbit aorta, and the effect is inhibited most by H_1 receptor antagonists such as mepyramine. Histamine stimulates smooth muscle contraction in this organ. The phosphatidylinositol pathway (section 9.1) activates calcium uptake into the cell so that muscle contraction may be stimulated (Villalobos-Molina and Garcia-Sainz, 1983; White *et al.*, 1993).

The antagonism of histamine is competitive and reversible. The H_1 antagonists are generally lipophilic and demonstrate variations on the following structure:

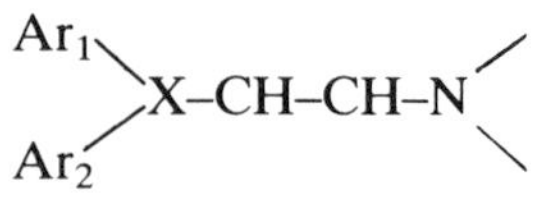

$$\begin{matrix} Ar_1 \\ Ar_2 \end{matrix} \rangle X{-}CH{-}CH{-}N \langle$$

where Ar is usually an aryl group and X is a nitrogen or carbon atom of a C—O—ether group (Garrison, 1990).

9.10.2 H_2 receptor antagonists

The rational development of the drug cimetidine, including the importance of physicochemical factors such as partition coefficient and ionization constant, makes an interesting story (Ganellin, 1978). The structure of histamine was taken as the starting point, and it was shown that 2-methylhistamine was a more selective agonist for the H_1 receptor, while 4-methylhistamine favoured the H_2 receptor. After separation of agonist and antagonist activities, burimamide was derived as the first true H_2 receptor blocker. Optimization of the H_2 receptor blocking activity resulted in the synthesis of metiamide. Various structural changes to eliminate unwanted side-effects finally led to the development of cimetidine.

Histamine activates adenylate cyclase in brain and this has been shown to

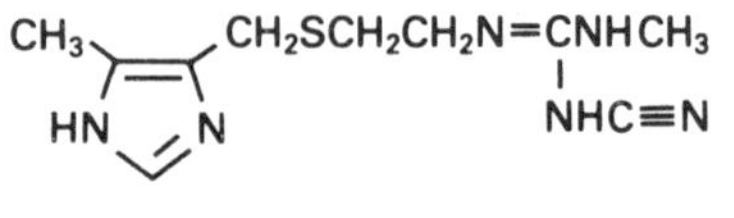

CH_3 (imidazole ring, HN, N) $CH_2SCH_2CH_2N{=}C(NHCH_3)NHC{\equiv}N$

Cimetidine

be mediated by H_2 receptors. Cimetidine antagonizes th
genates of the dorsal hippocampus of guinea-pig
8·9 × 10^{-7} M (Kanof and Greengard, 1979). This comp
of 7·9 × 10^{-7} M against histamine stimulation of the ra
right atrium of the guinea-pig *in vitro*, i.e. the posi
(Ganellin, 1978). The values for other H_2 antagonists, burima...
ide, also agree, thus confirming the connection between adenylate cyc...
activation and binding to the H_2 receptor. It should be noted that cimetidine is a compound that is much more hydrophilic than an H_1 receptor antagonist and thus does not cross the blood–brain barrier to react with brain receptors, and so the correlation obtained above is all the more convincing.

$CH_2CH_2CH_2CH_2NHC(=S)NHCH_3$ (imidazole ring: HN, N)

Burimamide

CH_3 ; $CH_2SCH_2CH_2NHC(=S)NHCH_3$ (imidazole ring: HN, N)

Metiamide

Tiotidine binds to H_2 receptors in homogenates of guinea-pig cerebral cortex with a binding constant of 1·7 × 10^{-8} M. That the H_2 receptor is involved in the binding is suggested by the correlation of binding with the antagonism of the chronotropic effect in guinea-pig heart. In addition, antagonism of histamine-induced adenylate cyclase activity in guinea-pig gastric mucosa also correlates very well with receptor binding for a number of antagonists including tiotidine (Gajtkowski *et al.*, 1983).

Cimetidine was put on the market in the mid-1970s for the treatment of gastric acid secretion in peptic ulcer patients, and effected a revolutionary improvement in the chemotherapy of ulcers. Most patients suffering from ulcers hypersecrete gastric acid, possibly as a result of stress or for some other reason. Gastric acid secretion is stimulated by food intake and by a variety of agents, known as secretagogues, e.g. histamine or caffeine, and by stimulation of the vagus nerve. Cimetidine will also block secretion induced by muscarinic agonists and by gastrin non-competitively. This is because gastrin activates the release of histamine from parietal cells (Sachs and Wallmark, 1989). Cimetidine thereby inhibits both basal and stimulated secretion. The drug both relieves the pain and allows the ulcer to heal.

Relapses are, however, quite common after treatment has stopped. In some cases the ulcer may not have healed completely, although the patient may have no symptoms. It has been remarked that the problem is not how to heal an ulcer but how to keep it healed (Thomas and Misiewicz, 1984). If the ulcer is healed initially but the predisposing factor is not addressed, then it is not surprising that the ulcer may recur. Cimetidine may be used prophylactically as a maintenance therapy to prevent relapse. Pepsin secretion is also reduced by the drug.

After further synthetic optimization of H_2 receptor blocking activity, another more potent receptor blocker, ranitidine, joined cimetidine on the market.

CHNO$_2$ ‖ CH$_2$SCH$_2$CH$_2$NHCNHCH$_3$, O, CH$_2$N(CH$_3$)$_2$ — Ranitidine

N, CH$_2$SCH$_2$CH$_2$NHCNHCH$_3$ ‖ NC≡N, NH, S, C—NH$_2$, NH — Tiotidine

Structurally the need for a small heterocyclic ring has been met by a furan, thus indicating that an imidazole, reminiscent of histamine, is not essential (moreover, tiotidine has a thiazole ring). Ranitidine is now regarded as the frontline therapy for suppression of gastric acid secretion (Grant *et al.*, 1989).

Questions

1. Name three receptors that activate (a) adenylate cyclase or (b) phospholipase C.
2. What special structural features can you identify for G-protein-coupled receptors?
3. Name two receptors that are linked to ion channels. How do they differ structurally from G-protein coupled receptors?
4. In the β-adrenergic receptors, which amino acid residues are believed to bind the agonists and where in the receptor are they sited? How do catecholamine and serotonin receptors differ in the transmembrane domain?
5. What protein is used as the model template for assigning structure to the 7-transmembrane receptors? How valid is this extrapolation?
6. What role do guanine nucleotides play in signal transduction?
7. What is meant by the terms (a) affinity (b) efficacy (c) partial agonists?
8. Which receptors mediate the action of the benzodiazepines? How does this interaction take place?
9. How does agonist binding to the G-protein-coupled receptors differ from that of antagonists?

References

Aapro, M.S., 1991, *Drugs,* **42**, 551–68.
Ahlquist, R.P., 1948, *Am. J. Physiol.,* **153**, 586–600.
Alexander, B.S. and Wood, M.D., 1987, *J. Pharm. Pharmacol.,* **39**, 664–6.
Allan, A.M. and Harris, R.A., 1986, *Molec. Pharmacol.,* **29**, 497–506.
Allerton, C.A., Boden, P.R. and Hill, R.G., 1989, *Brit. J. Pharmacol.,* **96**, 29–38.
Anden, N.E. and Stock, G., 1973, *J. Pharm. Pharmacol.,* **25**, 346–8.
Anon, 1991, *Trends in Pharmacol. Sci.,* **12**, 18.
Apperley, G.H., Daly, M.J. and Levy, G.P., 1976, *Brit. J. Pharmacol.,* **57**, 235–46.
Arrang, J.-W., Garbarg, M., Lancelot, J.-C., Lecomte, J.-M., Pollard, H., Robba, M., Schunack, W. and Schwartz, J.-C., 1987, *Nature,* **327**, 117–23.

Awouters, F., Niemegeers, C.J.E. and Janssen, P.A.J., 1983, *Annu. Rev. Pharmacol. Toxicol.*, **23**, 279–301.
Baldessarini, R.J., 1979, *Postgrad. Med.*, **65**, 123–8.
Barchfeld, C.C. and Medzihradsky, F., 1984, *Biochem. Biophys. Res. Commun.*, **121**, 641–8.
Barnes, J.M., Barnes, N.M., Costall, B. and Naylor, R.J., 1991, *Pharmaceut. J.*, **246**, 112–14.
Barnes, J.M., Barnes, N.M. and Cooper, S.J., 1992, *Neuroscience and Biobehavioural Reviews*, **16**, 107–13.
Bass, C. and Kerwin, R., 1989, *Brit. Med. J.*, **298**, 345–6.
Berridge, M.J., 1993, *Nature*, **361**, 315–25.
Bianchine, J., 1985, In: *The Pharmacological Basis of Therapeutics*, 7th Ed., A.G. Gilman, L.S. Goodman, T.W. Rall and F. Murad, Eds (New York: Macmillan) Ch. 21.
Birnbaumer, L., Abramowtiz, J. and Brown, A.M., 1990, *Biochem. Biophys. Acta*, **1031**, 163–224.
Blier, P., de Montigny, C. and Chaput, Y., 1990, *J. Clin. Psychiat.*, **51**, 4, Suppl, 14–21.
Borison, R.L., Hitri, A., Blowers, A.J. and Diamond, B.I., 1983, *Clin. Neuropharmacol.*, **6**, 137–50.
Bowen, W.D., Moses, E.L., Tolentino, P.J. and Walker, J.M., 1990, *Eur. J. Pharmacol.*, **177**, 111–18.
Bowery, N., 1989, *Trends in Pharmacol. Sci.*, **10**, 401–7.
Braestrup, C., Albrechtsen, R. and Squires, R.F., 1977, *Nature*, **269**, 702–704.
Braestrup, C., Nielsen, M., Krogsgaard-Larsen, P. and Falch, E., 1980, *Nature*, **280**, 331–3.
Brittain, R.T., Drew, G.M. and Levy, G.P., 1982, *Brit. J. Pharmacol.*, **77**, 105–14.
Brogden, R.N. and Sorkin, E.M., 1990, *Drugs*, **40**, 903–49.
Brogden, R.N. and Faulds, D., 1991, *Drugs*, **42**, 895–912.
Brogden, R.N., Speight, T.M. and Avery, G.S., 1973, *Drugs*, **5**, 6–91.
Brown, J.R. and Arbuthnott, G.W., 1983, *Neuroscience*, **10**, 349–55.
Buckner, C.B. and Abel, P., 1974, *J. Pharmacol. Exp. Ther.*, **189**, 616–25.
Bunney, B.S., Walters, J.R., Roth, R.H. and Aghajanian, G.K., 1973, *J. Pharmacol. Exp. Ther.*, **185**, 560–71.
Butchers, P.R., Vardey, C.J. and Johnson, M., 1991, *Brit. J. Pharmacol.*, **104**, 672–6.
Bylund, D., 1992, *FASEB J.*, **6**, 832–9.
Calne D.B., 1984, *New Engl. J. Med.*, **310**, 523–4.
Campbell, W.C., Fisher, M.H., Stepley, E.D., Albers-Schonberg, G. and Jacob, T.A., 1983, *Science*, **221**, 823–8.
Carlsson, A., 1984, *Adv. Biochem. Psychopharmacol.*, **39**, 213–21.
Carmine, A.A. and Brogden, R.N., 1985, *Drugs*, **30**, 85–126.
Caulfield, M.P. and Straughan, D.W., 1983, *Trends in Neurosci.*, **6**, 73–5.
Childers, S.R., 1991, *Life Sci.*, **48**, 1991–2003.
Chretien, M., Benjannet, S., Dragon, N., Seidah, N.G. and Lis, M., 1976, *Biochem. Biophys. Res. Commun.*, **72**, 472–8.
Christmas, A.J., Coulson, C.J., Maxwell, D.R. and Riddell, D., 1972, *Brit. J. Pharmacol.*, **45**, 490–503.
Coleman, R.A., Johnson, M., Nials, A.T. and Sumner, M.J., 1990, *Brit. J. Pharmacol.*, **99**, (Suppl), 121P.
Coppen, A., 1967, *Brit. J. Psychiat.*, **113**, 1237–64.
Coulson, C.J., 1986, *Med. Hypothes.*, **19**, 243–56.
Creese, I., Schneider, R. and Snyder, S.H., 1977, *Eur. J. Pharmacol.*, **46**, 377–81.
Cullum, V.A., Farmer, J.B., Jack, D. and Levy, G.P., 1969, *Brit. J. Pharmacol.*, **35**, 141–51.
Daum, P.R., Hill, S.J. and Young, J.M., 1982, *Brit. J. Pharmacol.*, **77**, 347–57.

Dougall, I.G., Harper, D., Jackson, D.M. and Leff, P., 1991, *Brit. J. Pharmacol.,* **104**, 1057-61.

Duce, I.R. and Scott, R.H., 1985, *Brit. J. Pharmacol.,* **85**, 395-401.

Egerton, J.R., Ostlind, D.A., Blair, L.S., Eary, C.H., Suhayda, D., Cifelli, S., Riek, R.F. and Campbell, W.F., 1979, *Antimicrob. Ag. Chemother.,* **15**, 372-8.

Ellenbroek, B.A., Artz, M.T. and Cools, A.R., 1991, *Eur. J. Pharmacol.,* **196**, 103-108.

Ericsson, C.D., 1990, *Amer. J. Med.,* **88**(S6A), 105-45.

Evans, S.M., Galdes, A. and Gall, M., 1991, *Pharmacology, Biochemistry and Behaviour,* **40**, 1033-40.

Fisher, M.H. and Mrozik, H., 1992, *Annu. Rev. Pharmacol. Toxicol.,* **32**, 537-53.

Fitton, A. and Heel, R.C., 1990, *Drugs,* **40**, 722-47.

Fitton, A., Faulds, D. and Goa, K.L., 1992, *Drugs,* **43**, 561-96.

Fowler, C.J., Mantle, T.J. and Tipton, K.F., 1982, *Biochem. Pharmacol.,* **31**, 3555-61.

Frederickson, R.C.A., 1977, *Life Sci.,* **21**, 23-42.

Gajtkowski, G.A., Norris, D.B., Rising, T.J. and Wood, T.P., 1983, *Nature,* **304**, 65-7.

Ganellin, G.R., 1978, *J. Appl. Chem. Biotechnol.,* **28**, 182-200.

Gantz, I., DelValle, J., Wang, L.-D., Tashiro, T., Munzert, G., Guo, Y.-G., Konda, Y. and Yamada, T., 1992, *J. Biol. Chem.,* **267**, 20840-3.

Garrison, J.C., 1990, Ch. 23 in *The Pharmacological Basis of Therapeutics*, 8th Ed. Eds, A.G. Gilman, T.W. Rall, A.S. Nies and P. Taylor. New York: Pergamon.

Gerlach, M., Riederer, P. and Youdim, M.B., 1992, *Eur. J. Pharmacol.,* **226**, 97-108.

Gilman, A.G., 1984, *Cell,* **36**, 577-9.

Goa, K.L., McTavish, D. and Clissold, S.P., 1991, *Drugs,* **43**, 640-58.

Goldstein, A., Fischli, W., Lowney, L.I., Hunkapilla, M. and Hood, L., 1977, *Proc. Nat. Acad. Sci. (U.S.A.),* **78**, 7219-23.

Goodwin, F.K. and Post, R.M., 1983, *Brit. J. Clin. Pharmacol.,* **15**, 393S-405S.

Goyal, R.K., 1988, *Life Sci.,* **43**, 2209-20.

Grant, S.M., Langtry, H.D. and Brogden, R.N., 1989, *Drugs,* **37**, 801-70.

Grossman, A., Moult, P.J.A., Gaillard, R.C., Delitalia, G., Toff, W.D., Rees, L.H. and Besser, G.M., 1981, *Clin. Endocrinol.,* **14**, 41-7.

Haefely, W., Burkard, W.P., Cesura, A.M., Kettler, R., Lorez, H.P., Martin, J.R., Richards, J.G., Scherschlicht, R. and de Prada, M., 1992, *Psychopharmacol.,* **106**, S6-14.

Hall, H. and Ogren, S.-O., 1984, *Life Sci.,* **34**, 597-605.

Hamik, A., Oksenberg, D., Fischette, C. and Peroutka, S.J., 1990, *Biol. Psychiat.,* **28**, 99-109.

Harrington, M.A., Zhong, P., Garlow, S.J. and Ciaranello, R.D., 1992, *J. Clin. Psychiat.,* **53**, 10, Suppl, 8-27.

Havard, C.W.H. and Fonseca, V., 1990, *Drugs,* **39**, 66-73.

Hegde, S.S., Ricci, A., Amenta, F. and Lokhandwala, F., 1989, *J. Pharmacol. Exp. Ther.,* **251**, 1237-45.

Hesketh, P.J. and Gandara, D.R., 1991, *J. Natl. Cancer Inst.,* **83**, 613-20.

Hill, S.J., 1990, *Pharmacol. Rev.,* **42**, 45-83.

Holcslaw, T.L. and Beck, T.R., 1990, *Am. J. Hypertens.,* **3**, 120S-5S.

Hughes, J., Smith, T.W., Kosterlitz, H.W., Fothergill, L.A., Morgan, B.A. and Morris, H.R., 1975, *Nature,* **258**, 577-9.

Hughes, S., Higgs, N.B. and Turnberg, L.A., 1982, *Gut,* **23**, 974-9.

Hulme, E.C., Birdsall, N.J.M. and Buckley, N.J., 1990, *Annu. Rev. Pharmacol. Toxicol.,* **30**, 633-73.

Humphrey, P.P.A. and Feniuk, W., 1991, *Trends in Pharmacol. Sci.,* **12**, 444-6.

Jack, D., 1991, *Brit. J. Clin. Pharmacol.,* **31**, 501-14.

Jaffe, J.H. and Martin, W.R., 1985, in *The Pharmacological Basis of Therapeutics*, 7th Ed., A.G. Gilman, L.S. Goodman, T.W. Rall and F. Murad, Eds. (New York: Macmillan) Ch. 22.

Johnston, J.P., 1968, *Biochem. Pharmacol.,* **17**, 1285-97.

Julins, D., Huang, K.N., Livelli,T.J., Axel, R. and Jessel,T.M., 1990, *Proc. Nat. Acad. Sci., U.S.A.*, **87**, 928-32.
Kakidani, H., Fumitani, Y., Takahashi, H., Noda, M., Morimoto, Y., Hirose, T., Asai, M., Inayama, S., Nakanishi, S. and Numa, S., 1982, *Nature,* **298**, 245-9.
Kanof, P.D. and Greengard, P., 1979, *J. Pharmacol. Exp. Therapeut.,* **209**, 87-96.
Kass, I.S., Stretton, A.O.W. and Wang, C.C., 1984, *Molec. Biochem. Parasitol.,* **13**, 213-25.
Kendall, 1992, *Biochem. Soc. Trans.,* **20**, 109-13.
Kendall, D.A. and Nahorski, S.R., 1985, *J. Pharmacol. Exp. Ther.,* **233**, 473-9.
Knoll, J. and Magyar, K., 1972, *Adv. Biochem. Psychopharmacol.,* **5**, 393-408.
Kobilka, B., 1992, *Annu. Rev. Neurosci.,* **15**, 87-114.
Kosterlitz, H.W., 1985, *Proc. Roy. Soc. Lond.,* Series B, **225**, 27-40.
Lambert, D.G., Burford, N.T. and Nahorski, S.R., 1992, *Biochem. Soc. Trans.,* **20**, 130-5.
Lance, J.W., Lambert, G.A., Goadsby, P.J. and Zagami, A.S., 1989, *Cephalalgia,* **9** (S9), 7-13.
Lefkovitz, R.J., 1975, *Biochem. Pharmacol.,* **24**, 583-90.
Leonard, B.E., 1992, *Drugs,* **43**, Suppl 2, 3-10.
LeWitt, P.A., 1991, *Acta Neurol. Scand. Suppl.,* **136**, 79-86.
Leysen, J.E., de Chaffoy de Courcelles, D., De Clerck, F., Niemegeers, C.J.E. and Van Neuten, J.M., 1984, *Neuropharmacology,* **23**, 1493-1501.
Li, C.., Chung, D. and Doneen, B.A., 1976, *Biochem. Biophys. Res. Commun.,* **72**, 1542-7.
Lieberman, A.N. and Goldstein, M., 1985, *Pharmacol. Rev.,* **37**, 217-27.
Limbird, L.E., Speck, J.L. and Smith, S.K., 1982, *Molec. Pharmacol.,* **21**, 609-17.
Lipinski, J.F., Cohen, B.M., Zubenko, G.S. and Waternaux, C.M., 1987, *Life Sci.,* **40**, 1947-63.
Low, M.G., Ferguson, M.A.J., Futerman, A.H. and Silman, I., 1986, *Trends in Biochem. Sci.,* **11**, 212-15.
Lucki, I., 1991, *J. Clin. Psychiat.,* **52**, S12, 24-31.
McPherson, G.A. and Summers, R.J., 1982, *Brit. J. Pharmacol.,* **77**, 177-84.
Manschrek, T.C., 1981, *New Engl. J. Med.,* **305**, 1628-31.
Maricq, A.N., Peterson, A.S., Brake, A.J., Myers, R.M. and Julius, D., 1991, *Science,* **254**, 432-7.
Martin, R.J., 1987, *Biochem. Soc. Trans.,* **15**, 61-5.
Matsui, H., Imafuki, J., Asakura, M., Tsukamoto, T., Ino, M., Saitoh, N., Miyamura, S. and Hasegawa, K., 1984, *Biochem. Pharmacol.,* **33**, 3311-14.
Maycock, A.L., Abeles, R.H., Salach, H.I. and Singer, T.P., 1976, *Biochemistry,* **15**, 114-25.
Meltzer, H.Y., 1992, *Brit. J. Psych.,* Suppl, 22-9.
Minneman, K.P., Hegstrand, L.R. and Molinoff, P.B., 1979, *Molec. Pharmacol.,* **15**, 21-33.
Napoliello, M.J. and Domantay, A.G., 1991, *Brit. J. Psychiat.,* **159**, S12, 40-44.
Pearce, J.M.S., 1984, *Brit. Med. J.,* **288**, 1777-8.
Pickar, D., Owen, R.R., Litman, R.E., Konicki, P.E., Gutierrez, R. and Rapaport, M.H., 1992, *Arch. Gen. Psych.,* **49**, 345-53.
Pinder, R.M., Brogden, R.N., Speight, T.M. *et al.*, 1977, *Drugs,* **1**, 321-52.
Pong, S.S. and Wang, C.C., 1982, *J. Neurochem.,* **38**, 375-9.
Powell, J.F., 1991, *Biochem. Soc. Trans.,* **19**, 199-201.
Prichard, B.N.C., 1992, *J. Cardiovasc. Pharmacol.,* **19**, Suppl 1, S1-S4.
Rasmussen, H., 1986, *New Engl. J. Med.,* **314**, 1094-101.
Reid, R.L. and Yen, S.S.C., 1983, *Clin. Obstet. Gynaecol.,* **26**, 710-18.
Rodbell, P., 1980, *Nature,* **284**, 17-22.
Rosenberg, T.L., 1975, *Adv. Enzymol.,* **43**, 103-218.
Rudd, P. and Blaschke, T.F., 1985, In: *The Pharmacological Basis of Therapeutics,* 7th

Ed., Eds A.G. Gilman, L.S. Goodman, T.W. Rall and F. Murad (New York: Macmillan) Ch. 32.
Sachs, G. and Wallmark, B., 1989, *Scand. J. Gastroenterol.*, **24**, Suppl 166, 3-11.
Sanger, D.J., 1985, *Life Sci.*, **36**, 1503-13.
Saverese, T.M. and Fraser, C.M., 1992, *Biochem. J.*, **283**, 1-19.
Saxena, P.R. and Ferrari, M., 1989, *Trends in Pharmacol. Sci.*, **10**, 200-4.
Scadding, G.K. and Havard, C.W.H., 1981, *Brit. Med. J.*, **283**, 1008-12.
Schiller, L.R., Santa Ana, C.A., Morawski, S.G. and Fordtran, J.S., 1984, *Gastroenterol.*, **86**, 1475-80.
Schmidt, M.J., Fuller, R.W. and Wong, D.T., 1988, *Brit. J. Psychiat.*, **153**, Suppl. 3, 40-6.
Schoeffter, P. and Hoyer, D., 1989, *Naunyn-Schmiedeberg's Arch Pharmacol.*, **340**, 135-8.
Schoffelmeer, A.N.M., Van Vliet, B.J., De Vries, T.J., Heijna, M.H. and Mulder, A.H., 1992, *Biochem. Soc. Trans.*, **20**, 449-53.
Shapiro, G. and Enz, A., 1992, *Drugs of the Future*, **17**, 489-501.
Shih, J.C., 1991, *Neuropsychopharmacol.*, **4**, 1-7.
Shih, J.C., Yang, W., Chen, K. and Gallaher, T., 1991, *Pharmacol. Biochem. and Behaviour*, **40**, 1053-8.
Sibley, D.R. and Monsma, F.J., 1992, *Trends in Pharmacol. Sci.*, **13**, 61-9.
Sieghart, W., 1989, *Trends in Pharmacol. Sci.*, **10**, 407-11.
Silverman, R.B., 1983, *J. Biol. Chem.*, **258**, 14766-9.
Silverman, R.B., 1991, *Biochem. Soc. Trans.*, **19**, 201-6.
Snyder, S.H., 1981, *Am. J. Psychiat.*, **138**, 460-3.
Snyder, S.H., 1986, *Nature*, **323**, 292-3.
Soreq, H., Gnatt, A., Lowenstein, Y. and Neville, L.F., 1992, *Trends in Biochem. Sci.*, **17**, 353-8.
Speight, T.M. and Avery, G.S., 1972, *Drugs*, **3**, 159-203.
Speth, R.I., Wastek, G.J., Johnson, P.C. and Yamamura, H.I., 1978, *Life Sci.*, **22**, 859-66.
Sternweis, P.C. and Smrcka, A.V., 1992, *Trends in Biochem. Sci.*, **17**, 502-6.
Taussig, R., Quarmby, L.M. and Gilman, A.G., 1993, *J. Biol. Chem.*, **268**, 9-12.
Tetrud, J.W. and Langston, J.W., 1989, *Science*, **249**, 519-20.
Thomas, J.M. and Misiewicz, G., 1984, *Clin. Gastroenterol.*, **13**, 501-41.
Vauholden, R., Lameire, N. and Ringoir, S., 1985, *Eur. J. Clin. Pharmacol.*, **28**, 125-30.
Vicini, S., 1991, *Neuropsychopharmacol.*, **4**, 9-15.
Villalobos-Molina, R. and Garcia-Sainz, J.A., 1983, *Eur. J. Pharmacol.*, **90**, 457-9.
Vincent, A., 1991, *Biochem. Soc. Trans.*, **19**, 180-3.
Wakade, A.R., Malhotra, R.K. and Wakade, T.D., 1986, *Nature*, **321**, 698-700.
Weinstein, D., Schenker, J.G., Gloger, I., Slonim, J.H., DeGroot, N., Hochberg, A.A. and Folman, R., 1981, *Fed. Eur. Biochem. Soc. Lett.*, **126**, 29-32.
White, T.E., Dickenson, J.M. and Hill, S.J., 1993, *Brit. J. Pharmacol.*, **108**, 196-203.
Wieland, H.A., Luddens, H. and Seeburg, P.H., 1992, *J. Biol. Chem.*, **267**, 1426-30.
Williams, M. and Yarbrough, G.G., 1979, *Eur. J. Pharmacol.*, **56**, 273-6.
Wilson, I.B., Harrison, M.A. and Ginsburg, S., 1961, *J. Biol. Chem.*, **236**, 1498-500.
Wojcik, W.J. and Holopainen, I., 1992, *Neuropsychopharmacol.*, **6**, 201-14.
Wright, D.J., 1987, *Biochem. Soc. Trans.*, **15**, 65-7.
Yamashita, M., Fukui, H., Sugama, K., Horio, Y., Ito, S., Mizuguchi, H. and Wada, H., 1991, *Proc. Nat. Acad. Sci., (U.S.A.)*, **88**, 11515-19.
Zavecz, J.H., Jackson, T.E., Limp, G.L. and Yellin, T.O., 1982, *Eur. J. Pharmacol.*, **78**, 375-7.

Chapter 10

Membrane-active agents

10.1 Introduction

This chapter considers those examples of drugs which act directly on the cell membrane, whether mammalian or fungal. The processes concerned govern the transport of ions across the cell membrane in either direction, and are either channels through which ions pass, or require the hydrolysis of ATP to drive various ions across the membrane. Alternatively, the drug may disrupt the membrane of a pathogenic organism in such a way that it becomes highly permeable to small molecules and consequently loses its effectiveness as a semi-permeable membrane. These effects may all be regarded as direct; indirect effects, whereby a drug binds to a receptor which subsequently modulates membrane permeability, are discussed in Chapter 9.

In order to understand more about how drugs can interact with membranes, it is important to know how present theories of membrane structure relate to transport of ligands and ions across membranes, whether carried out by enzyme or by ion channel. Furthermore, some understanding of the action potential (potential difference) that exists across the membrane is also important for an understanding of the mechanism of action of local anaesthetics and anti-arrhythmic agents that act via blockade of sodium channels, and those anti-hypertensive drugs that interfere with calcium ion channels.

10.1.1 Membrane structure

Our present view of membrane structure in eukaryotic cells has been outlined by Singer and Nicolson (1972) and is depicted in Fig. 10.1. The major matrix of the membrane comprises a double leaflet of phospholipid molecules. The polar headgroups of the two layers face out towards the aqueous environment of the cytoplasm and the extracellular medium respectively, while the non-polar fatty acyl hydrocarbon chains (two per molecule) are orientated towards the middle of the membrane. Cholesterol in mammalian cell membranes and ergosterol in fungal cell membranes are inserted into the lipid bilayer between phospholipid molecules. The 3-hydroxyl group common to both of the sterols is orientated towards the aqueous environments and interacts with the polar headgroups of the phospholipids, while the non-polar sterol skeleton and

Table 10.1 Drugs and their targets discussed in Chapter 10

Section	*Target*	*Drug*	*Therapeutic Use*
10.2	Sodium channel	Lidocaine Lidocaine Amiloride, Triamterene	Local anaesthetic Anti-arrhythmic Diuretic
10.3	Calcium channel	Verapamil Nifedipine	Angina
10.4	Coupled sodium/ chloride channels	Ethacrynic acid Frusemide	Diuretics
10.5	Potassium channels	Lemakalin Nicorandil	Coronary vasodilator
10.6	Na/K ATPase H/K ATPase	Digoxin Omeprazole	Cardiac failure Anti-ulcer
10.7	Membrane stabilizer	Cromoglycate	Anti-allergic
10.8	Calcineurin	Cyclosporin	Immune suppressant
10.9	Ergosterol	Amphotericin	Antifungal

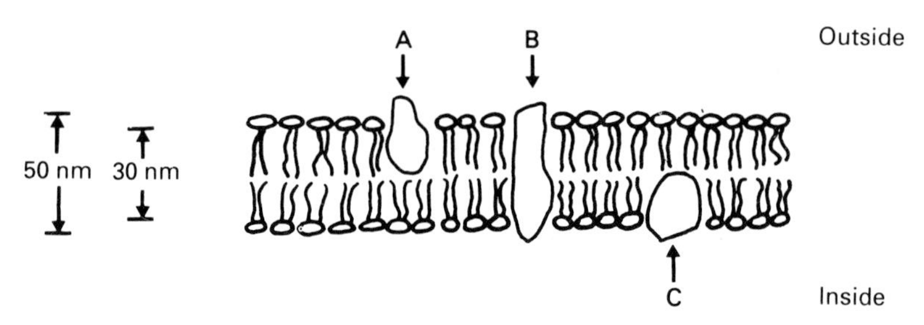

In the lipid bilayer, phospholipid molecules are shown with their polar head groups as small circles, and the non-polar tails as wavy lines. The leaflet is held together by the non-polar interaction between the fatty acyl side-chains. A is an integral protein bound in the extracellular surface of the membrane, while B is a transmembrane protein. C is an integral protein facing the cytoplasm. Both A and C penetrate only one-half of the bilayer.

Figure 10.1 Membrane structure.

hydrocarbon tail are positioned so that they interact with the hydrocarbon chains of the phospholipid fatty acyl groups.

Proteins are also present in the membrane; the receptors for some hormones span the membrane, unlike adenylate cyclase which is positioned on the inside. Some transmembrane proteins, e.g. Na^+,K^+-ATPase, are the transport carriers for ions, while there are other protein carriers for small molecules such as glucose. These proteins have been shown to operate on both sides of the membrane. Thus for Na^+,K^+-ATPase, the sodium and ATP binding sites are on the cytoplasmic side of the membrane while the potassium and ouabain binding sites are situated on the extracellular side. As noted in Chapter 9, the GABA receptor contains a channel for chloride ions.

Membrane fluidity has a marked effect on the transport of small molecules across the membranes - the more fluid the membrane the more permeable it is. Fluidity is controlled to some extent by the type of fatty acid present in the phospholipids: saturated fatty acids decrease and unsaturated fatty acids increase the fluidity. Cholesterol also has the effect of reducing fluidity in areas that are rich in unsaturated fatty acids and *vice versa* in areas high in saturated fatty acids. The net effect is to produce regions of markedly differing permeability within the same membrane.

With regard to the mechanism of transport, Scarborough (1985) has suggested that when membrane-bound proteins bind ligands, they undergo a conformational change rather like the operation of a hinge so that the ligand is embedded within the protein. Scarborough views the open or ligand-free state as having a cleft (state 1 in Fig. 10.2) within which the ligand binds (state 2). This binding induces a conformational change which opens an aqueous diffusion pathway from the binding site towards the far side of the membrane (state 3). The ligand is then free to escape to the opposite side (state 4). The conformation then changes back to the original to allow the process to begin again.

10.1.2 Dynamics of the heart beat

From a biochemical point of view the above description might be considered sufficient for an understanding of membrane events. The cell membrane has, however, been shown to support a considerable potential difference; 70-90 mV in cardiac cells and 200 mV in cells of the filamentous fungus *Neuros-*

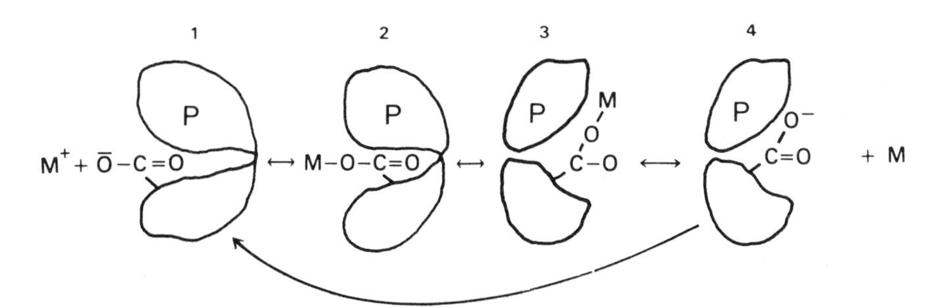

P, the transmembrane protein that acts as a carrier.
M, the ligand to be transported; in this case a cation, and the carboxylate ion the binding site — probably the aspartate side-chain.
1–4, the different states of the transport system.

Figure 10.2 Membrane transport.

pora crassa (the inside of the membrane is negative with respect to the outside). The consequences that flow from changes in the electrophysiological state of the membrane are crucial for an understanding of the mechanism of action of three of the types of drug that are discussed in this chapter which modulate the transport of cations: local anaesthetics and type 1 anti-arrhythmic agents, both of which block the sodium channel, and calcium channel blockers that are also used for arrhythmias and for high blood pressure. In order to understand more about how these drugs exert their effects, an outline is presented here of the generation of the action potential in a pacemaker cell that controls the beating of the heart, specifically in the Purkinje fibres (Kumana and Hamer, 1979, described in Muhiddin and Turner, 1985; Fig. 10.3).

The sodium channel initiates the process in phase 0 by allowing a rapid transport of sodium ion into the cell to occur, thus causing the membrane potential to depolarize from a potential of −70 mV to approximately +25 mV. These channels are then shut off or inactivated and cannot open again until the membrane has been repolarized. Next in phase 1 there is a sharp fall of about 20 mV probably due to chloride ion entry. In phase 2, calcium begins to enter through the slow calcium channels (called slow because the time constant for inactivation is 50 ms compared with 0·5 ms for the 'fast' sodium channel). Eventually chloride and calcium channels close and the major repolarization of the membrane is effected by a large outflow of potassium in phase 3. The potassium channels are inactivated when the membrane potential has reached about −80−−90 mV, the resting potential.

A second outward potassium channel, known as the pacemaker current, continues to operate for a short time. Eventually, the point is reached at which the inward sodium and calcium background current, which do not change with time, produce a net inflow of current. The overall effect in phase 4,

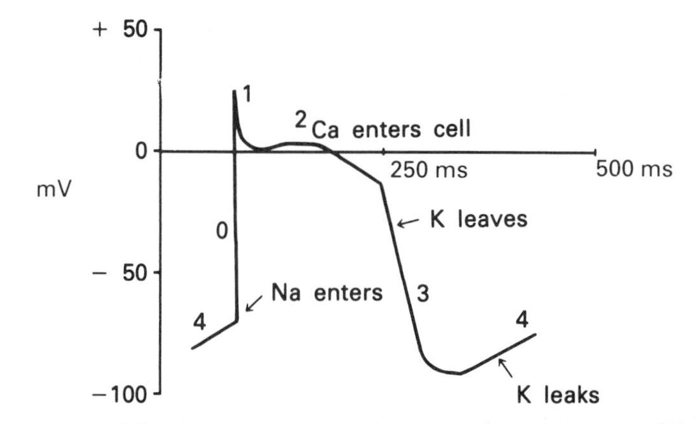

The phases of the normal action potential are shown; 0, depolarization, 1,2,3, repolarization and the diastolic phase (4).

Figure 10.3 Diagram of myocardial action potential.

therefore, is that of gradual depolarization as the background repolarizing current gradually falls away. When the membrane potential has dropped to −70 mV, the reactivated sodium channel is triggered and the process repeats itself. It usually takes about 400 ms from start to finish in a pacemaker cell, which acts, as its name suggests, to initiate the process by spontaneously generating an action potential - a property known as automaticity. A group of pacemaker cells act together to initiate an impulse which spreads throughout the rest of the heart. Although the details of ionic involvement may differ in other cells in the heart, the requirement for fast sodium and slow calcium channels is constant.

Arrhythmias, or abnormal beating of the heart, can be caused either by faults in the mechanism for initiating the electrical impulse or by disturbances in the conduction of an impulse. Abnormal automaticity occurs when the membrane depolarizes too soon before phases 3 and 4 have fully run their course, and an extra impulse is generated that is propagated throughout the heart. The result can be almost random beating of the heart. This is often referred to as after-depolarization and, if it occurs during phase 3 (early after-depolarization), is due to the increased influx of a cation, either sodium or calcium, or reduced efflux of potassium. The membrane potential has not fallen low enough to activate the fast sodium channels. After-depolarizations may also occur that are delayed until full repolarization has been reached (Bigger and Hoffman, 1990).

Alternatively, other cells that are not normally pacemaker cells may take it upon themselves to initiate an impulse - a form of automaticity that arises in the wrong place. This may be a consequence of disease affecting the responsiveness in that the sodium channel is oversensitive to stimulation. In arrhythmia caused by either situation, a drug that can partially block the initial sodium current is of value in steadying the heart beat and returning the process to normal. A partial blockade of the calcium channel is also found to be of value in some cases, particularly after depolarization.

The conduction of an impulse can also lead to arrhythmias if it re-enters a part of the heart that has already been excited. This normally requires that the rate of conduction be slowed, usually as a consequence of disease, and that a block in conduction occurs which can divert the impulse back into a region of the heart that it has already traversed (Wit, 1985).

10.2 The sodium channel

The mechanism of transmission of nerve impulses, and the propagation of electrical impulses in the heart that accompany beating, both require a sharp and rapid change in the trans-membrane potential difference. This is achieved by means of a swift transport of sodium cations into the cell through what is known as the sodium channel. Estimates of 10^7 ions per second or an electrical conductance of 2–8 pS (the unit of conductance is Siemens, S) per site have been made (reviewed in Catterall, 1980). This rate is too fast to be associated with a mobile carrier and the cation must pass independently through the selective pore. It is not, however, open at all times and requires repolarization of the membrane to take place before it is reactivated. After a rapid transient influx of cations has taken place, the channel is closed by a mechanism that is not yet understood.

At least two types of drug interact with the electrically excitable sodium channel to block it; local anaesthetics, where the transmission of nerve impulses must be blocked, and anti-arrhythmic drugs that are needed to restore the beating of the heart to normal. Our understanding of the structure of the sodium channel has increased greatly in recent years (reviewed in Catterall, 1988).

The sodium channel from mammalian brain consists of three subunits: α (260 kDa), β_1 (36 kDa) and β_2 (33 kDa). The β_2 subunit is linked to the α subunit by disulphide bonds; all three subunits are heavily glycosylated at the surface which faces the extracellular space, while the α subunit contains a site for phosphorylation by cyclic-AMP-dependent kinase on the cytoplasmic face. The α subunit contains the voltage-gated channel, while the other subunits stabilize the channel in the functional state.

The α subunit consists of three cytoplasmic and four extracellular loops separated by four homologous transmembrane domains - each with six segments that traverse the membrane in the form of α-helices. The amino and carboxy termini lie within the cell. The ion channel is believed to lie in an almost symmetrical square array 0·3 to 0·5 nm in size, formed by the four transmembrane domains.

The voltage dependence of channel opening is linked to conformational changes that result from the movement of charges (voltage sensors) when the transmembrane potential changes. Positively charged voltage sensors have been found in the fourth segment (S_4) of each transmembrane domain. At rest all the positive charges in S_4 are paired with negative ones and the transmembrane segment is held in that position by the negative membrane potential. Depolarization reduces the forces holding the charges together. The S_4 helix slides and rotates in such a way as to produce an unpaired negative charge on the inner surface and a positive one on the outside surface to give a net charge transfer of +1 (Catterall, 1988; Grant and Wendt, 1992).

10.2.1 Sodium channel blockers - procaine as a local anaesthetic

Local anaesthetics block both the generation and the transmission of sensory nerve impulses by binding to sodium channels in the nerve cell membrane. Normally, a slight depolarization of the membrane opens the sodium channel to allow a rapid but transient inward current. Local anaesthetics raise the threshold for electrical excitation of the membrane to such an extent that the permeability of the membrane becomes so low as to result in a block of nerve conductance. Small nerve fibres are blocked more rapidly than thick ones, possibly because the former are demyelinated, unlike the latter. All the commonly used local anaesthetics contain a secondary or tertiary amine with a pK_a between 8 and 9, so that at pH 7·4 between 5 and 20 per cent of the drug will be in the form of a neutral amine (pK_a is the pH at which 50 per cent ionization occurs. As pH is a logarithmic scale, at one unit either side of the pK_a, only 5 per cent of one component is present.) The activity of the drug increases with increasing lipophilicity, whereas the cationic form binds to the sodium channel. Increasing lipophilicity probably helps the drug to diffuse into the cell in the neutral form (Postma and Catterall, 1984). The drugs act more rapidly when added to the cytoplasmic side of the axonal membranes, rather than the extracellular side, suggesting that the binding site is nearer the cytoplasm. Affinity of the drugs for the channel depends on the voltage, and the channels are blocked more rapidly and completely when the channels are repeatedly activated (use-dependent inactivation or block). The cationic form binds more rapidly to the open form of the channel because it can gain access more readily but more tightly to the inactivated channel. In the resting state the channel is not sensitive to the drug.

The drugs have a strictly limited life in the circulation since they are all esters and are hydrolysed by plasma esterases. This ensures that one of the major requirements of a local anaesthetic is met, namely a transient and reversible anaesthesia. In principle, however, it is possible to see that such drugs could be of use as anti-arrhythmics if their duration of action were sufficient; in procainamide, an amide link replaces the ester group in procaine and the reduced hydrolysis by esterases allowed the drug to be used as an anti-arrhythmic. Lidocaine is an exception in being used both as a local anaesthetic and anti-arrhythmic, but in the latter case it is usually during acute myocardial infarction or cardiac surgery where a long duration of action is not required.

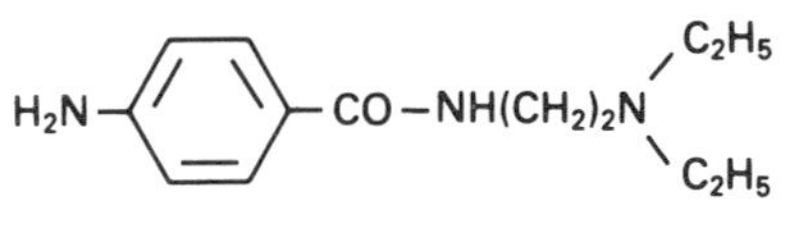

$H_2N-C_6H_4-CO-NH(CH_2)_2N(C_2H_5)_2$

Procainamide

$(CH_3)_2C_6H_3-NH-CO-CH_2-N(C_2H_5)_2$

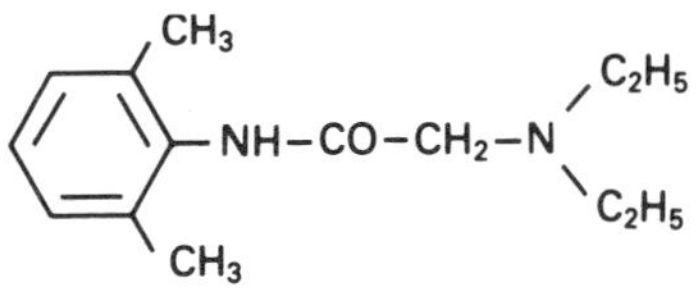

Lidocaine

Local anaesthetics find a use in dentistry and a variety of surgical procedures to reduce the pain suffered by the patients, without causing unconsciousness. Examples of this procedure include the injection of a local anaesthetic to block a particular nerve such as the median or ulnar at the elbow to allow surgery at the wrist (nerve-block anaesthesia); infiltration anaesthesia where a particular tissue is to be removed and surface anaesthesia where an operation is to be carried out on the mucous membranes of, for example, nose, mouth and throat (Ritchie and Green, 1985).

10.2.2 Anti-arrhythmic agents - lidocaine as an anti-arrhythmic

Anti-arrhythmic drugs may be classified into five groups (Muhiddin and Turner, 1985); sodium channel blockers such as lidocaine and procainamide belong to class 1; sympatholytic agents such as β-blockers (Chapter 9) form class 2; class 3 is composed of agents such as amiodarone that prolong the repolarization phase; calcium channel antagonists form class 4 (section 10.3.1) and chloride channel antagonists form class 5 (section 10.4). It is the agents of class 1 with which we are concerned in this section.

Class 1 anti-arrhythmic drugs block sodium channels in both nerve and cardiac cells, but their efficacy is greater on the latter by a factor of between 10 and 200, and so local anaesthetic action is likely to make only a very modest contribution while the drugs are being used for heart conditions. Three states of the cardiac sodium channel are postulated: resting, activated (open) or inactivated (closed) after use. These different states of the channel have different affinities for the drugs. Furthermore, the idea is confirmed that drug binding to the channel shuts off conductance. Repolarization of the membrane closely parallels the activation of the channel in that when the membrane potential has returned to the resting state, the channels are reactivated. The class 1 anti-arrhythmics have, as their main effect, the reduction of the maximum rate of depolarization. In addition, they may raise the threshold of excitability, making the cardiac cell less sensitive to being triggered. These drugs may also prolong the period during which no action potential can be initiated, known as the refractory period. They produce no change in the resting potential (Muhiddin and Turner, 1985).

Sodium channel blockers (group 1) vary somewhat in their mode of action and are classified into three groups based on the effect they have on the duration of the action potential. Those that prolong the action potential, e.g. disopyramide, are grouped in 1A. Those that shorten it, such as lidocaine, are in group 1B, while group 1C have no effect. Lidocaine binds to the inactivated channel and so the degree of block increases as the membrane remains depolarized; the drug reduces the chance of the channel opening during depolarization (Grant, 1992).

The concept of a use-dependent block, noted above for local anaesthetics, also holds good here - if the rate of heart beat is sufficiently fast the effect of the drug increases with every beat. This has been termed the 'modulated receptor hypothesis' (Hondeghem and Katzung, 1984). As with local anaesthetics (section 10.2.1) the neutral form of the drug can penetrate the channel via the lipid membrane and ionize in the channel. This is particularly favoured if the channel is closed (Hille, 1977).

Disopyramide binds rapidly to the open channel and little further block occurs after the channel has closed. This drug is useful for both atrial and ventricular arrhythmias. In contrast, lidocaine is only useful for ventricular arrhythmias where there is time for block to develop.

The sodium channel blockers also vary in the speed with which they dissociate from the channel, and this fortuitously follows their classification. Lidocaine dissociates rapidly with a rate constant of less than 1 second, whereas disopyramide is between 1 and 10 seconds (type 1C drugs have a rate constant greater than 10 seconds). The slower the rate of dissociation, the greater the prolongation of slower conduction (Grant, 1992). This may have caused the threefold increase in sudden death noted for the 1C drugs on the Cardiac Arrhythmia Suppression Trial (CAST, 1989). If the conduction is slowed too much the impulse may re-enter the circuit and produce arrhythmias. This effect may lie behind the increase in mortality. Furthermore, patients were selected who were at a moderately low risk of dying so the drug treatment was more dangerous than placebo (Grant and Wendt, 1992).

10.2.3 Diuretics - amiloride and triamterene

The type of sodium channel discussed in the previous two sections is that present in tissue which is electrically excitable. In epithelial tissue that is not electrically excitable, such as that lining the surface of the kidney or cornea, there are three other types of sodium channel which are characterized by insensitivity to toxins, such as tetrodotoxin even at high concentrations, and by a slower rate of ion transport than the fast sodium channels of the heart. One group of these channels is characterized by a high trans-epithelial potential difference and resistance; these are known as 'tight' epithelia and occur in the toad urinary bladder. Another group, which includes the proximal tubule of the kidney (Fig. 10.4), have a very low or negligible trans-epithelial potential and are able to transport large volumes of isotonic fluid (leaky epithelia). Intermediate between these extremes are the channels with moderate trans-epithelial potential difference such as those in the cells of the distal portion of the kidney tubule (reviewed in Cuthbert, 1977) and occur in the hatched area for amiloride and triamterene action (see below) in Fig. 10.4.

The role of these sodium channels in the kidney is to reabsorb sodium ion which, together with chloride ion and other solutes, has been filtered from the blood in the glomerular portion of the kidney (Fig. 10.4). At the same time, in order to maintain equal osmotic pressures either side of the kidney cell membrane, fluid is absorbed.

The transport of sodium across the tight epithelia is reasonably well understood in that sodium passes across the membrane from the bladder into the cell, passively down an electrochemical gradient. Sodium pumps located on the other side of the cell (known as the basolateral) remove the sodium by an active transport process across the membrane coupled to an enzyme hydrolysing ATP. The sodium ion flow may help to carry chloride ions in addition, usually against the electrochemical gradient of the latter. Transport systems in leaky and intermediate epithelia are less well understood at present although it is probable that an electroneutral sodium exchange for protons may take place as in the renal proximal tubule.

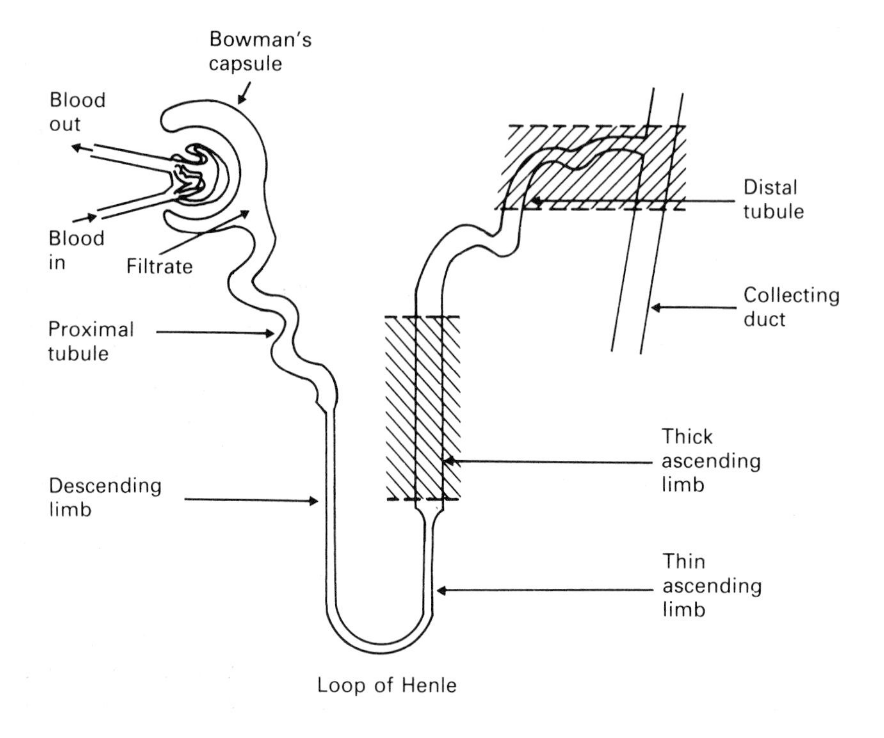

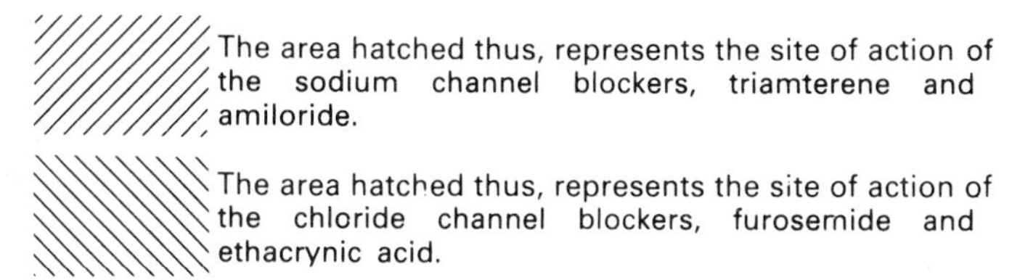

Figure 10.4 Site of action of sodium and chloride channel blocking diuretics.

Amiloride and triamterene are both commonly used diaminopyrimidine diuretics which block the sodium channel in the distal portion of the kidney tubule (Fig. 10.4). Amiloride has been shown to bind to the exterior surface of the cell membrane instantaneously and reversibly. The drug thereby prevents sodium transport and increases the electrical resistance of the toad bladder that is used as a model of a tight epithelium. The reduction in the negative charge of the membrane (depolarization) is blocked, and consequently the secretion of potassium ion into the lumen of the tubule, which normally serves to maintain the membrane potential, does not occur (reviewed in Kokko, 1984; Lant, 1985a).

Amiloride blocks sodium transport in tight epithelia with an IC_{50} less than 10^{-6} M (see the comprehensive review by Benos *et al.*, 1992). The predominant form of the drug at neutral pH is cationic with the positive charge resonating in the amidinium moiety (pK_a 8·7). As noted for local anaesthetics, however, the proportion of the uncharged form will be appreciable and this may be the form that crosses lipophilic membranes.

The binding site for amiloride is not known at present, nor is the kinetics of its interaction with sodium defined as competitive, non-competitive and mixed inhibition have all been noted in different experiments. The evidence suggests that amiloride acts as a molecular cork by physically plugging the channel, rather than by binding outside and causing it to block by a conformational change. The molecular weight of the channel is 730 kDa and it is composed of six non-identical subunits ranging from 40 to 315 kDa in molecular weight. The smallest subunit is apparently the G-protein α_{1-3} which renders the channel sensitive to inhibitory control by atrial natriuretic hormone. The membrane-spanning 150 kDa subunit binds amiloride on its extracellular surface and may form the channel by itself (Benos *et al.*, 1992).

Amiloride exerts its effects on the sodium channel and not on the outward potassium channel, which indicates that the two alkali metal cations are transported by different channels - not surprising in view of their different sizes. Amiloride and triamterene do not give rise to excessive potassium secretion (spironolactone acts similarly; Chapter 6) - thus preventing hypokalaemia. The drugs' major action is to prevent sodium (and concomitantly chloride ion) reabsorption together with fluid. The urine is thereby increased in volume, sodium and bicarbonate ion excretion is increased (the pH of the urine rises) and there is a fall in potassium excretion.

There are some conditions where there is an excess of water (oedema) in certain peripheral tissues of the body and also the lung, e.g. after heart failure, when it is therapeutically advantageous to block the reabsorption of fluid. This can be carried out by blocking the re-uptake in the kidney of sodium ion through specific channels, as with amiloride and triamterene, or chloride ion in a similar fashion by furosemide and ethacrynic acid (section 10.4), although the specific agents operate at different sites in the kidney tubule. These drugs all help to reduce excess fluid in the tissues, and thereby reduce the workload of the heart - clearly an important contribution to the therapy of heart disease.

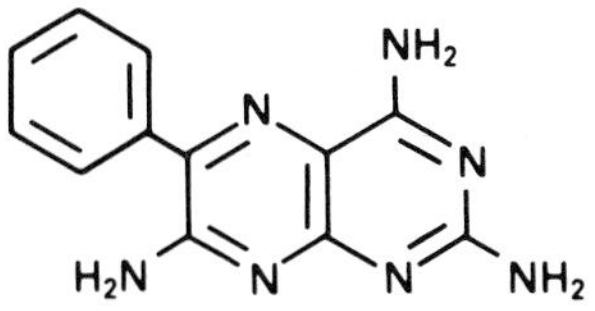

Triamterene

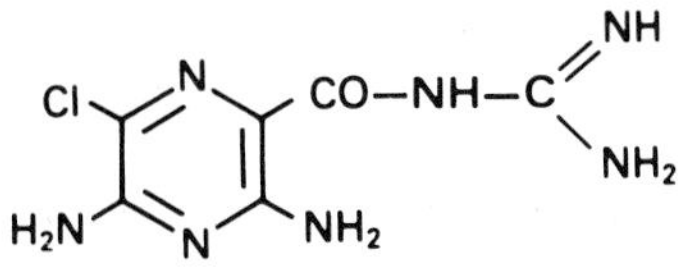

Amiloride

10.3 *The calcium channel*

Calcium is indispensable for the initiation of a number of essential cellular processes, notably muscle contraction (e.g. in the heart), secretion of neurotransmitters and hormones from neurone and endocrine cells, and rhythmic firing of heart and nerve cells. Some of the calcium required for this activation may be found intracellularly, bound to endoplasmic reticulum, mitochondria and to the calcium binding proteins such as calmodulin. The quantity of calcium available in these sites is sufficient to support only a short burst of activity, however, and for longer sustained activity calcium has to be drawn into the cell from outside through specific ion channels (reviewed in Droogmans *et al.*, 1985), except for the platelet which can sustain the release of ADP from intracellular stores of calcium.

In the heart, the depolarizing effect of calcium entry has two effects: it controls the conductance in the sinoatrial and atrioventricular (AV) nodes (a small mass of conducting tissue that links atrium and ventricle via the bundle of His and forms an important part of the conducting system of the heart). In other parts of the heart, where the sodium current is responsible for the action potential, the major role of calcium is to mediate excitation–contraction coupling of the heart muscle. Calcium binds to troponin, thus relieving the inhibition of contraction, and releasing actin and myosin to cause contraction.

Calcium channels are in a continuous state of flux, opening and closing repeatedly, typically remaining open for 1 ms and then closing for between 1 and 100 ms. The channels may be opened by the binding of a neurotransmitter or hormone to the cell surface, in which case they are termed receptor-dependent. Neurotransmitters activate the channel not by increasing the number of ions flowing through per unit time (conductance), but rather by increasing the length of time that the channel remains open (and reducing the time that it is shut) (reviewed in Schramm and Towart, 1985). Catecholamines, such as noradrenaline, usually act through cyclic AMP and its attendant protein kinases to phosphorylate a membrane protein to open the channel while, in contrast, phosphatases dephosphorylate the protein during inactivation (see Glossman *et al.*, 1982).

Alternatively, other calcium channels may be influenced by changes in the membrane potential and these are known as voltage-dependent channels. Using the patch-clamp technique, whereby a single cell can be maintained at a given potential difference across its membrane, the kinetics of opening and closing (known as 'gating'), conductance and sensitivity to various drugs can be determined.

Recent studies indicate that there are three types of voltage-operated calcium channels in sensory neurones and two types in cardiac cells (reviewed in Spedding and Paoletti, 1992). They differ primarily in that different reductions in the membrane potential are required to effect their gating. As the membrane is depolarized a channel is opened (called T for transient) that shows a moderate current and rapid decay. Stronger depolarization opens a channel termed L (for longer lasting) as it decays only slowly, while another channel termed N (neither T nor L) is also opened by strong depolarization and grows more strongly with increasing depolarization. The most prominent calcium channel types in the heart are T and L. The L type, which is the target for the calcium channel blockers is pentameric with subunits of α_1 that contains the pore, α_2, β, γ and δ. Although the calcium channel has two extra subunits, it otherwise resembles the sodium channel, in that two of the subunits are linked by disulphide bonds to an α subunit and the subunits are extensively glycosylated. The respective α subunits show an extensive homology, possibly caused by gene duplication. There are phosphorylation sites on the α_1 and β subunits which respond to β-adrenergic stimulation and mediate channel opening via cyclic-AMP-dependent kinases. The sodium channels, however, operate a faster conduction close to the diffusion limit and generate a more rapidly rising action potential than the calcium channels, originally known as the slow channels (Catterall, 1988; Katz, 1992).

The calcium channel subunits are made up of four potential membrane spanning domains separated by cytoplasmic and extracellular loops. As with the sodium channel, there are six homologous transmembrane stretches of α-helix in each domain. The fourth (S_4) has been proposed as the voltage sensor responding to depolarization as it contains a string of positively charged arginine and lysine residues.

As with all biologically active molecules, mechanisms must exist for terminating their action. In the case of calcium it clearly cannot be metabolized and so it has to be removed from the cell cytoplasm. One mechanism is via the calcium pump - a calcium-calmodulin-activated membrane enzyme which uses the energy from the hydrolysis of ATP to drive calcium ions out of the cell. This enzyme is activated as soon as the cellular level of free calcium ion rises much above the resting level of approximately 2×10^{-7} M. The maximum activation of the contractile proteins in smooth muscle cells, for example, is achieved at 10^{-5} M, and this level also induces secretion in hormone-producing cells that are activated by calcium ions.

Another cellular mechanism for removing calcium from the cytoplasm is by the sodium-calcium exchange, whereby calcium is pumped out of the cell in return for sodium entry. Cardiac glycosides exert their pharmacological effects on this exchange (section 10.5.1). The extracellular level of calcium ion is approximately 10^{-3} M and the gradient is maintained to a large extent by the very low permeability of the cell membrane to the doubly charged alkaline earth metal.

10.3.1 Calcium channel antagonists - verapamil, nifedipine for hypertension and heart failure

In addition to the patch-clamp technique noted above, pharmacological intervention has led to great advances in our understanding of the calcium channel. There are three main families of calcium channel blockers (or calcium antagonists) in clinical use: phenylalkylamines exemplified by verapamil, 1,4-dihydropyridines of which nifedipine is a major contender and benzothiazepines such as diltiazem. These ligands include a number of drugs used for the control of angina pectoris, arrhythmias and high blood pressure.

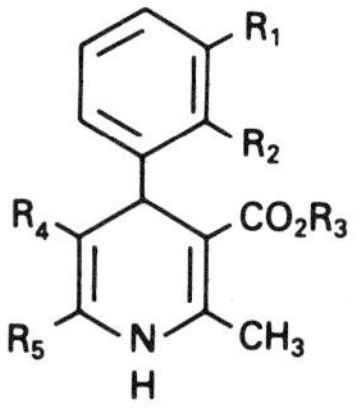

	R_1	R_2	R_3	R_4	R_5
Nifedipine	H	NO_2	CH_3	CO_2CH_3	CH_3
Nimodipine	NO_2	H	$CH(CH_3)_2$	$CO_2(CH_2)_2OCH_3$	H
Bay k 8644	H	CF_3	CH_3	NO_2	CH_3

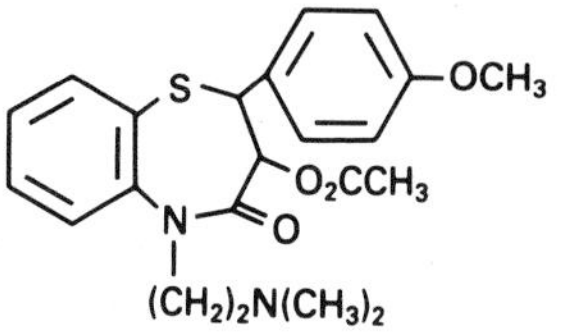

Diltiazem

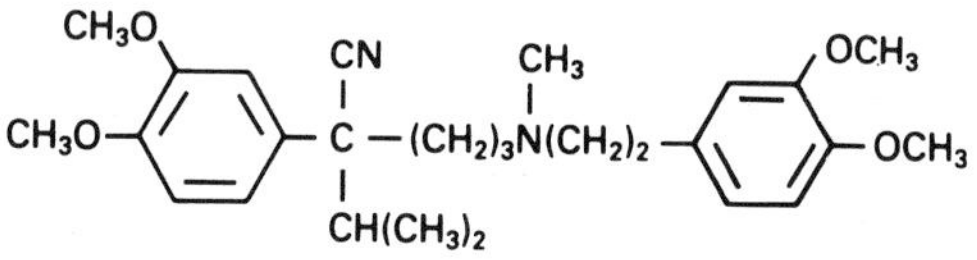

Verapamil

The earlier concept that the channel could be either open, closed and resting, or closed and inactivated has been modified in the light of measurements on a single channel. The channel can remain open for either short or long periods. The effect of the calcium channel blockers is, however, not to act as molecular plugs but rather to reduce the probability that the channel will be in the open state long. In addition, the ligands bind more effectively to the channel when the membrane is depolarized.

These three classes of drug vary, however, in their effects on the (slow) calcium channel. Although nifedipine reduces the inward calcium current, it does not affect the rate of recovery of the calcium channel. Verapamil, on the other hand, as well as reducing the inward flow of calcium also reduces the rate of recovery of the channel. Furthermore, verapamil and diltiazem also block more effectively as the frequency of stimulation increases (use-dependent block).

Although there is a much greater understanding of the molecular architecture of the calcium channel than in earlier years, we still do not know much about the specific binding sites of the classes of antagonists. Recently, however, a potential verapamil-binding site on channels from skeletal muscle has been identified on the cytoplasmic face of the channel. Verapamil and analogues behave in a fashion similar to the local anaesthetic blockers of the sodium channel in that they are weak bases with pK_a in the region of 8·5. There is sufficient drug in the uncharged form at neutral pH to enter the cell and block the open state of the channel by approaching from the cytoplasmic side. A putative binding site composed of aspartic acid residues has been proposed on segment 6 near the mouth of the channel (Striessnig *et al.*, 1991).

Chirality plays a part in the action of these drugs in that the *S*-(−)-form of verapamil is six to ten times more active than the *R*-(+)-form; diltiazem is most active as the 2*S*,3*S* form. In fact, ion channels have been described as chiral receptors (Kwon and Triggle, 1991).

The effects of calcium entry blockade are more readily seen on the whole heart. Verapamil was first shown by Fleckenstein's group to mimic the effects of calcium withdrawal, as the drug reduced (a) calcium-dependent contraction without affecting sodium-dependent aspects of the action potential and (b) calcium-dependent myofibrillar ATPase activation and oxygen consumption of the beating heart (Fleckenstein-Grun *et al.*, 1984).

The situation in which these effects may be particularly useful is when the oxygen available to part of the heart is greatly reduced as, for example, after a myocardial infarction. In this condition a blockage occurs in a branch of an artery supplying the heart, cutting off part of its blood supply. The area involved rapidly uses up the available oxygen and becomes hypoxic. It is believed that catecholamines released as a result will greatly increase calcium uptake into the myocardial cell, sending the muscular apparatus into a contractile spasm and using up ATP in the proces. Mitochondria try to take up calcium from the cytoplasm and use up more ATP. Eventually the calcium load damages the structure of the organelle and uncouples oxidative phosphorylation. The loss of ATP ultimately shuts off the plasma membrane calcium pump, thus allowing calcium to accumulate further and increase the damage. Calcium channel blockers, although less effective against neurotransmitter-operated channels may still be of use in a cardioprotective fashion (Kendall and Horton, 1985), and indeed are already used before cardiac surgery (Fleckenstein-Grun *et al.*, 1984).

At present, however, the main uses of calcium channel antagonists are for the treatment of angina pectoris, hypertension and arrhythmias (Schramm and Towart, 1985). Angina pectoris is characterized by dull chest pain brought on by conditions of stress such as over-exercise, over-indulgence, smoking etc. and is the result of insufficient oxygen for the oxidation of substrates in the heart muscle. Calcium antagonists are able to reduce the oxygen demand of the heart muscle by reducing the amount of muscular activity and are particularly useful for the treatment of angina pectoris.

Another type of angina, known as variant angina, is characterized by a reduction in blood flow in the heart rather than lack of oxygen. Calcium channel blockers are of use in this condition because they are particularly effective at relaxing cardiac smooth muscle in the coronary arteries, thereby increasing the diameter of the artery. As a consequence, the effort the heart has to make to drive the blood through the vessels is reduced and the blood pressure falls. In addition, calcium channel antagonists relax peripheral blood vessels, thereby reducing blood pressure, and can be used to treat both acute hypertensive crises and chronic hypertension.

The type of arrhythmia in which there is a change in the normal beating of the heart by a group of cardiac cells beating in a disorganized fashion (when pacemaker groups of cells arise in the 'wrong' part of the heart) responds to verapamil, but not nifedipine (Spedding, 1985). In some cases these arrhythmias can arise from a partial depolarization of the membrane causing the excitation process to depend on the slow calcium channel rather than a fast sodium channel (Fleckenstein-Grun *et al.*, 1984).

10.4 Coupled sodium-chloride ion channels - furosemide and ethacrynic acid as diuretics

As noted in the section on sodium ion channels (section 10.2.3), chloride is reabsorbed in the kidney by active transport systems of which that in the ascending limb of Henle's loop is particularly active (Fig. 10.4). The transport system has been proposed as a sodium-chloride co-transport with sodium entering the cell down its electrochemical potential and chloride moving against its electrochemical gradient. The chloride ion exits across the opposite border of the cell down a favourable potential difference (Frizzell *et al.*, 1979). In earlier years it was thought that chloride uptake alone was the target of the drugs discussed in this section and so much of the work has been directed to the transport of chloride but not sodium (hence the emphasis on chloride in the studies reported below). Nevertheless, it is now clear that the co-transport of the ions is the major target.

Two structurally unrelated diuretics, furosemide and ethacrynic acid, among others, act to inhibit this coupled transport and are known as 'loop' diuretics. Up to 25 per cent inhibition of chloride and sodium reabsorption is possible with these agents which is a far higher proportion than that obtainable with other diuretics - hence the term 'high ceiling' is used to describe these diuretics. Perhaps not surprisingly, such a powerful effect is characterized by swift onset in about 15 minutes and rapid cessation in 4-6 hours (Kokko, 1984; Lant, 1985a).

Furosemide acts to inhibit active chloride transport from within the tubular lumen, i.e. the fluid side not the intracellular side (Odlind, 1984). The drug reaches the lumen by being transported from the proximal tubule cell through a non-specific channel that secretes organic acids, thus using the normal kidney transport pathways to reach its site of action (Kokko, 1984; Lant, 1985a). Furosemide demonstrates competitive kinetics with chloride ion uptake and non-competitive with sodium ion which suggests, but does not prove, that the drug may bind at or near the chloride uptake sites (Ludens, 1982). Coupled transport exhibits saturation kinetics as would be expected of an active transport system. Furthermore, ouabain blocks active transport of chloride which suggests that the process is linked to a Na^+,K^+-ATPase as this drug is a specific inhibitor (section 10.5). It is impossible, at present, to be more precise because no definitive receptor has yet been identified for furosemide.

Frog cornea, as another epithelial system, is believed to contain similar coupled transport systems to the kidney, although the cornea secretes chloride whereas absorption occurs in the ascending limb of Henle. Electrochemical studies have confirmed the selective inhibitory effect of furosemide on chloride uptake and have shown that the drug increases the electrochemical gradient of chloride across the apical membrane of frog cornea. It prevents equalization of chloride concentrations either side of the membrane, leading to hyperpolarization of the apical membrane and depolarization of the basolateral membrane (Patarca *et al.*, 1983). The net effect is to reduce the permeability of the membrane to chloride transport.

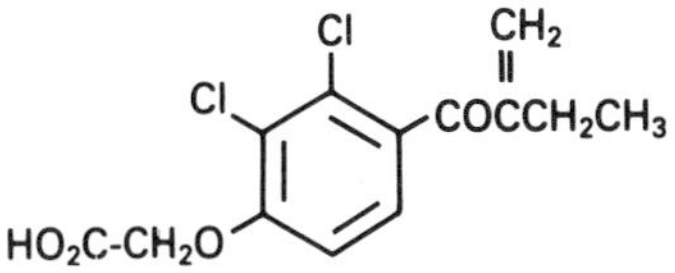

Ethacrynic acid

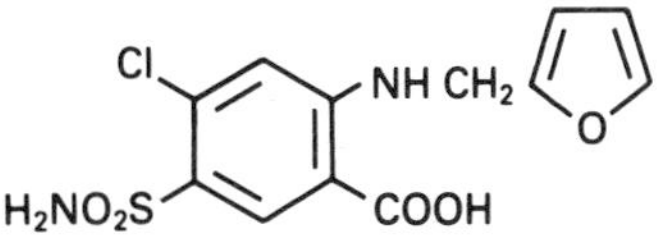

Furosemide

The receptor for ethacrynic acid has also not yet been identified. It is likely, however, bearing in mind the differences in structure between furosemide and ethacrynic acid, that the receptors for the two drugs differ. Ethacrynic acid is effective at blocking chloride transport across the epithelial membrane of the ascending loop of Henle with a concomitant decrease in potential difference. A decrease is found in this instance because the potential difference across the luminal membrane is maintained by the chloride transport (Burg and Green, 1973). Interestingly, the cysteine adduct of ethacrynic acid, which appears to be a normal excretion product, is far more effective than the drug itself; this adduct may therefore be the active form of the drug *in vivo*.

The contention that the loop of Henle is the sole site of action of these drugs has been challenged by Caffruny and Itskovitz (1982) who demonstrated that both drugs can act to block chloride transport in the proximal tubule, but the importance of this observation is not entirely clear. Obviously, there is much that we do not yet understand in the precise molecular mode of action of the sodium-chloride channel blocking diuretics.

10.5 Membrane-bound ATPases

There are a variety of enzymes in eukaryotic and prokaryotic systems that use the energy of hydrolysis of ATP to drive the transport of protons and other cations across membranes that are otherwise impermeable to these ions. These enzymes are all dependent upon magnesium, presumably because the true substrate is an ATP-magnesium complex. One group that is part of the process of oxidative phosphorylation is characterized by the reverse reaction of ATP synthesis coupled to oxidation and is found in the membranes of mitochondria, bacteria, chloroplasts etc. The mitochondrial enzyme is crucial to the chemiosmotic principle of Mitchell which states that phosphorylation of ADP is driven by the energy of a proton gradient across the mitochondrial membrane. These enzymes are often referred to as F_0/F_1-ATPases because they consist of two portions: one is responsible both for anchoring the enzyme in the membrane (F_1) and for the translocation of hydrogen ion, while F_0 is the binding site for the interconversion of adenine nucleotides. General inhibitors of these enzymes are oligomycin which acts on F_0 and blocks proton transport and dicyclohexylcarbodiimide (DCCD), which reacts with a carboxyl group crucial for the enzyme reaction (reviewed in Pedersen, 1982).

Another group of ATPases, bound to the plasma membrane of the cell, is concerned with the pumping of cations against a concentration gradient out of the cell; notable examples are: (a) the enzyme that catalyses the exchange of sodium for potassium in the cardiac cell membrane particularly, although it is also found in other tissues, (b) the calcium pump which drives calcium out of secretory cells, among others and (c) the proton pump which exchanges protons for potassium ion in the parietal cells of the stomach and thereby produces the low gastric pH. In general, the cation pumps consist of at least one polypeptide with molecular weight in the region of 100 kDa which usually acts as a channel for the cation transport.

Oligomycin does not inhibit these enzymes but DCCD does, again because there is a carboxyl group in a glutamate side-chain which is essential for activity. A phosphate-enzyme intermediate is an essential step in the mechanism, the phosphate being attached to the active site aspartate. Vanadate is a characteristic inhibitor of these enzymes, probably because it can replace the phosphate in the intermediate, forming a vanadyl-enzyme intermediate. A considerable degree of homology between cation pumps has been discovered, which suggests that they may have evolved from a common ancestor (Serrano *et al.*, 1986).

Mitchell has suggested a charge relay type mechanism in which the transport of a proton in one direction can be balanced by the transport of an ion such as potassium in the opposite direction. The aspartyl-phosphate conveys a proton outwards with the phosphate group, while potassium is translocated inwards together with another phosphate on the same aspartyl group. The aspartyl-phosphate group is assumed to rotate and go from a high energy to a low energy conformation (reviewed in Pedersen, 1982). Not all the cation pumps are electroneutral, however, since the Na^+,K^+-ATPase pumps three sodium ions out of the cell in exchange for the entry of two potassium ions for every ATP molecule hydrolysed, resulting in a change of potential difference across the membrane.

10.5.1 Sodium-potassium-ATPase - cardiac glycosides for heart failure

Heart cells, in common with most eukaryotic cells, need to maintain gradients of the alkali metal cations - potassium levels must be kept high inside the cell and sodium low, in order to maintain the cell's integrity. To do this, the cell requires a pump to eject sodium ions from the cell in exchange for potassium ions. This will counterbalance the effect of sodium influx and potassium efflux through the respective ion channels during the early part of the action potential which accompanies a heart beat. The transmembrane ATPase that effects this exchange has been purified from a number of sources, and, when inserted into lipid vesicles, can catalyse the exchange transport of alkali metal cations at a rate approaching that seen *in vivo* (Jorgensen, 1982).

The enzyme is made up of two subunits: the α subunit of molecular weight about 100 kDa and the β subunit of about 40 kDa, although minor variations in the α subunit may be found between tissues to a sufficient extent to warrant the designation of isozymes (Lytton, 1985). The enzyme spans the membrane with the catalytic and sodium activator sites facing the cytoplasm and the potassium binding site on the extracellular surface. All these sites are located on the α subunit: no functional role has yet been described for the β subunit. Both the amino and carboxyl termini of the α subunit face the cytoplasm and the polypeptide chain traverses the lipid bilayer several times in between. The mode of transport of the cations is not known, although it has been suggested that the ionic channel may pass either through the α subunit or between adjacent α subunits (Jorgensen, 1982).

In the presence of sodium ions and Mg–ATP on the intracellular side of the enzyme, the carboxyl side-chain of an essential aspartate group is phosphorylated to yield a high-energy phosphate. A conformational change takes place as ADP is released ($E_1P \rightarrow E_2P$; where E_1 and E_2 represent different conformational forms of the enzyme). Potassium ions enter the α subunit from the other side of the membrane and bind to E_2P causing it to lose the phosphate group. K^+E_2 breaks down to free potassium and enzyme to complete the catalytic cycle.

The conformational change induced by ATP phosphorylation is potentiated by sodium and dephosphorylation is activated by potassium. In separated and purified enzyme preparations, sodium ion stabilizes E_1 and potassium E_2. ATP binds with high affinity to E_1Na in a hydrophobic pocket and only weakly to E_2K in an open structure. Magnesium ion is required for activity as, with most 3enzymes that use ATP, Mg–ATP is the true substrate. This rather complicated reaction scheme may be envisaged as follows (after Jorgensen, 1982): $Na^+{}_i$, $K^+{}_i$ represent cytoplasmic cations; $Na^+{}_e$, $K^+{}_e$ extracellular.

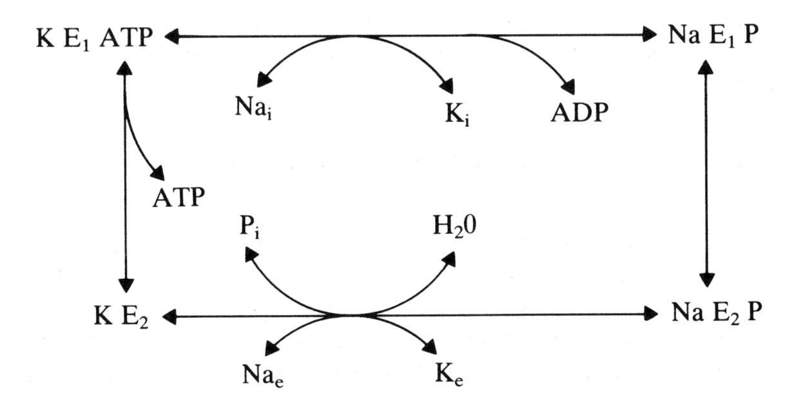

Cardiac glycosides are used in the treatment of congestive heart failure, and appear to act by inhibiting the sodium pump (Hansen, 1984). They are composed of a large steroid-like structure with an unsaturated lactone ring attached to position 17 and a sugar, or chain of sugars, linked to position 3. All these parts of the molecule are essential for activity. These drugs were discovered as part of the extract of foxgloves, digitalis, which was recommended for the treatment of heart failure as long ago as 1785. Digitoxin is the major constituent of digitalis, while digoxin is also present to a lesser extent. The original preparations from foxgloves have now been replaced by chemical preparations because the dose is much easier to control and is not subject to variable composition. The cardiac glycosides are still very useful drugs for the treatment of cardiac failure despite occasionally serious incidents of kidney toxicity. Digoxin is probably the most widely used agent (Crozier and Ikram, 1992).

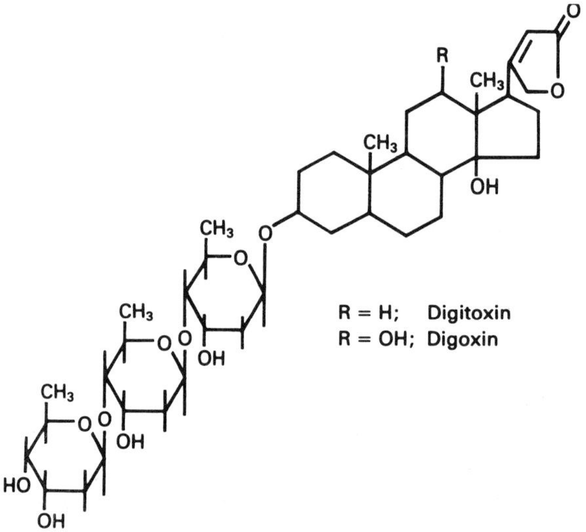

R = H; Digitoxin
R = OH; Digoxin

Experimentally, ouabain is probably the most studied of the cardiac glycosides and it has been shown to bind tightly but reversibly to the α subunit of the enzyme on the extracellular face with an Ic_{50} of 10^{-7} M. The affinity of the drug for the isozyme α^{+} of brain and kidney is much less with an Ic_{50} of 10^{-4} M (Sweadner, 1979). The enzyme undergoes a large conformational change on ouabain binding which can be partially antagonized in a non-competitive fashion by potassium, suggesting that the latter binds at a different site. Ouabain binds preferentially to the E_2P conformation of the enzyme but the affinity of that form is reduced for the drug if it is complexed with potassium. Ouabain–enzyme complexes will also bind ATP but with greatly reduced affinity (Hansen, 1984).

Ouabain is thus able to block the translocation of alkali metal cations in and out of the cell at concentrations of about 10^{-6} M, and the rate of ouabain binding parallels inhibition of rate of pumping. If the pump is inhibited, intracellular sodium levels rise. Consequently, calcium levels also rise as a result of a transmembrane sodium–calcium exchange. The raised calcium level activates the heart to contract more forcefully (the positive inotropic effect). There have been suggestions, however, that this positive inotropic effect can be dissociated from pump inhibition, since after the cardiac glycoside has been washed out of the heart the enzyme was still significantly inhibited whereas the positive inotropic effect was no longer present. One explanation may be that another unknown mechanism operates in addition to pump blockade (reviewed in Godfraind, 1984). On the other hand, the degree of enzyme inhibition required to show an observable effect on the pumping of the heart may be very high.

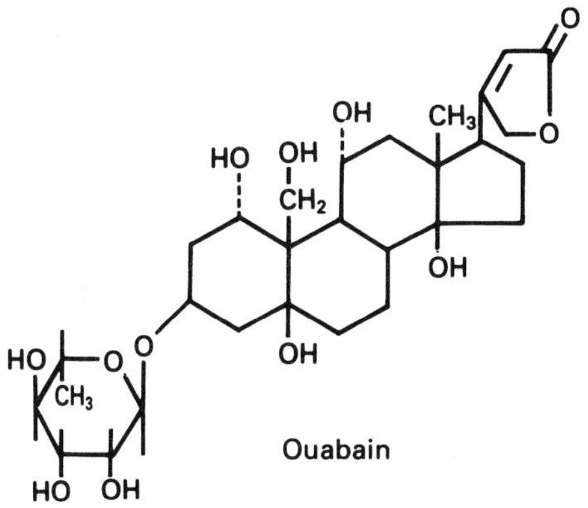

Ouabain

Other studies by Brown and Erdmann (1984), however, have cast doubt on this hypothesis. They obtained biphasic Scatchard plots and suggested that there were two sets of binding sites for ouabain on rat and guinea-pig cardiac membranes – a high affinity, low capacity site, the occupancy of which was associated with the positive inotropic effect, and a low affinity, high capacity site which was apparently the sodium pump. The high affinity site was not identified. With cat and human cardiac membranes the position was much less clear-cut with only a very slight curvature of the Scatchard plots. Bearing in mind that biphasic Scatchard plots can arise for reasons other than multiple sets of binding sites, the detailed mode of action of the cardiac glycosides is still a subject for debate.

The action of the cardiac glycosides is particularly valuable in congestive heart failure which is characterized by, among other things, an enlarged heart and a high venous pressure as a consequence of blood flow congestion. The increased contraction forces the blood out of the heart and it reduces in size. As a consequence, the sympathetic activity is reduced and the venous pressure falls back towards normal. In addition, the drugs slow the ventricular rate of beating by rendering this part of the heart less sensitive to electrical stimuli known as a refractory state. Digoxin may not be as effective as the angiotensin-converting enzyme inhibitors captopril and enalapril, however, nor as free from serious side-effects and the latter are preferred for mild to moderate congestive heart failure (Crozier and Ikram, 1992).

10.5.2 Potassium-hydrogen-ATPase - omeprazole for ulcers

Recently, an enzyme has been discovered which pumps protons out of the parietal cells that release acid into the stomach and is believed to be the terminal link in the chain of acid secretion. This proton pump is the target of a new type of anti-ulcer drug that is exemplified by omeprazole, a substituted benzimidazole.

The enzyme appears to be unique to the parietal cell of the gastric mucosa, and is the last link in the chain of the transport of hydrogen ions into the small channels (canaliculi) that eventually lead to the surface of the cell leading into the gut. These canaliculi form a highly acidic space in the cell and exhibit a convoluted membraneous structure with short microvilli projections. The pump is found in a particular type of vesicle which is isolated from the microsomal fraction of gastric parietal cells. This fraction is probably derived from the plasma membrane and vesicles which have a smooth surface. Normally, there are a large number of these vesicles (about 0·2 μm in diameter) in the cytoplasm of the resting cell. When acid secretion is stimulated by histamine, for example, these vesicles coalesce with the canaliculi and allow protons to be transported to the cell exterior (see Sachs, 1986 for review).

Potassium ion and ATP are required for the action of the proton pump, in fact the enzyme is a proton-stimulated ATPase which uses the energy of ATP hydrolysis to drive the exchange of protons for potassium across the vesicle membrane. ATP can thus generate a pH gradient across the canalicular membrane in parietal cells. The following scheme has been proposed for the overall reaction (Wallmark, 1989):

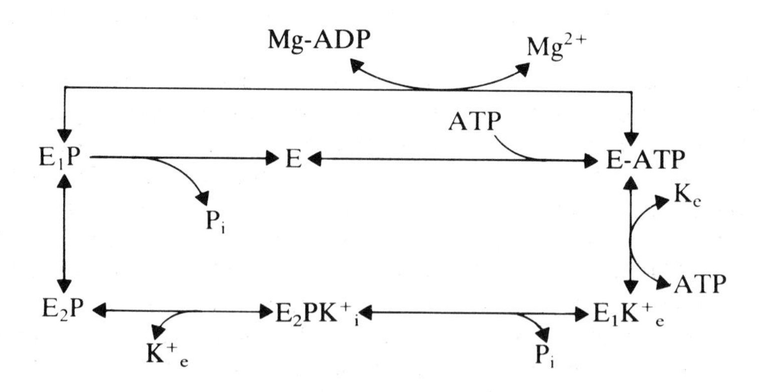

$K^+{}_i$, internal potassium; $K^+{}_e$, external potassium

Studies on the cloned enzyme show that it is a single 1033-amino-acid polypeptide with a molecular weight of 110 kDa, sharing 60 per cent homology with the Na^+,K^+-ATPase on the cardiac cell membrane, discussed in the previous section. This similarity extends to the catalytic cycle where a phosphorylated form is the obligatory intermediate in the reaction, indeed the phosphorylated forms of both enzymes are active whereas the dephosphorylated forms are not. Magnesium ion is also required for activity, as is common for enzymes where ATP is a co-substrate and it is likely that the Mg–ATP complex is the true substrate. The following scheme has been proposed for the overall reaction (Wallmark, 1989).

The mechanism of protein pumping is outlined by Sachs and Wallmark (1989). Mg–ATP binds to the cytosolic face of the enzyme and a hydrated hydrogen ion (H_3O^+) binds to the ion-binding region. Phosphate is transferred from the ATP to the carboxyl group of an aspartate residue to give the E_1 form. The enzyme changes conformation to E_2P to allow the hydronium ion to face the outer luminal side; the affinity for the hydronium ion is sharply reduced and it exchanges for the potassium ion. The phosphoenzyme breaks down rapidly to give the E_2K form which converts to the E_1K thus undergoing the reverse conformational change to present K^+ to the cytosolic face. Potassium dissociates, Mg–ATP binds and the whole process can recommence. Potassium and chloride ion are co-transported out through the cell membrane and thus recycling occurs. The pump maintains a pH gradient with the acid outside, cytosol neutral, and is electroneutral.

Omeprazole is taken up from the circulation into the acidic canaliculus of the parietal cell where it becomes protonated. Omeprazole is a pro-drug because it is stable at neutral pH, but breaks down in the presence of acid to form a tetracyclic sulphenamide (Fig. 10.5). The acidic lumen has a sufficiently low pH for this to happen. Sulphydryl groups on the luminal side of the enzyme make a nucleophilic attack on the sulphur atom, opening the ring to give a covalent disulphide bond. Complete inhibition occurs when two moles of inhibitor are bound per mole of active site. This reaction can be inhibited by pH neutralization and also by sulphydryl reagents such as mercaptoethanol (Wallmark, 1989).

Gastric mucosal ATPase is the final step in the induction of acid secretion by various secretagogues; histamine is the most potent and acts through adenylate cyclase and the formation of cyclic AMP; acetylcholine stimulates the uptake of calcium ion into the cytoplasm and has no effect on cyclic nucleotide levels, while the peptide gastrin, which is the physiological mediator of gastric acid secretion in response to feeding, appears to release calcium from intracellular stores. In addition, both acetylcholine and gastrin induce the release of histamine from neighbouring paracrine cells (Sachs and Wallmark, 1989).

Clearly, it is possible to inhibit the secretion of acid by (a) blocking the histamine receptor, as demonstrated by the anti-ulcer drugs cimetidine and ranitidine and (b) the acetylcholine (muscarinic) receptor, e.g. the drug pirenzepine. Furthermore, the histamine H_2, receptor blockers inhibit gastrin and acetylcholine induced secretion because these secretagogues also release histamine (section 9.8.2). The proton pump, however, is the last link in the

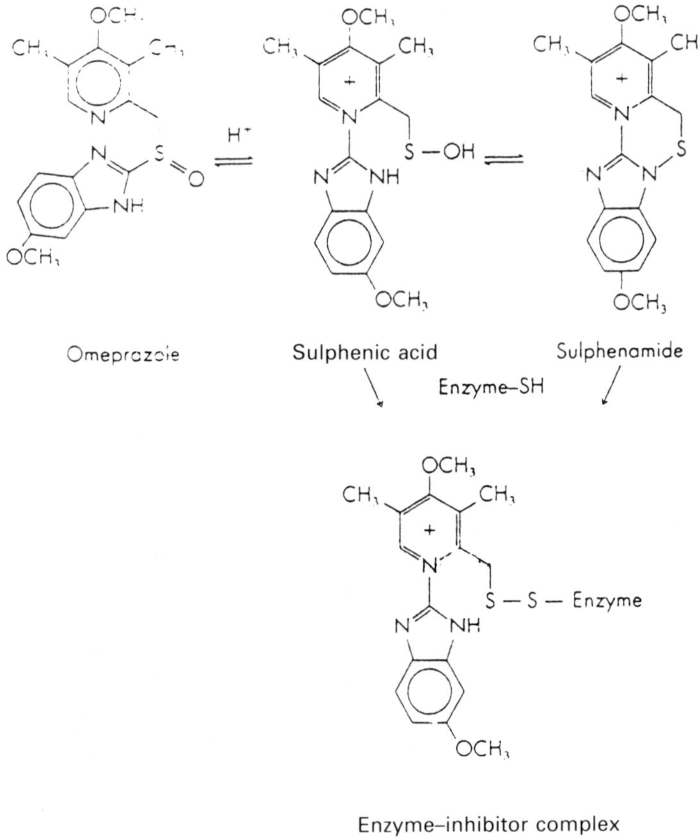

Figure 10.5 Chemical reactions of omeprazole that lead to inhibition of H^+,K^+-ATPase.

chain of proton secretion and blockers of this may be expected to be at least as effective as blockers of either of the two receptors described above. In addition, the expected blockade of both histamine, dibutyryl cyclic AMP and pentagastrin induced secretion in human gastric glands *in vitro* and *in vivo* was found by Olbe *et al.* (1986). Pentagastrin is a synthetic pentapeptide containing the C-terminal tetrapeptide amide moiety of gastrin and is often used in tests to measure inhibition of acid secretion as it has appreciable secretory activity.

In clinical trials, omeprazole has been shown to heal both gastric and duodenal ulcers more rapidly and completely than cimetidine or ranitidine; in some cases which were resistant to prolonged treatment with H_2 receptor blockers, omeprazole was effective. As we might expect from its mode of action producing a type of irreversible inhibition, less drug and less frequent administration is required for omeprazole. Other gastric-acid-linked disorders such as Zollinger–Ellison syndrome and reflux oesophagitis are also responsive to omeprazole but not to H_2 receptor blockers (McTavish *et al.*, 1991). Omeprazole is another drug that has made a pharmacological breakthrough and has also taught us much about the workings of a previously little-known enzyme.

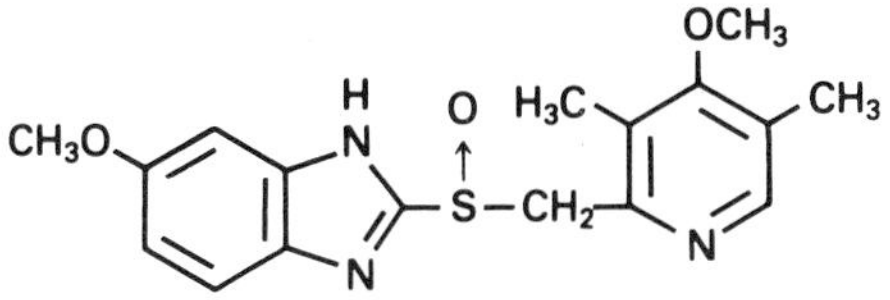

Omeprazole

10.6 Cromoglycate - calcium antagonist or membrane stabilizer?

Asthma is characterized by episodes of difficulty in breathing, varying from the mild to extremely severe due to widespread narrowing of the airways in the lung which can be due to a number of factors: the lining of the airway may swell, there may be mucus in the airway, a spasm in the muscle in the walls of the airway (Kauffman, 1984). In man, there is an initial acute phase of the condition followed by two late phases, ranging from 17 to 72 hours later.

Many cells are involved in the pathogenesis of asthma. It seems likely that the response of the mast cell to an allergen or other stimulus, such as exercise or cold air, controls the acute phase by releasing chemical mediators such as histamine and leukotrienes. The late phases may result from activation and subsequent invasion of the lining of the airways by macrophages, eosinophils and T lymphocytes. The airways are abnormally sensitive to stimuli which may be a result of the presence of a large number of eosinophils, a type of white blood cell that contains granules that readily stain with the red fluorescent dye eosin (Barnes, 1992).

As a result, asthma is now regarded as a special type of inflammatory disease, such that the anti-inflammatory steroids are now given much earlier in the treatment instead of being retained for the really serious cases (Barnes, 1992). Cromoglycate is still widely used 25 years after it was launched, although its mode of action is, even now, not understood (Kuzemko, 1989).

10.6.1 Cromoglycate action

The story of the research project that eventually led to cromoglycate makes very interesting reading, not only with regard to the science, but also the politics of the pharmaceutical industry (Kauffman, 1984). Khellin, a chromone (benzopyrone) obtained from the plant *Ammi visnagi*, was the model for synthesis of chromone 2-carboxylic acids because it possessed smooth muscle relaxant properties. This was believed to be an advantage in relieving the bronchoconstriction of asthma. A doctor, Roger Altounyan, who worked for the company and who was himself a chronic asthmatic, tested the compounds on himself by breathing in a soup of guinea-pig hair, producing an attack of asthma by antigenic challenge.

Eventually, by chance, an agent was found that did not possess bronchodilator activity but still protected against antigen challenge. The following synthetic efforts resulted in the classic situation where the first batch of a compound showed great activity but succeeding batches did not. A highly active impurity was suspected and, on this occasion, the agent was isolated and found to be a bis (or doubled-up) chromone carboxylic acid, a breakthrough which led directly to cromoglycate. During the course of this work the project was officially closed down but work still carried on clandestinely!

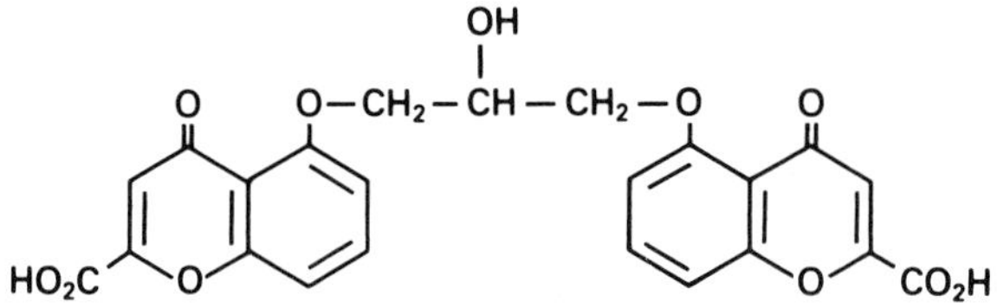

Cromoglycate

Cromoglycate is very hydrophilic, being a dicarboxylic acid with pK_as both in the region of 2. It is therefore a dianion at neutral pH and would not be expected to pass through the cell membrane easily if at all. Indeed, the drug may not need to enter the mast cell to inhibit degranulation since it is active when linked covalently to polyacrylamide beads (Mazurek *et al.*, 1980). Cromoglycate is only active when given by inhaled aerosol, and the rate of passage across the lung membranes governs its rate of disappearance from the circulation.

Mast cells obtained from different tissues vary greatly in sensitivity to cromoglycate inhibition of the anti-allergic reaction for which the IC_{50} for liberation of histamine can vary from the high concentrations of 10^{-4}–10^{-3} M for the human lung mast cell (Holgate, 1989) to $5{\cdot}5 \times 10^{-7}$ M for the mast cell from rat peritoneum (Moqbel *et al.*, 1988). It is difficult to understand how such a high figure for the 'correct' tissue in humans can relate to the disease process but, nevertheless, the rise in circulating levels of histamine in an asthma attack are inhibited (Howarth *et al.*, 1985).

Another mechanism must be at work, however, because the drug blocks not only allergen-induced, 'extrinsic', asthma but also asthma in non-allergic models, i.e. induced by exercise or cold air ('intrinsic'). Furthermore, cromoglycate does not relax bronchial or smooth muscle *in vitro*, or acutely *in vivo*, whereas after long-term administration, bronchoconstriction induced by stimulants such as histamine or exercise is greatly reduced (Holgate, 1989).

The late reaction may be governed by the recruitment of neutrophils, eosinophils and possibly macrophages which can exert an inflammatory effect. Here cromoglycate can be shown to be very active, i.e. at levels below 10^{-7} M, in inhibiting the stimulation of eosinophils by formyl-methionyl-leucyl-alanine, a bacterial peptide which activates macrophages to kill bacteria (Kay *et al.*, 1987).

It is interesting to speculate that cromoglycate may owe its action in some way to potentiation of the nitric oxide pathway. Nitric oxide has been implicated as one possible neurotransmitter in the nerve system in the lung which inhibits allergic bronchoconstriction and reduces arterial plasma histamine (Lammers *et al.*, 1992). Macrophages are activated by cytokines to produce nitric oxide, with the help of calmodulin, which stimulates cyclic GMP production (Moncada *et al.*, 1991), and the most efficient inhibitors of the allergic reaction are compounds which potentiate cyclic GMP by inhibiting its hydrolysis rather than cyclic AMP hydrolysis (Coulson *et al.*, 1977).

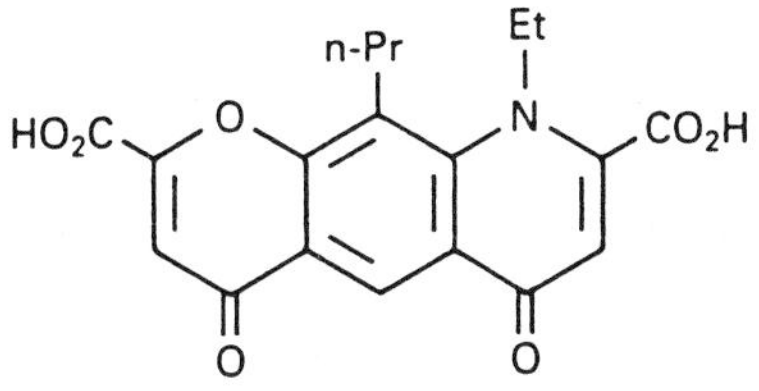

Nedocromil

Nedocromil, an analogue of cromoglycate, is the only agent to reach the market subsequently, despite the very great effort expended in recent years trying to find new anti-allergic agents based on the *in vitro* mast cell degranulation test. Many new compounds were discovered that were more active than cromoglycate *in vitro* but show little or no advantage *in vivo*; indeed many were inactive. Although in some cases this has been attributed, amongst other things, to serum binding and different pharmacokinetic profiles, the lack of correlation in other cases is likely to be due to an inappropriate *in vitro* test. As Roger Altounyan observed, there is no substitute for testing in man (Kauffman, 1984).

10.7 Cyclosporin

There have been a number of examples in the field of drug discovery where a condition has been treated by a drug without its mechanism of action being known. Cyclosporin is one such drug. Although its mechanism of action has not been worked out, the involvement of the calcium-binding phosphatase calcineurin, that can regulate calcium channels in the brain, suggested cyclosporin's inclusion in this chapter. Cyclosporin has played a major part in the recent increase in transplant surgery for organ and bone marrow by suppressing the immune response at the level of the T lymphocyte – sufficiently well to allow a great improvement in the survival rate for kidney transplants (Schreiber, 1991; Walsh *et al.*, 1992).

Cyclosporin is an 11-membered cyclic peptide with the methyl amide between residues 9 and 10 in the *cis* position; all the other methyl amides are *trans*. It was originally discovered as a metabolite of the fungus *Tolypocladium inflatun Gams* and had weak anti-fungal activity. Cyclosporin interferes with

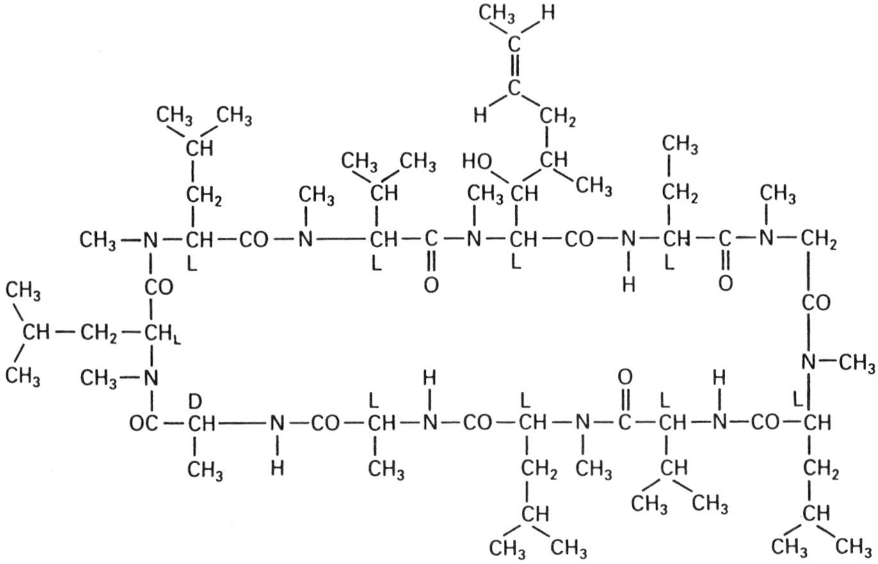

```
                                                        CH3  H
                                                           \/
                                                           C
                                                           ||
                                                           C
                                                          / \
           CH3  CH3                                      H   CH2
             \ /                                             |           CH3
             CH                    CH3  CH3    HO           CH           |
             |                       \ /         \         /  \          |
             CH2          CH3         CH          CH3 CH       CH3       CH2         CH3
             |            |           |             |   |                |           |
   CH3—N—CH—CO—N————————CH—C—N—CH—CO—N—CH—C—N—CH2
       |  L                          L   ||       L         |   L   ||        |
CH3    CO                                O                  H       O         |
 \     |                                                                      CO
  CH—CH2—CHL                                                                  |
 /      |                                                                     N—CH3
CH3  CH3—N                           H                   O        H           |
         |  D                   L    |         L         ||   L   |          L|
         OC—CH————N—CO—CH—N—CO—CH—N—C—CH—N—CO—CH
              |        |       |                 |    |        |              |
              CH3      H       CH3               CH2  CH3      CH             CH2
                                                 |            /  \            |
                                                 CH         CH3  CH3          CH
                                                /  \                         /  \
                                              CH3  CH3                     CH3  CH3
```

Cyclosporin

the activation of T lymphocytes. T cells need activation by an antigen or cytokines before they can destroy virally infected cells and stimulate macrophages to kill bacteria and aberrant cells. Activation is a complex series of changes including the synthesis of interleukins, principally interleukin 2, and other cytokines that induce proliferation and differentiation. Cyclosporin blocks interleukin 2 gene expression in a T cell after antigen presentation (Kronke *et al.*, 1984; reviewed in Walsh *et al.*, 1992).

Cyclosporin binds to a cytoplasmic receptor known as a cyclophilin, found in many tissues and eukaryotic cells, catalysing the folding of proteins. Cyclophilin has the enzyme activity of peptidyl-proline *cis-trans* isomerase. Cyclophilins may help newly formed proteins to fold, particularly as proline cannot participate in an α-helix and is often found at a turning point in protein structure. Cyclosporin initially binds in the *cis* form found in solution ($K_D = 4 \times 10^{-6}$ M) but isomerizes in the complex into the *trans* form ($K_D = 4 \times 10^{-8}$ M) thus mimicking the normal substrates of cyclophilin. This activity, however, is not directly responsible for the immunosuppressive action because (a) most of the cyclophilin in the cell is not bound by cyclosporin when T cell activation is inhibited, and (b) analogues of cyclosporin have been synthesized which have cyclophilin binding activity but no immunosuppressive action (Liu *et al.*, 1992).

The active agent appears to be the drug–protein complex. The intracellular target is even less obvious, being the calcium-activated protein, calcineurin (Clipstone and Crabtree, 1992). Calcineurin is a phosphatase that removes phosphate groups from serine and threonine. It is a heterodimer with subunits 61 and 19 kDa; the smaller subunit binds calcium with similar calcium-binding sites to calmodulin. Calcineurin was originally discovered in mammalian brain, hence the name, where it regulates voltage-activated calcium channels by dephosphorylating a membrane-associated protein and inactivating the channel (Armstrong, 1989). It is not yet clear how calcineurin interferes with T cell activation, but cyclosporin has already greatly increased our molecular understanding of the immune process.

10.8 Potassium channel opening

The discovery of new information by novel drugs has shown up again in the field of potassium channels. Nicorandil and cromokalin (or its optically active isomer, lemakalin) open potassium channels and induce smooth muscle relaxation. This has led to potential use in hypertension, asthma and incontinence. Relaxation of arterial smooth muscle leads to anti-hypertensive effect and beneficial reduction in plasma lipids and effects on coronary heart disease. The activity resides in the (−)-isomers of these drugs, disparate though they are in structure (Longman and Hamilton, 1992).

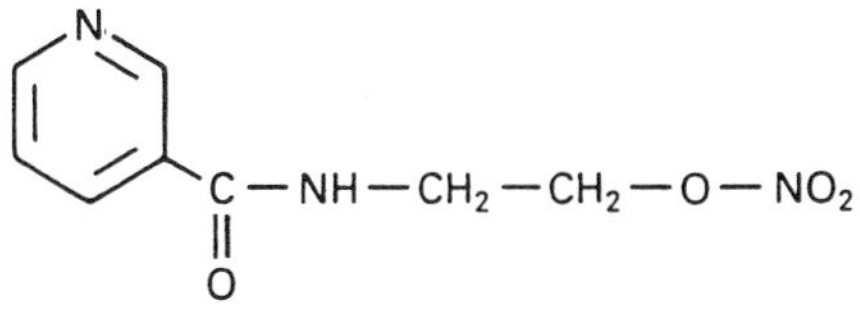

Nicorandil

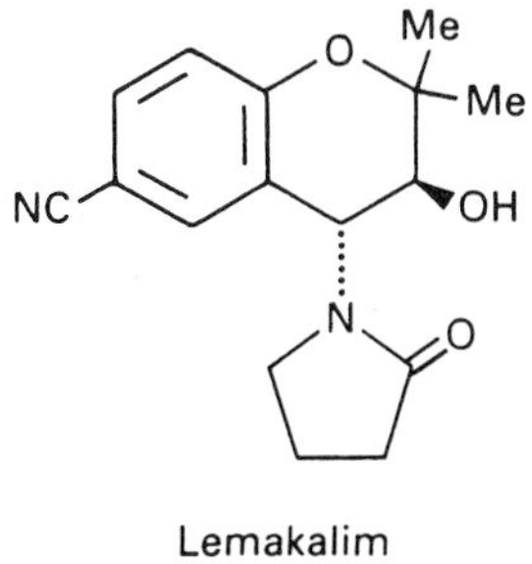

Lemakalim

Although there is little mechanistic information available at the molecular level, the inhibition of potassium efflux from the cell results in hyperpolarization. Calcium does not enter, is not released from intracellular stores, and the tissue relaxes. The hyperpolarization can reach as high as −90 mV which is the equilibrium level of potassium. Once this level is reached no further loss of potassium can occur.

ATP regulates the calcium channel; at higher levels of ATP the channel is more difficult to open. As the ATP level falls, the channel opens more easily and consequently the drugs become more effective. Glibenclamide, the anti-diabetic drug (Chapter 10) which is known to block ATP-sensitive K^+ channels, antagonizes these effects.

Nicorandil is unusual by having two mechanisms of action. As well as potassium channel activation, nicorandil is a nitrate and induces relaxation by stimulating cyclic GMP formation (Kukovetz *et al.*, 1992).

10.9 Sterol ligands - polyene antibiotics

Fungal infections have become progressively a more serious problem in recent years. This has been attributed partly to our mastery of bacterial infections, leaving a vacuum that pathogenic fungi take the opportunity to fill, and partly to the fact that a large number of hospital patients - particularly those with cancer and transplant patients - have a partially suppressed immune system as a consequence of drug treatment. Thus there has been a great increase in the number of systemic infections by fungi, particularly opportunist infections such as thrush, caused by *Candida albicans*, and they now present a serious medical problem. It has been known for a long time, however, that *Candida* infections can be lethal. Gooday (1977) notes that a paper of 1665 records at the height of the plague: 'The diseases and casualties this week, London, Sept. 12 to the 19; Consumption, 129 . . . Fever, 332 . . . Plague, 6544 . . . Thrush, 6'.

For nearly 30 years, systemic fungal infections have been treated with amphotericin B as the drug of choice (indeed for much of that period the only drug) because it is very active against a wide range of pathogenic organisms, is fungicidal not fungistatic and does not induce resistance (reviewed in Medoff *et al.*, 1983). The drug's major drawback, and it is a serious one, is kidney toxicity; in about 80 per cent of the patients treated, the glomerular filtration rate falls to about 40 per cent of normal. The effect is usually reversible when treatment is withdrawn, but in some cases there is some residual impairment of kidney function. Amphotericin B, however, was for many years the only polyene antibiotic sufficiently safe to be used systemically in man. Nystatin, a close analogue, may be given orally for *Candida* infections of the mouth or intestinal tract, but since it is almost completely non-absorbed it is not used for systemic infections.

Polyene antibiotics are natural products produced by the Actinomycetes (notably the Streptomycetes) and are characterized by a macrolide ring closed by an internal ester or lactone. As their name implies, polyenes have a system of conjugated double bonds (normally between 4 and 7) in the *trans* configuration, which gives rise to a characteristic ultraviolet absorption spectrum with a number of peaks in the visible and near ultraviolet. Amphotericin B is a heptene (seven double bonds) and, additionally, contains an amino sugar (mycosamine) attached to the ring. The ring size can range from 12 to 37 carbon atoms and is substituted by a number of hydroxyl groups. The combination of several polar hydroxyl groups and a carboxyl, on the one hand, and several hydrophobic double bonds on the other, gives a dual character to the molecule, particularly as these groups reside on opposite sides of the ring (Norman *et al.*, 1976).

10.9.1 Amphotericin action

The specific target of the polyenes is the sterol in the plasma membrane of the eukaryotic cell; ergosterol in fungi and cholesterol in protozoal and mammalian cells. The evidence for this may be summarized as follows:

1. Bacteria do not contain sterol and are resistant to amphotericin. Protozoa, fungi and mammalian cells are all sensitive. The mycoplasma, *Acholeplasma laidlawii*, can grow whether sterol is present in the culture medium or not and incorporates it into the membrane if it is present. The microorganism is sensitive to the polyene only when sterol is present in the membrane.
2. Free sterols in the culture medium antagonize the effect of the drug on microorganisms, such as *Saccharomyces cerevisiae* and *Candida albicans*.
3. A complex between amphotericin and sterols which can be demonstrated by various techniques, such as ultraviolet difference spectroscopy, e.s.r. and circular dichroism, is sufficiently tightly bound to explain the antimicrobial action.
4. Fungal cells die as a result of damage to the membrane where the drug-sterol complexes are found. The membrane begins to leak small ions, notably ammonium and potassium.

The binding of polyenes to sterol can be demonstrated in model membrane systems such as vesicles composed of sterol and phosphatidylcholine. The binding constant for the amphotericin B-vesicle complex is $9{\cdot}7 \times 10^{-7}$ M when cholesterol is used and $1{\cdot}52 \times 10^{-6}$ M with ergosterol. Although this suggests that the drug might interfere more with animal cell membranes, the number of binding sites is important - ergosterol containing vesicles have a larger number of binding sites for the drug than those made with cholesterol, giving a tenfold factor in favour of the former in terms of product of binding sites and binding constant (Witzke and Bittman, 1984). Furthermore, circular dichroism studies suggest that there is more than one conformer of the drug present in membrane vesicles at higher drug concentrations but one conformer predominates at lower concentrations whichever sterol is present (Vertut-Croquin *et al.*, 1983).

The fact that cholesterol binds polyenes tightly (indeed amphotericin is very unusual in binding more effectively to ergosterol) probably explains why these compounds are toxic to mammals. There is no direct correlation between anti-*Candida* activity and binding affinity in model vesicles for a small group of four polyenes including amphotericin B, but the group may be too small to expect correlations.

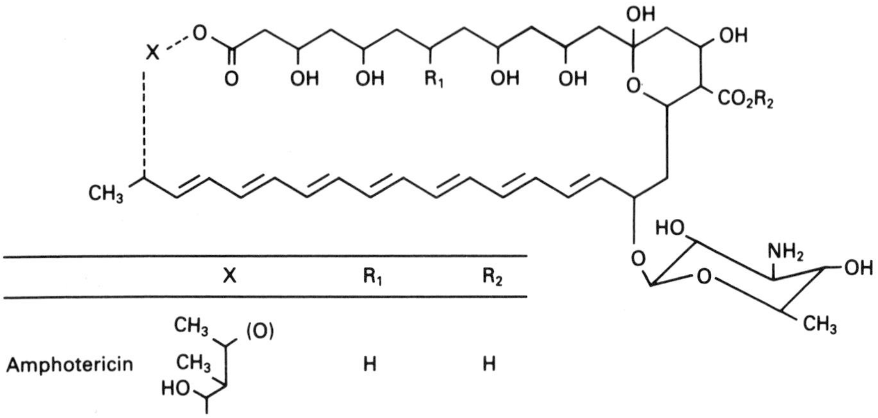

	X	R_1	R_2
Amphotericin	CH_3 (O) CH_3 HO	H	H

Alternatively, the nature of the phospholipid in the vesicles may be important as phosphatidylcholine faces the extracellular side of the plasma membrane in that its polar head groups are pointing in that direction while phosphatidylethanolamine and phosphatidylinositol face the cytoplasm, and so a different composition in the vesicles might give a better correlation. It is also possible that the leakiness induced by binding may not be directly related to the extent of binding, but rather to the particular complexes made by the polyenes and how they disrupt the membrane architecture (Witzke and Bittman, 1984).

A number of features on the sterol nucleus are essential for the interaction to take place, namely a 3β-hydroxyl group and a Δ-22 double bond together with a cholestane skeleton. An intact polyene is also required - if the ring is broken no interaction will take place (Norman *et al.*, 1976).

Membrane fluidity plays a key role in the action of the polyenes. They can induce permeability in the 'solid' or crystalline form of the membrane in the absence of cholesterol and above the transition temperature when the membrane is in the liquid form, amphotericin can still have a small effect, but the addition of cholesterol markedly increases it (reviewed in Medoff *et al.*, 1983).

The precise molecular nature of the interaction of polyenes with membrane components, and how this induces permeability, is still obscure even for model membranes. Electron microscope pictures of erythrocyte membranes treated with amphotericin show a series of pits and protuberances but no obvious pores, and so it is difficult to accept the original view that the polyene–sterol complex forms aqueous channels or pores in the membrane through which the ions pass. An alternative hypothesis is that the complex may alter the local membrane fluidity by removing the sterol from its interaction with membrane phospholipids, and thereby induce leakage of cell constituents. Particularly sensitive points may occur where there is a change in the membrane from solid to liquid phase (reviewed in Medoff *et al.*, 1983). At present the situation is still unresolved.

Amphotericin at low concentrations *in vivo* can stimulate cell growth - indeed macrophages are stimulated to kill microorganisms, and the drug can markedly increase the number of antibody-forming cells in the mouse spleen and lymph node. Not surprisingly, therefore, the drug has a stimulatory effect on the immune system *in vivo*, but the clinical relevance of this finding is as yet unknown. In view of the immune suppression of many of the patients with fungal infections this property could be very useful.

In conclusion, therefore, amphotericin B is still the drug of choice for life-threatening fungal infections by primary pathogens such as *Histoplasma capsulatum* and *Paracoccidioides brasiliensis* as well as opportunist pathogens such as *Candida albicans*. The problem of the toxicity of amphotericin may be countered by the concomitant use of flucytosine which can allow the dose of amphotericin to be reduced (Lyman and Walsh, 1992). Therapy with amphotericin, however, is not always straightforward because, in addition to toxicity, there can be a lack of clinical response even when the organism is susceptible to the drug *in vitro*. In some cases this is because the drug is unable to penetrate in sufficient quantity the particular part of the body which harbours the pathogen. This is the case in infections of the central nervous system with *Coccidioides* species which cause meningitis (Medoff *et al.*, 1983).

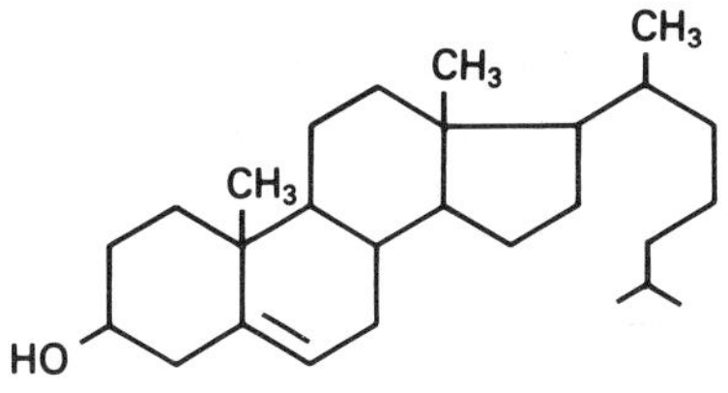

Cholesterol

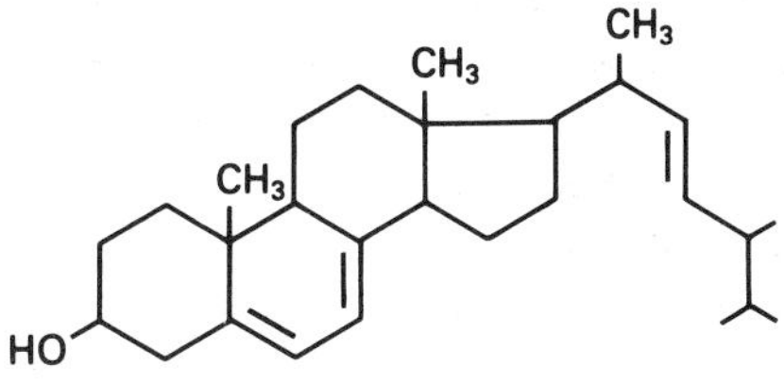

Ergosterol

Questions

1. Name two ions that pass through channels in the mammalian cell membrane without requiring the energy of ATP. Name two drugs that interfere with the passage of these ions. What are the drugs used for?
2. Why is the energy of ATP required to drive ion pumps? Name two classes of drug that block ion pumps. How do they do this?
3. Explain these terms (a) depolarization (b) action potential (c) arrhythmia.
4. How do local anaesthetics interfere with the sodium channel?
5. How do the sodium and calcium channels differ in structure?
6. What classes of calcium channel blockers are available? How does their mechanism of action differ?

7. For what cellular processes is calcium required?
8. Name one major difference between the fungal and mammalian cell membranes. How is this exploited for the treatment of fungal infection?
9. How might amphotericin be packaged to make it less toxic?
10. On what test does cromoglycate show the greatest activity? How does this relate to the pathogenesis of asthma?

References

Armstrong, D.L., 1989, *Trends in Neurosciences,* **12**, 117-21.
Barnes, P.J., 1992, *J. Int. Med.,* **231**, 453-61.
Benos, D.J., Cunningham, S., Baker, R.R., Beason, K.B., Oh, Y. and Smith, P.R., 1992, *Rev. Physiol. Biochem. Pharmacol.,* **120**, 31-113.
Bigger, J.T. and Hoffman, B.F., 1990, Ch. 35 in the *Pharmacological Basis of Therapeutics,* 8th edn (eds A.G. Gilman, T.W. Rall, A.S. Nies and P. Taylor). New York: Pergamon Press.
Brown, L. and Erdmann, E., 1984, *Basic Res. Cardiol.,* (Suppl.), 50-5.
Burg, M. and Green, N., 1973, *Kidney Int.,* **4**, 301-308.
Caffruny, E.J. and Itskovitz, H.D., 1982, *J. Pharmacol. Exp. Ther.,* **223**, 105-109.
CAST investigators, 1989, *N. Engl. J. Med.,* **321**, 406-12.
Catterall, W.A., 1980, *Annu. Rev. Pharmacol. Toxicol.,* **20**, 14-43.
Catterall, W.A., 1988, *Science,* **242**, 50-61.
Clipstone, N.A. and Crabtree, G.R., 1992, *Nature,* **357**, 695-7.
Coulson, C.J., Ford, R.E., Marshall, S., Walker, J.L., Wooldridge, K.R.H., Bowden, K. and Coombs, T.J., 1977, *Nature,* **265**, 545-7.
Crozier, I. and Ikram, H., 1992, *Drugs,* **43**, 637-50.
Cuthbert, A.W., 1977, *Progr. Med. Chem.,* **14**, 1-50.
Droogmans, G., Himpens, B. and Casteels, R., 1985, *Experientia,* **41**, 895-900.
Fleckenstein-Grun, G., Frey, M. and Fleckenstein, A., 1984, *Trends Pharmacol. Sci.,* **5**, 283-6.
Frizzell, R.A., Field, M. and Schultz, S.G., 1979, *Am. J. Physiol.,* **236**, F1-8.
Glossman, H., Ferry, B.R., Lubbecke, F., Mewes, R. and Hoffman, F., 1982, *Trends Pharmacol. Sci.,* **3**, 431-7.
Gooday, G.W., 1977, *J. Gen. Microbiol.,* **99**, 1-11.
Godfraind, T., 1984, *Eur. Heart J.,* **5**, Suppl. F, 303-308.
Grant, A.O., 1992, *Amer. Heart J.,* **123**, 1130-6.
Grant, A.O. and Wendt, D.J., 1992, *Trends in Pharmacol. Sci.,* **13**, 352-8.
Hansen, O., 1984, *Pharmacol. Rev.,* **36**, 143-63.
Hille, B., 1977, *J. Gen. Physiol.,* **69**, 497-515.
Holgate, S.T., 1989, *Resp. Med.,* **83** (Suppl), 25-31.
Hondeghem, L.M. and Katzung, B.G., 1984, *Annu. Rev. Pharmacol. Toxicol.,* **24**, 387-423.
Howarth, P.H., Durham, S.R., Lee, T.H., Kay, A.B., Church, M.K. and Holgate, S.T., 1985, *Am. Rev. Respir. Dis.,* **132**, 986.
Jorgensen, P.L., 1982, *Biochem. Biophys. Acta,* **694**, 27-68.
Katz, A.M., 1992, *Amer. J. Cardiol.,* **69**, 17E-22E.
Kauffman, G.B., 1984, *Chem. Educ.,* **21**, 42-5.
Kay, A.B., Walsh, G.M., Moqbel, R., McDonald, A.J., Nagakura, A., Carroll, M.P. and Richerson, H.B., 1987, *J. Allergy Clin. Immunol.,* **80**, 1-8.
Kendall, M.J. and Horton, R.C., 1985, *J. Royal Coll. Phys. Lond.,* **19**, 85-9.
Kokko, J.P., 1984, *Am. J. Med.,* **77**, 5A, 11-17.

Kronke, M., Leonard, W.J., Depper, J.M., Arya, S.K., Wong-Staal, F., Gallo, R.C., Waldmann, T.A. and Greene, W.C., 1984, *Proc. Nat. Acad. Sci. (U.S.A.),* **81**, 5214-18.
Kukovetz, W.R., Holzmann, S. and Poch, G., 1992, *J. Cardiovasc. Pharmacol.,* **20**, Suppl 3, S3-S7.
Kumana, C. and Hamer, J., 1979, In: *Drugs for Heart Disease*, Ed. J. Hamer (London: Chapman and Hall), p. 61.
Kuzemko, J.A., 1989, *Respir. Med.,* **83**, (Suppl), 11-16.
Kwon, Y.W. and Triggle, D.J., 1991, *Chirality,* **3**, 393-404.
Lammers, J.-W.J., Barnes, P.J. and Chung, K.F., 1992, *Eur. J. Respir.,* **5**, 239-46.
Lant, A., 1985a, *Drugs,* **29**, 57-87.
Lant, A., 1985b, *Drugs,* **29**, 162-88.
Liu, J., Albers, M.W., Wandless, T.J., Luan, S., Alberg, D.G., Belshaw, P.J., Cohen, P., Mackintosh, C., Klee, C.B. and Schreiber, S.L., 1992, *Biochemistry,* **31**, 3896-901.
Longman, S. and Hamilton, T.C., 1992, *Med. Res. Rev.,* **12**, 73-148.
Ludens, J.W., 1982, *J. Pharmacol. Exp. Ther.,* **223**, 25-9.
Lyman, C.A. and Walsh, T.J., 1992, *Drugs,* **44**, 9-35.
Lytton, J., 1985, *Biochem. Biophys. Res. Commun.,* **132**, 764-9.
Mazurek, N., Berger, G. and Pecht, I., 1980, *Nature,* **286**, 722-3.
McTavish, D., Buckley, M.M. and Heel, R.C., 1991, *Drugs,* **42**, 138-70.
Medoff, G., Brajtburg, J., Kobayashi, G.S. and Bolard, J., 1983, *Annu. Rev. Pharmacol. Toxicol.,* **23**, 303-30.
Moncada, S., Palmer, R.M.J. and Higgs, E.A., 1991, *Pharmacol. Revs.,* **43**, 109-42.
Moqbel, R., Cromwell, O., Walsh, G.M., Wardlaw, A.J., Kurlak, A.K. and Kay, A.B., 1988, *Allergy,* **43**, 268-76.
Muhiddin, K.A. and Turner, P., 1985, *Postgrad. Med. J.,* **61**, 665-78.
Norman, A.W., Spielvogel, A.M. and Wong, R.G., 1976, *Adv. Lipid. Res.,* **14**, 127-70.
Odlind, B., 1984, *Acta Pharmacol. Toxicol.,* **54**, Suppl. 1, 5-15.
Olbe, L., Lind, T., Cederberg, C. and Ekenved, G., 1986, *Scand. J. Gastroenterol.,* **21**, S118, 105-7.
Patarca, R., Candia, O.A. and Reinach, P.S., 1983, *Am. J. Physiol.,* **245**, F660-9.
Pedersen, P., 1982, *Ann. N.Y. Acad. Sci.,* **401**, 1-20.
Postma, S.W. and Catterall, W.A., 1984, *Molec. Pharmacol.,* **25**, 219-27.
Ritchie, J.M. and Green, N.M., 1985, In: *The Pharmacological Basis of Therapeutics*, 7th Ed, eds A.G. Gilman, L.S. Goodman, T.W. Rall and F. Murad (New York: Macmillan) Ch. 15.
Sachs, G., 1986, *Scand. J. Gastroenterol,* **21**, 1-10.
Sachs, G. and Wallmark, B., 1989, *Scand. J. Gastroenterol.,* **24**, (Suppl 166), 3-11.
Scarborough, G.A., 1985, *Microbiol. Rev.,* **49**, 214-31.
Schramm, M. and Towart, R., 1985, *Life Sci.,* **37**, 1843-60.
Schreiber, S.L., 1991, *Science,* **251**, 283-7.
Serrano, R., Kielland-Brandt, M.C. and Fink, G.R., 1986, *Nature,* **319**, 689-93.
Singer, S.J. and Nicolson, G.I., 1972, *Science,* **175**, 720-31.
Spedding, M., 1985, *Trends Pharmacol. Sci.,* **6**, 109-14.
Spedding, M. and Pauletti, R., 1992, *Pharmacol. Rev.,* **44**, 363-76.
Striessnig, J., Glossmann, H. and Catterall, W.A., 1991, *Proc. Natl. Acad. Sci. (U.S.A.),* **87**, 9108-102.
Sweadner, K.J., 1979, *J. Biol. Chem.,* **254**, 6060-67.
Vertut-Croquin, A., Bolard, J., Chabbert, M. and Gary-Bobo, C., 1983, *J. Biol. Chem.,* **22**, 2939-44.
Wallmark, B., 1989, *Scand. J. Gastroenterol.,* **24**, (Suppl 166), 12-18.
Walsh, C.T., Zydowsky, L.D. and McKeon, F.D., 1992, *J. Biol. Chem.,* **267**, 13115-18.
Wit, A.L., 1985, *Clin. Physiol. Biochem.,* **3**, 127-34.
Witzke, N.M. and Bittman, R., 1984, *Biochemistry,* **23**, 1668-74.

Chapter 11

Microtubule assembly

11.1 Introduction

Microtubules are cytoplasmic organelles which play a crucial role in the process of cell division (mitosis) in eukaryotic cells as part of the mitotic spindle. Furthermore, various other activities including intracellular transport and secretion are mediated by these organelles. Indeed, the secretion of cytoplasmic granules of hormonal mediators of the asthmatic process (see section 10.6) also requires the involvement of microtubules (Kaliner, 1977). Interestingly, the cilia of the protozoon *Tetrahymena pyriformis*, that the organism uses for swimming, are mainly composed of microtubules.

During the stage of mitosis known as prophase the chromosome doublets appear, linked by a body known as a centromere. The two centrioles, normally located near the nucleus, migrate to opposite ends of the cell. Around each centriole a system of radiating fibres, known as the aster, develops. In the next phase, metaphase, the centrioles organize microtubules into the mitotic spindle which links the centrioles. The centromeres become attached to a spindle fibre and migrate to the median position between the poles. The loose ends of the chromosomes are orientated in a random fashion but the centromeres of each chromosome lie exactly in a plane called the equatorial plate. In anaphase which follows, the centromeres duplicate themselves. The two resulting centromeres move apart towards the poles, still attached to the spindle fibre, with their attached chromosome 'dragging behind'. Subsequently, in telophase a nuclear membrane begins to encircle the uncoiling chromosomes, while a furrow appears at the equator, eventually deepening and splitting the daughter cells apart.

The spindle is composed of a number of microtubules which have the general structure of long hollow cylinders with inside and outside diameters of 15 and 25 nm, respectively. Microtubules are composed of the protein tubulin which itself is constructed of two closely related subunits (40 per cent sequence homology) with molecular weights in the region of 50 kDa. The functional form of tubulin is an α/β heterodimer of sedimentation coefficient ($s_{20,w}$) of 5·8S (reviewed in Correia, 1991).

Tubulin from brain, the most prolific source, can be assembled *in vitro* into microtubules by raising the temperature from 4 to 37°C. The polymerization may be measured by an increase in turbidity. Usual requirements are: (a) the presence of a family of associated proteins, known not surprisingly as microtubule-associated proteins (or MAPs) which assist the process in a way that is not understood, (b) GTP which is hydrolysed to GDP and (c) magnesium ion. Tubulin will assemble in the absence of MAPs but usually requires a high magnesium concentration or the presence of a polycation. Calcium, on the other hand, causes the microtubules to depolymerize, probably by binding to the wall of the microtubule and inducing an increase in the rate of subunit dissociation from the ends of the microtubules (Weisenberg and Deery, 1981). Reduction of the temperature to 10°C also causes microtubules to disassemble, and repeated cycles of assembly and disassembly are usually used to purify tubulin.

Tubulin contains two GTP binding sites per dimer. One of these sites, the E site, will rapidly exchange GTP for GDP in the presence of magnesium ion, undergoing a conformational change in the process. GTP and GDP stabilize different conformations of tubulin. GTP bound at the E site is hydrolysed when microtubules are formed and is linked to the addition of a tubulin dimer to the growing microtubule. The other site (N site) is not involved in catalysis and is always occupied by non-exchangeable GTP.

In the absence of MAPs, the process of assembly requires a sufficiently large polymeric aggregate before assembly will begin. A threshold concentration of tubulin is required for this stage. Once the reaction is initiated, 5·8S dimers add sequentially until the polymer reaches its final length within 10–20 minutes at 37°C. The process terminates if the tubulin dimer level falls below the critical threshold value. Nevertheless, the process is dynamic with tubulin dimers joining and leaving the polymer - provided there is sufficient GTP available. *In vivo*, the polymer is in the form of a cylinder 30 nm in diameter composed of 13 rods lying parallel to each other to form a cylindrical sheet (Correia, 1991). Growth or elongation occurs at both ends of the cylinder albeit at different rates. The faster rate is at the plus end while the minus end is anchored to a microtubule-organizing centre such as the centrosome. Subunit disassembly occurs through loss of dimers when GDP bind at the E site (Fig. 11.1).

Various drugs interfere with the process of assembly. In general, they all have similar actions *in vitro* in that they block mitosis. Their uses, however, differ widely in a medical sense from anti-fungal agents like griseofulvin, anthelminthic drugs such as the benzimidazole family, to the vinca alkaloids, vincristine and vinblastine, which are used as anti-tumour agents. Colchicine, one of the best known of the microtubule ligands is used for the treatment of gout, but there is some doubt as to whether its therapeutic efficacy relies only, if at all, on inhibition of microtubule polymerization.

11.2 Colchicine for gout

Colchicine is an alkaloid derived from the autumn crocus (*Colchicum autumnale*) and other *Colchicum* species that grow in Colchis in Asia Minor

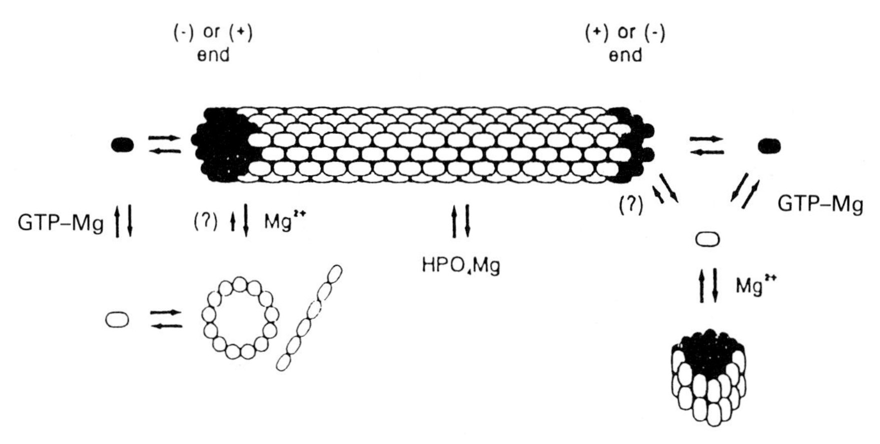

Figure 11.1 An idealized mechanism of microtubule assembly/disassembly. GTP-tubulin heterodimers with GTP at the E-site are represented by dark ovals. GDP-tubulin heterodimers are represented by light ovals. Microtubule growth occurs at both ends *in vitro*. *In vivo* the slow growing end, the minus end, may be anchored at MTOCs. GTP hydrolysis occurs upon assembly and, after the release of HPO_4Mg, generates a microtubule that is stabilized by a GTP cap, a layer of GTP-tubulin subunits that prevent catastrophic disassembly. Disassembly may occur by the loss of oligomers. Subunits may also form nonmicrotubule polymers, typically double walled rings. (Correia, 1991)

and it is probably one of the most studied of the microtubule ligands. The precise site at which the drug binds has yet to be determined and both subunits have been implicated by genetic data. This could either mean a site at the interface between the α and β subunits or that changes in one subunit can alter the conformation of the other (Correia, 1991; Levy *et al.*, 1991).

The drug binds to the protein in a time-dependent fashion which resembles an irreversible process in that a conformational change takes place in the protein. The colchicine bound to the tubulin is, consequently, not in equilibrium with free drug, i.e. it is non-exchangeable. The process is not irreversible in practice because colchicine gradually dissociates from the complex with a half-life at 37°C of approximately 12 hours (Wilson, 1970). Estimates of the binding constant vary between 0·1 to $3{\cdot}0 \times 10^{-6}$ M, depending on how much time has been allowed for equilibration. The reason for this apparent irreversibility is the large activation energy required for binding (E = 100·4 kJ/mol) because of a major conformational change – hence the dissociation is also very slow.

Colchicine has three rings denoted A, B and C, and all three play a part in binding. The trimethoxybenzyl A ring interacts hydrophobically and with hydrogen bonding, while the tropolone C ring provides stacking interactions. The slow kinetics of the irreversible colchicine binding requires distortion of the B ring. Removing the B ring increases the reaction rate and lowers the activation barrier (Correia, 1991).

The drug inhibits the assembly of microtubules by binding to soluble tubulin, forming a complex that adds to microtubule ends and sharply reduces the affinity of the microtubule for fresh tubulin dimers. The tubulin addition rate to the growing microtubule is thereby greatly reduced (Margolis *et al.*, 1980). Colchicine acts, in fact, as a type of chain terminator.

Furthermore, colchicine stimulates the GTPase activity of tubulin, perhaps as a result of drug binding to a site which is responsible for inducing polymerization-dependent GTPase activity. Colchicine, therefore, induces a conformational change in the absence of microtubule assembly that mimics assembly by leading to increased GTPase activity. Vinblastine, an indole alkaloid, inhibits this activation of GTPase activity by colchicine (David-Pfeuty *et al.*, 1978).

Colchicine is used in the treatment of acute gouty arthritis to reduce pain and inflammation. The swellings on joints and tendon sheaths composed mainly of sodium urate crystals (tophi), characteristic of gout, are reduced in size and eventually disappear. Polymorphonuclear leukocytes ingest the crystals and produce a glycoprotein which acts as a chemotactic factor and may be the cause of gouty arthritis by recruiting granulocytes into the inflamed area which then release hydrolytic enzymes. Colchicine suppresses the release of this factor. The role of microtubules in this is not clear, although colchicine interferes with the microtubules in the neutrophil and disassociates microtubules and the mitotic spindle. One problem is the activity of trimethylcolchicinic acid, which is almost as active as colchicine itself against gout but is no inhibitor of microtubule assembly (Wallace, 1975).

Trimethylcolchicinic acid could be converted *in vivo* to colchicine (or an analogue which can bind to microtubules). This could give rise to low levels of colchicine in the circulation, which are as high as is needed for colchicine action *in vivo* (namely 5×10^{-8} M) but still difficult to detect (Wallace, 1975). Although this concentration is below the apparent binding constant for tubulin–colchicine complex formation, the existence of a pseudo-irreversible complex suggests that time-dependent inactivation could occur. Low levels of the drug would then have the desired effect. This biotransformation has not been confirmed, however, and the question remains unresolved.

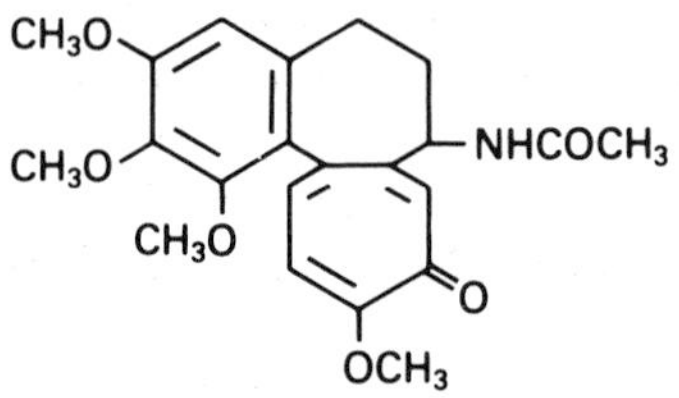

CH_3O, CH_3O, CH_3O, $NHCOCH_3$, O, OCH_3

Colchicine

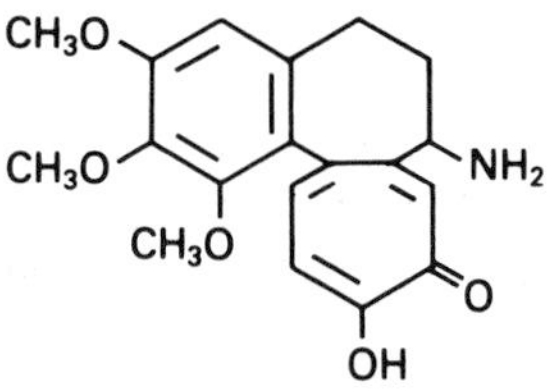

CH_3O, CH_3O, CH_3O, NH_2, O, OH

Trimethylcolchicinic acid

A second difficulty arises from the fact that colchicine is only a weak anti-inflammatory agent in conditions induced by substances other than urate crystals, and so it is possible that there is a particular step in crystal-induced inflammation which is more important for that type of inflammation than it is for any other. The release of a neutrophil-derived chemotactic factor has been proposed in this context (Spilberg *et al.*, 1979). Whether microtubules can be implicated in this secretory process is not entirely clear, although they are known to be involved in a number of other secretory processes such as the release of histamine from human lung tissue after antigen challenge (Kaliner, 1977). In conclusion, inhibition of microtubule assembly remains the only plausible explanation for the mode of action of colchicine in the absence of any other convincing hypothesis (Levy *et al.*, 1991).

11.3 Vinca alkaloids as anti-tumour agents

Vincristine and vinblastine are two members of the family of indole alkaloids derived from the periwinkle (*Catharanthus* or *Vinca rosea* – hence the name vinca alkaloids). These drugs bind to tubulin in a different fashion and at a different site for colchicine. The considerable difficulties, however, in measuring binding constants accurately in this system have been noted by Correia (1991).

There are two binding sites for vinblastine per tubulin dimer, instead of one for colchicine. Moreover, vinblastine will bind instantaneously and reversibly, unlike colchicine, to binding sites on the microtubule, in the absence of soluble dimer, with a binding constant of $1{\cdot}9 \times 10^{-6}$ M. The number of high-affinity binding sites that is available on the microtubule, however, is far less than the total number possible if one site on each individual subunit were freely accessible (Wilson *et al.*, 1982).

Inhibition of *de novo* polymerization has yielded IC_{50} of 4·3, and $3{\cdot}2 \times 10^{-7}$ M for vinblastine and vincristine respectively (Owellen *et al.*, 1976). If the experiment is carried out by measuring the inhibition of addition of tubulin dimers to microtubules at steady state, known as 'freewheeling', in contrast to polymerization *de novo*, a figure of $1{\cdot}38 \times 10^{-7}$ M for half-maximal inhibition is obtained for vinblastine (Wilson *et al.*, 1982). Under these 'freewheeling' conditions, 1·16 molecules of vinblastine are detected per microtubule, implying that the addition of one vinblastine molecule at the assembly end of a microtubule is enough to have a very marked effect on the polymerization process. This is referred to as 'substoichiometric poisoning' since the concentration of drug is well below that of the tubulin (although greater than that of the microtubule ends). No effect on depolymerization is noted at these concentrations.

At higher concentrations (i.e. $>10^{-6}$ M), the drug binds to a greater number of sites on the microtubule and appears to loosen its structure. As a consequence, splaying and peeling of protofilaments at microtubule ends occurs together with active depolymerization (Wilson *et al.*, 1982).

Since freewheeling is more sensitive to drug inhibition than is *de novo* polymerization, the ability of four vinca alkaloids to inhibit freewheeling has been compared with their ability to inhibit cell proliferation in tumour cell cultures *in vitro*. Although all four drugs were very potent in both tests, no correlation was found between orders of potency on the two tests (Jordan *et al.*, 1985).

It was suggested that this may be because the relative potency of the alkaloids may vary with respect to tumour type studied. Alternatively, it may be suggested that four compounds are hardly enough on which to base firm conclusions. In the cell, vinblastine blocks mitosis and causes metaphase arrest.

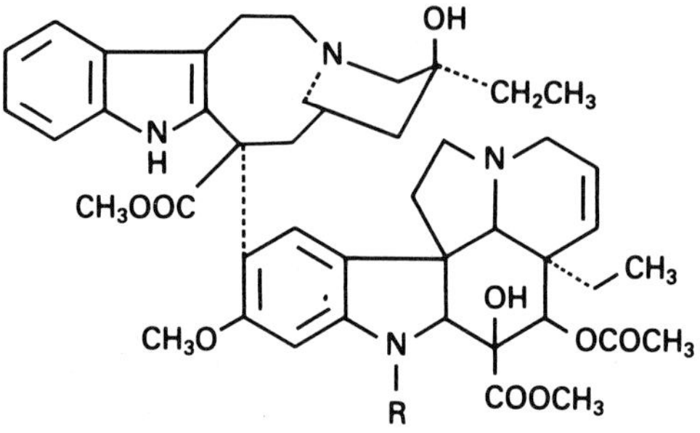

Vinblastine: R = CH_3
Vincristine: R = CHO

The drug binds to microtubules causing disassociation and disruption of the mitotic apparatus. Chromosomes cannot be segregated correctly, indeed they aggregate in unusual groupings such as balls or stars, and the cell dies (Calabresi and Chabner, 1990).

Vinblastine and vincristine are used normally in combination with other anti-tumour agents for the treatment of Hodgkin's and non-Hodgkin's lymphomas, breast carcinoma, cancers of the head and neck and others. One curious feature is that cross-resistance between the vinca alkaloids is not observed (see Calabresi and Chabner, 1990).

11.4 Griseofulvin as an anti-fungal agent

Griseofulvin is a secondary metabolite elaborated by the fungus, *Penicillium griseofulvium dierckx*. The action of griseofulvin is to arrest the target fungal cell in metaphase (Gull and Trinci, 1973). As a consequence, multinucleate hyphal cells are formed, frequently together with a Y-shaped metaphase equatorial plate. The microtubules maintain their normal morphological appearance but are disorientated within the spindle (Grisham *et al.*, 1973). The overall effect is thus antimitotic, although the site of action differs from that of colchicine. Griseofulvin does not prevent the binding of colchicine to tubulin nor does it affect the ability of the vinca alkaloids to stabilize the binding of colchicine to tubulin (Wilson, 1970).

Originally it was thought that the drug bound to the MAPs (Roobol *et al.*, 1977), thereby inhibiting microtubule assembly both in rate and extent (Roobol *et al.*, 1976). More recently, however, radioactive griseofulvin has been found to bind directly to tubulin dimer (0·83 mol/mol). Initiation of polymerization requires microtubule-associated proteins but subsequent elongation does not, and griseofulvin blocked this later stage. Moreover, griseofulvin also induces depolymerization of preformed microtubules at 37°C (reviewed in Kerridge, 1986).

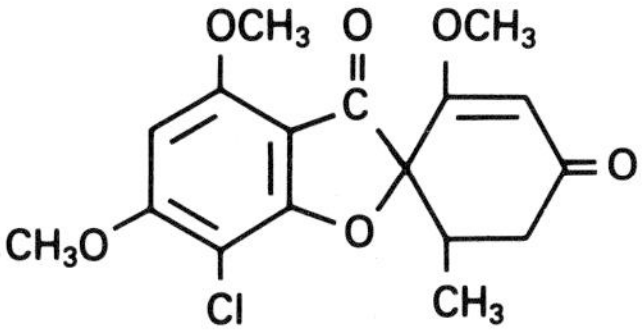

Griseofulvin

Keates (1981) has drawn attention to the need to carry out studies *in vitro* with griseofulvin at low concentrations since the solubility limit is 3×10^{-5} M in aqueous buffers. Attempts to introduce higher levels into solution may induce artifacts by precipitating proteins out of solution. Keates finds that griseofulvin inhibits both polymerization and depolymerization with an IC_{50} of $1{\cdot}25 \times 10^{-5}$ M, but the equilibrium position is unchanged.

Griseofulvin is fungistatic *in vitro* for a number of fungi that cause skin infections (dermatophytes, such as *Trichophyton, Micromonospora* species) and has been used for many years to treat infections such as athlete's foot (caused by *Trichophyton mentagrophytes*). The drug owes its efficacy partly to the fact that it is concentrated in the skin in keratin precursor cells, binding particularly to keratin so that the fungus is unable to gain a foothold on new hair and nails (Gull and Trinci, 1973). Secondly, many dermatophytes concentrate the drug particularly if they are sensitive to it (El-Nakeeb and Lampen, 1965).

11.5 Benzimidazoles as anthelminthics

The benzimidazole family of drugs has long been used to treat helminthic infections of sheep, cattle, goats etc. Various nematodes infect man and mebendazole is considered to be front-line treatment for several infections including hookworm (*Ancylostoma duodenale*), roundworm (*Ascaris lumbricoides*) and whipworm (*Trichuris trichiura*). Mebendazole is reasonably effective and non-toxic - a very important consideration in anti-parasitic chemotherapy, where many of the drugs available are toxic (Webster, 1990). Most of the experimental work has been done with nematodes that do not infect man.

Several benzimidazoles inhibit the assembly of microtubules from a nematode, *Ascaridia galli*, which infects chickens, with IC_{50} of 5×10^{-6} M and 6×10^{-6} M for mebendazole and oxfendazole respectively. Some but not all of the drugs bind to mammalian tubulin at similar concentrations, although thiabendazole and oxfendazole had IC_{50} values in excess of 10^{-4} M. Electron microscopy of microtubules, polymerized under benzimidazole inhibition, showed a reduction in both the number and the length of microtubules formed (Dawson *et al.*, 1984). Correlation of *in vitro* with *in vivo* results is not possible for all the compounds, perhaps because of pharmacokinetic and/or metabolic considerations. Nevertheless, evidence from benzimidazole-resistant helminths, such as *Trichostrongylus colubriformis*, suggests that tubulin has been altered to a form which shows less drug affinity. Furthermore, resistant organisms show no change in the appearance of their cytoplasmic microtubules when treated with thiabendazole, unlike intestinal cells from the drug-sensitive organisms, that lose almost all their organelles (Sangster *et al.*, 1985).

	R_1	R_2
Mebendazole:	C_6H_5–C(=O)–	$-NHCOOCH_3$
Oxfendazole:	C_6H_5–S(=O)–	$-NHCOOCH_3$
Thiabendazole:	H	thiazol-4-yl
Flubendazole:	F–C_6H_4–C(=O)–	$-NHCOOCH_3$

Antagonism of colchicine binding can also be used as an index of microtubule ligand affinity for the benzimidazole drugs because they bind at the colchicine binding site, despite their obvious difference in structure. Mebendazole has been shown to interfere with colchicine for the binding site on tubulin from bovine brain with an inhibitor constant of $3{\cdot}6 \times 10^{-6}$ M. One binding site per tubulin dimer was found by Scatchard analysis (Laclette *et al.*, 1980) – see Appendix for discussion on Scatchard analysis. Furthermore, an IC_{50} of 1×10^{-6} M was found for albendazole by inhibition of [^{3}H]colchicine binding to tubulin from the intestinal cells of *Ascaris suum*, a nematode that infects pigs (Barrowman *et al.*, 1984). In confirmation of these findings, benzimidazole resistance results in lack of colchicine binding to microtubules (Sangster *et al.*, 1985).

The importance of microtubules to the organism appears to be in the secretion of acetylcholinesterase – regulating peristalsis in the host intestine. This allows the nematode to hang on to the stomach wall more efficiently and also delays the approach of lymphocytes. Mebendazole is a powerful inhibitor of cholinesterase release at submicromolar concentrations and causes the enzyme to accumulate inside the parasite. Other microtubule inhibitors such as colchicine are moderately effective at inhibiting enzyme release but vinblastine is ineffective, probably because it does not bind to nematode tubulin (Tekwani, 1992).

11.6 Taxol as an anti-tumour agent

The needles and bark of the common yew, *Taxus baccata*, yield an anti-tumour agent named, not surprisingly, taxol. This compound has been known for some time but only recently has it been developed for the treatment of lung, breast and ovarian cancers.

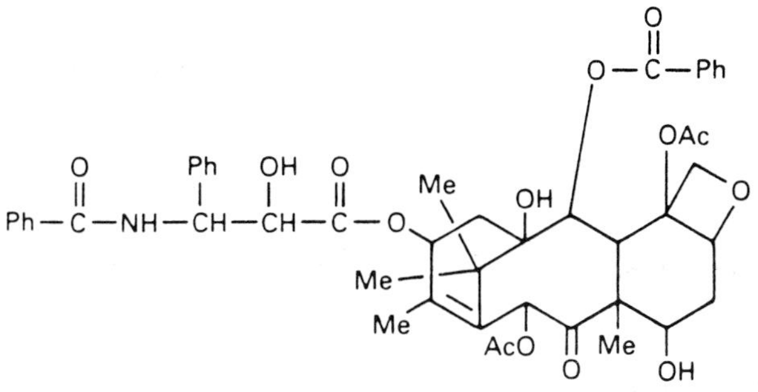

Taxol

Taxol is unusual amongst anti-mitotic agents in causing tubulin to associate into microtubules in the absence of GTP and MAPs. Taxol also instigates microtubule hydrolysis of GTP by assembled tubulin. There are more centres for growth, the microtubules are shorter and ribbon-like structures are formed (Schiff *et al.*, 1979; Ringel and Horwitz, 1991). The drug binds to microtubules with a binding constant near 1×10^{-6} M. The purification of brain microtubules by taxol is a sufficiently reliable process to form the basis of a publication (Wabbee and Collins, 1988).

Taxol binds at a different site from the other anti-mitotic drugs, colchicine, vinblastine and vincristine (Correia, 1991), and there is no cross-resistance between them. Taxol and its semi-synthetic analogue, taxotere, induce microtubule bundle formation in cells and thereby prevent cell division (Ringel and Horwitz, 1991).

The wide variety of structures that will bind to tubulin even at the same site argues for considerable heterogeneity on the protein surface, and would make a very interesting X-ray study. Furthermore, the variety of the uses of microtubule ligands is intriguing and suggests that other structural types may be found that bind to tubulin and show pharmacological activity.

Chapter 12

Hormonal modulators

12.1 Introduction

This chapter concerns the drugs that modulate the action of two types of hormone: insulin and analogues of gonadotrophin hormone releasing hormone (GnRH). Although we do not know the exact mechanism for the drugs that lower blood glucose or the detailed mechanism of action of insulin itself (e.g. the nature of the second messenger, if any), it is necessary to include the important area of anti-diabetic drugs. It is interesting to compare our limited knowledge of the mode of action of insulin, which has been investigated for several decades, with the way in which GnRH analogues, that have only been known for just over a decade, have broadened our knowledge of the relationship between hypothalamus, pituitary and gonads.

12.2 Diabetes mellitus

The condition of diabetes is still a very serious medical problem which can only partially be alleviated by pharmacological intervention. There are two broad categories of diabetes mellitus: one in which the patient is totally dependent on insulin injections because the β-cells of the pancreas which normally secrete insulin under the influence of glucose have been destroyed; and the other in which the patient has circulating levels of insulin but is resistant to the action of the hormone. The former, insulin-dependent, type is less frequent than the latter but is more serious and usually occurs early in life (hence the term 'juvenile onset' is used to describe it). It may result from viral infection - Coxsackie B virus has been implicated in at least one case (Yoon *et al.*, 1979). Non-insulin dependent diabetes of which several sub-types exist, is more often a disease of later life ('maturity onset') and is frequently associated with obesity.

The basic metabolic defect lies in the fact that glucose cannot enter cells, either because there is no insulin in the plasma to induce transport or because the insulin that is present (and its level may be higher than normal) is rendered ineffective, possibly by inadequate or insufficient receptors or by a post-receptor lesion. Consequently, excess glucose is found in the plasma (hyperglycaemia) and is excreted into the urine together with copious amounts of water (polyuria). The patient becomes dehydrated and drinks to assuage the thirst (polydipsia). Glucose cannot enter the appetite-regulating cells of the hypothalamus, so the patient always feels hungry and eats frequently (polyphagia).

The overall consequence is an increase in weight, leading to frank obesity in some cases; the production of vast quantities of sweet-tasting urine (mel is the Latin for honey and diabetes is related to the Greek work for syphon); and fatigue through the lack of available energy (Maurer, 1979).

The degree of severity of the condition may be studied by the administration of a glucose 'meal' (50 or 100 grams), otherwise known as a glucose tolerance test. In non-diabetics, the levels of blood glucose rise to a peak in 45 minutes and return to the starting level in about $1\frac{1}{2}$ hours. In those with mild disease, the glucose level rises much higher and takes up to 4 hours to return to the baseline, followed by an overshoot before returning to normal. Others with more serious disease show elevated glucose levels in the absence of food and very high levels of glucose for a long time after a glucose meal (Montgomery *et al.*, 1983).

If the condition is uncontrolled, the patient can relapse into a coma and die. This is partially a consequence of the dehydration and partly from metabolic acidosis. Because the cells cannot utilize glucose for the production of energy, they break down triglyceride fat instead. Acetyl-CoA levels are raised, leading to the production of acetone and acetoacetate and β-hydroxybutyrate, which are secreted into the blood by the liver. The latter two can be used by the brain, heart and skeletal muscle as a source of energy to a limited extent but then they begin to accumulate in the plasma. Acetoacetate and β-hydroxybutyrate are relatively strong acids, and eventually the buffering capacity of the blood is overcome, causing acidosis. The kidneys excrete these anions together with cations, such as sodium, and water, thus contributing to the dehydration, until eventually the patient goes into a coma as a result of the combination of acidosis and dehydration (Montgomery *et al.*, 1983).

12.2.1 The action of insulin

The action of insulin in cellular terms is responsible for the stimulation of glycogen and lipid synthesis and glucose oxidation concomitantly with the inhibition of glycogenolysis, lipolysis and gluconeogenesis. Any lesion in insulin action, whether it occurs at the receptor on the cell surface or subsequently, will cause increases in lipolysis and glycogenolysis, with elevation of glucose and acetyl-CoA levels leading to the formation of acetone etc. Extracellular levels of glucose will rise since the major route of glucose removal from the blood is into the peripheral muscle tissue (Pollet, 1983).

The intracellular effects of insulin are possibly initiated by means of an, as yet, unidentified second messenger (Malchoff *et al.*, 1987). Molecules of glucose carrier protein are transferred from the Golgi apparatus to the cell surface ready to transport glucose into the cell. The major series of metabolic changes noted above requires the phosphorylation status of the key enzymes of glycogen breakdown, lipolysis and lipogenesis to be altered:

Dephosphorylation	*Result*	*Insulin*
Glycogen synthase	Glycogen synthesis activated	+
Glycogen phosphorylase	Glycogen breakdown inhibited	+
Pyruvate dehydrogenase	TCA cycle activity increased	+
HMG-CoA reductase	Cholesterol synthesis activated	+
Phosphorylation	*Result*	
Glycogen synthase	Glycogen synthesis inhibited	−
Glycogen phosphorylase	Glycogen breakdown activated	−
ATP-citrate lyase	Lipogenesis activated	+
Acetyl-CoA carboxylase	Lipogenesis activated	+

HMG = hydroxmethylglutaryl
TCA = tricarboxylic acid

These events follow from insulin binding to its receptor. The insulin receptor is a glycoprotein heterodimer ($\alpha_2\beta_2$) of molecular weight 430 kDa, composed of two 125 kDa α subunits and two β subunits of 90 kDa. The receptor contains two functional binding sites and is stabilized by disulphide bonds between the subunits. The α subunits are extracellular, while the β subunits are transmembrane proteins that can autophosphorylate (ATP-linked) the hydroxyl of several tyrosine residues. Insulin binds to the α subunit and causes the expression of tyrosine kinase activity, together with receptor aggregation. The receptor can react with several cellular proteins to generate a cascade of phosphorylation and dephosphorylation on serine or threonine residues as noted above.

Not all of the actions of insulin may be mediated through autophosphorylation, as insulin receptor mutants without tyrosine kinase activity can still mediate some of the actions of insulin (Sung, 1992). For example, a specific guanine regulatory protein activates a specific phospholipase C, catalysing the formation of phosphatidylinositol-glycan and diacylglycerol (Romero *et al.*, 1988; Yip, 1992).

The binding of hormone to receptors in liver and muscle also causes the receptors to cluster in small groups followed by endocytosis of a receptor cluster: subsequent fusion with lysosomes leads to the degradation of insulin and the final event is recycling of the receptors.

12.2.2 Insulin therapy

From a therapeutic point of view, diabetes can be approached in a number of ways. Insulin-dependent diabetics, by definition, require insulin which in earlier years was extracted from animal pancreas (usually that of pigs or cows). In 1979, Goeddel *et al.* reported the expression of the human insulin gene in *E. coli*. They chemically synthesized the genes for the A chain (21 amino acids) and the B chain (30 amino acids) and inserted them into the plasmid pBR 322. The insulin genes were attached to the carboxy terminus of the gene for β-galactosidase so that the human insulin would be efficiently transcribed and translated as part of one long protein chain. The separate insulin chains were subsequently cleaved from the β-galactosidase using cyanogen bromide. The sulphydryl groups were converted to *S*-sulphonyl groups, the chains were mixed and were allowed to combine by first reducing the mixture followed by air oxidation. The presence of insulin was confirmed by radioimmunoassay. The production of sufficient quantities of insulin for human treatment is described by Johnson (1983) - an interesting story of the problems faced by the development of the first genetically engineered health-care agent to be produced on a commercial scale.

Long-acting insulin preparations are produced by the reaction of insulin with the basic protein protamine in the presence of zinc. When the preparation is injected subcutaneously in aqueous suspension, it dissolves slowly at the injection site and is thus absorbed slowly.

Periodic injections of insulin, however, tend to cause the blood level of glucose to oscillate wildly (Bressler, 1978) because the action of insulin is normally balanced by glucagon, whose role is to raise blood glucose levels by increasing liver glucose production and opposing liver glucose storage (as glycogen). The net result in the normal subject is to maintain levels of glucose close to the norm. The peaks in glucose level in the diabetic treated with periodic insulin injections are believed to contribute to the complications associated with diabetes. Recently, studies have been carried out with insulin infusion pumps taped on to the wall of the abdomen, such that the point of needle lies in the subcutaneous tissue. Insulin is then pumped through the syringe from a reservoir at a steady basal level, with additional doses just before meals (see Nathan, 1992).

In patients with non-insulin-dependent diabetes that is relatively mild, the patients are likely to be insulin-resistant in the sense that they have a reduced number of receptors. In this case the maximum response to the hormone is merely attained at higher concentrations. Treatment with a drug that lowers serum glucose levels (a hypoglycaemic agent) causes the number of receptors usually to return to normal.

With patients more seriously affected by the disease, down-regulation of the receptors is less likely to be of importance since the circulating insulin levels range from slightly elevated to below normal. A marked post-receptor defect is almost certainly present in these patients since an increased insulin level does not return glucose levels to normal. The nature of this defect is not known at present (Lockwood and Amatruda, 1983).

Glucagon is another peptide secreted by the pancreas, in this case by the α cells. It has been proposed that too much glucagon as well as too little insulin may be necessary for insulin-dependent diabetes to manifest (Unger and Orci, 1981b), since the actions of glucagon are antagonistic to those of insulin.

These actions are inhibition of cholesterol synthesis and stimulation of lipolysis, glycogenolysis and gluconeogenesis, with eventual hyperglycaemia. Also produced in the pancreas is the 14-amino acid peptide somatostatin which, amongst other actions, inhibits the release of both glucagon and insulin. Indeed, since the α cells of the pancreas produce glucagon, the β cells insulin and the δ cells somatostatin, an intrapancreal control mechanism has been proposed (Unger and Orci, 1981a; Fig. 12.1). Consistent with this suggestion, there is at least one report of combined treatment with insulin and somatostatin which has given better results than insulin alone (Raskin and Unger, 1978). Somatostatin probably works by suppressing glucagon release (Fig. 12.1).

Although this is a book about the uses of pharmaceuticals to combat disease, it is not the intention of the author to imply that drugs are the ultimate panacea for all illness. In no disease is this more relevant than in non-insulin-dependent diabetes. In many cases, dietary control with concomitant weight loss, coupled with exercise, will often control the condition. The rationale for this approach lies in the fact that many patients, particularly those who are obese, have elevated levels of insulin together with a lowered number of receptors, as has been seen in some diabetic animals (Roth *et al.*, 1975). A reduction in weight causes the number of receptors to rise and the level of circulating insulin to fall as is likely to be the consequence if the intake of mono- and disaccharides is restricted. Intake of polysaccharides apparently need not be restricted, presumably because these give rise to less available mono- and disaccharides. Saturated fat intake should also be lowered, in view of the greatly increased risk of heart disease with diabetes (American Diabetic Association, 1979). Weight reduction, even to a modest extent, greatly improves the responsiveness of obese subjects to insulin (Lockwood and Amatruda, 1983). Exercise is also

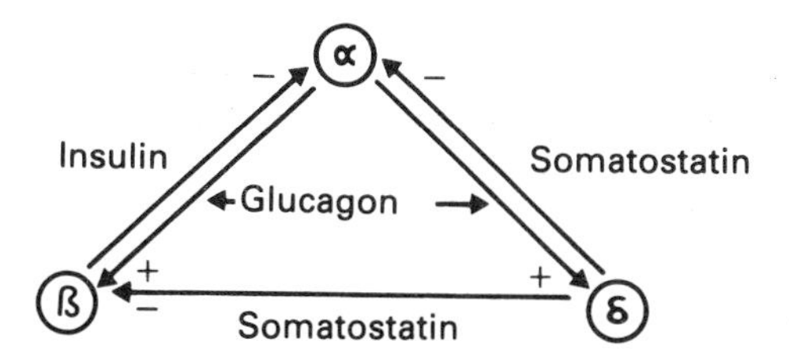

Arrows, an interaction between the hormone secreted from one cell type and the product of another. +, stimulation, −, inhibition.

Figure 12.1 The interactions between pancreatic hormones.

recommended, as it helps to activate the transport of glucose across the cell membranes even in the absence of insulin.

The main goal of therapy is to lower the serum glucose level, since not only does this reduce the symptoms of diabetes, but also because a number of serious complications associated with diabetes are related to excess serum glucose. One such complication is damage to the kidney by virtue of thickening of the basement membrane, which can lead to chronic renal failure. Retinal damage leading to oedema and haemorrhage makes blindness 25-times more common among diabetics. Cataracts also are more frequent in these patients. Atherosclerosis which predisposes to heart attacks, is another complication. Both types of diabetes lead to these sequelae, although non-insulin-dependent diabetes may result in them less frequently. The metabolic defect in insulin level or action is almost certainly the main cause of these complications. It is likely that hyperglycaemia is the cause of damage to kidney and retina, possibly because glucose glycosylates membrane proteins.

12.3 Sulphonylureas as hypoglycaemic agents

The circulating levels of glucose in non-insulin-dependent diabetes can be lowered by a family of drugs based on the sulphonylurea moiety. Tolbutamide and chlorpropamide were the earliest to be introduced over 30 years ago, while recently a second generation of drugs have been developed which include glibenclamide (glyburide) and glipizide. These drugs may, however, be regarded as only an adjunct to dietary control and exercise (Jackson and Bressler, 1981).

The effectiveness of these agents seems to depend on two actions. The first is an effect on the β cells of the pancreas to improve the secretion of insulin. The drugs need a functioning pancreas to be fully effective and are thus of no value in insulin-dependent disease. In resting β cells, the membrane potential is maintained by an efflux of potassium ions through an ATP-sensitive potassium channel. Sulfonylureas bind to a specific receptor which closes this channel. Potassium can no longer be pumped out of the cell and the membrane is depolarized. Voltage-sensitive calcium channels are activated, there is an influx of calcium and release of insulin, which requires cytosolic calcium ion (Garrino *et al.*, 1986), is stimulated. Glibenclamide binds the most tightly of the drugs to the membranes (K_i of $2{\cdot}7 \times 10^{-8}$ M) while tolbutamide is less tightly bound (7×10^{-6} M) (Longman and Hamilton, 1992). Drug action leads to an increase of both fasting and meal-stimulated insulin secretion (Lebovitz, 1992). In addition, these drugs suppress glucagon release in diabetics but not in normal subjects (Lebovitz, 1984).

The link between the binding to the β cell and hypoglycaemic activity has been demonstrated by Geisen *et al.* (1985), who used a rat β cell tumour to which ^{3}H-glibenclamide bound tightly, reversibly and in a saturable fashion with a K_d of 5×10^{-11} M. A number of analogues of glibenclamide were compared with respect to both inhibition of radioactive ligand binding and hypoglycaemic activity in rabbits. A good correlation was observed, although a num-

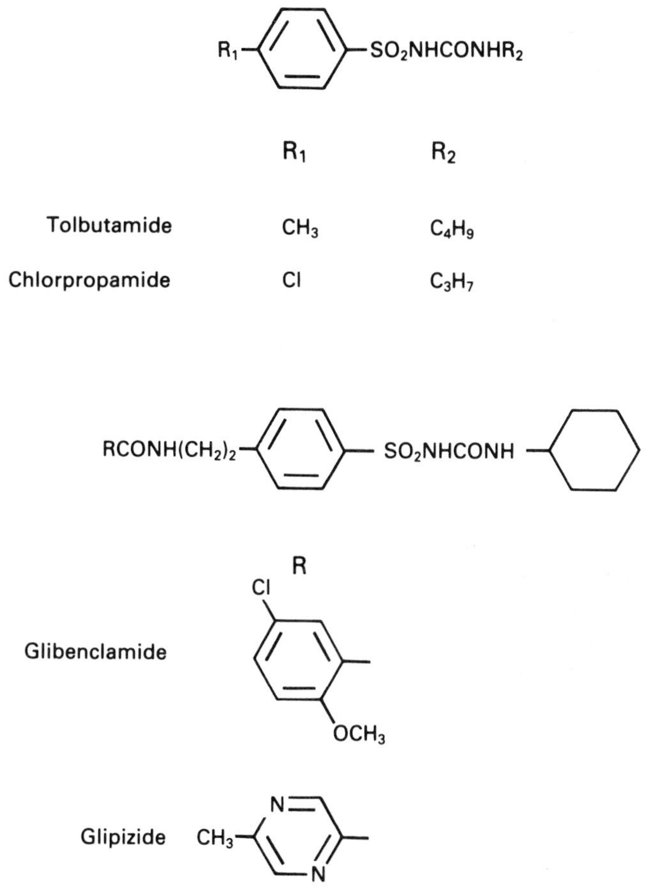

R_1–C₆H₄–$SO_2NHCONHR_2$

	R_1	R_2
Tolbutamide	CH_3	C_4H_9
Chlorpropamide	Cl	C_3H_7

$RCONH(CH_2)_2$–C₆H₄–$SO_2NHCONH$–C₆H₁₁

	R
Glibenclamide	Cl, OCH_3 (substituted phenyl)
Glipizide	CH_3 (pyrazinyl; N, N)

ber of compounds with binding constants between 10^{-7} and 10^{-8} M had to be omitted from the analysis because they failed to show sufficient hypoglycaemic activity - probably because they were not absorbed sufficiently or were too heavily serum-bound to be effective.

The second action of the hypoglycaemic agents is potentiation of the action of insulin. They increase the absorption of glucose engendered by a given dose of insulin; whether this is by increasing the number of insulin receptors through preventing down-regulation or by a post-receptor effect is not known (Lebovitz, 1984).

Chronic administration of these drugs eventually results in only normal (or even reduced) insulin secretion (with the possible exception of glipizide). In contrast, acute dosing raises insulin levels. It is likely, therefore, that the extra-pancreatic effects are more important in the long term, although in some cases the drugs may gradually lose their efficacy.

There is no clear indication that the second generation drugs are any more effective therapeutically than the first, although the former are more active at low concentrations. The number of failures with each type is almost identical, and so there seems to be little to choose between them (Kreisberg, 1985). They are nevertheless very useful for the treatment of the more severely hyperglycaemic patients for whom insulin is not yet necessary (Lebovitz, 1992).

12.4 Gonadotrophin-hormone-releasing hormone (GnRH) analogues

A growing area of pharmacological intervention has been opened up for a number of cases of hormonal aberrations and cancers that depend on sex steroid hormones for growth, by a greater understanding of the axis that links the hypothalamus to the gonads via the pituitary gland (Fig. 12.2). The hormones that carry the information from pituitary to gonads are collectively known as the gonadotrophins: luteinizing hormone (LH) and follicle-stimulating hormone (FSH) in females - LH is referred to as interstitial-cell-stimulating hormone (ICSH) in males.

Gonadotrophin-hormone-releasing hormone (GnRH or gonadorelin) is released from the hypothalamus and acts on the anterior portion of the pituitary gland to release gonadotrophins, although there are suggestions that FSH may also be released by another as yet undiscovered hormone. GnRH secretion is under the control of higher centres in a way that we do not yet understand, although it is clear that in women some form of 'biological clock' must operate in order to maintain the functioning of the menstrual cycle.

GnRH is released in short pulses of millisecond duration. This can be recognized by following the pulsation in plasma levels of LH, since LH release is

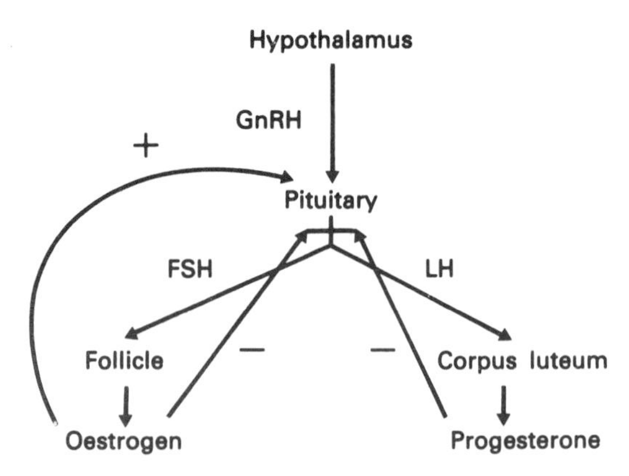

Figure 12.2 Hypothalamus-pituitary-gonad axis.

very sensitive to GnRH pulsation and the half-life of LH in plasma is shorter than the time between pulses. FSH release is much less sensitive to GnRH levels and does not pulse because the response is much slower and takes longer to die away (reviewed in Lincoln *et al.*, 1985).

In the female, FSH stimulates the growth of follicles in the ovary which are responsible for the production of the main oestrogen, the steroid hormone 17β-oestradiol (oestrone and oestriol are derived from oestradiol in the liver and are less active as oestrogens). By the time that ovulation is about to occur, various follicles are almost mature. There is then a surge in the level of LH which initiates ovulation in one or more follicles while in some way causing the others to regress. The follicle bursts and discharges the egg. The follicle then undergoes a re-organization of its cells developing into a yellow body known as the corpus luteum (a process therefore called luteinization) that secretes progesterone still under the influence of LH.

Progesterone prepares the female for pregnancy by stimulating the development of the lining of the uterus and inhibiting contraction of the smooth muscle of the uterine wall. It also prevents the development of a new follicle. Both progesterone and oestrogen act to close down FSH and LH release from the pituitary and possibly GnRH release from the hypothalamus in a negative feedback loop. Furthermore, oestrogen exerts a secondary, but positive, feedback effect on the pituitary in conjunction with GnRH which is believed to be responsible for the pre-ovulatory surge in LH production (Fig. 12.2). The use of GnRH analogues played a part in the discovery of these interactions (Schally, 1978; Asch *et al.*, 1983).

FSH and LH bind to membrane fractions from ovary tissue and exert their biochemical action through the stimulation of adenylate cyclase. Cleavage of the side-chain of cholesterol is induced, the rate-limiting step in steroid biosynthesis, in a fashion similar to the events in the adrenal cortex (Funkenstein *et al.*, 1983; Fig. 6.2). Oestrogen and progesterone production are thereby greatly increased (Fig. 12.3).

In the male, LH stimulates the Leydig cells of the testis to secrete testosterone. As these cells are also known as interstitial cells, LH is sometimes referred to as interstitial-cell-stimulating hormone or ICSH. FSH, however, is required only for the maturation of the spermatids into spermatozoa in some way that is not fully understood. Nevertheless, there is a similarity with the female in that testosterone exerts a negative feedback inhibition on LH secretion from the pituitary.

LH and FSH are glycoproteins of molecular weight about 16 kDa and are dimers of non-identical subunits α and β. The α subunits are nearly identical in the two hormones (and in thyroid-stimulating hormone), whereas the β subunits show some differences - although homology is still 80 per cent (Pierce and Parsons, 1981). GnRH is a decapeptide (see structures) that binds to pituitary cell membrane receptors belonging to the 7-transmembrane group with a binding constant of $3{\cdot}0 \times 10^{-9}$ M (Conn *et al.*, 1985). Occupation of the receptors causes them to aggregate, probably into dimers, and open a

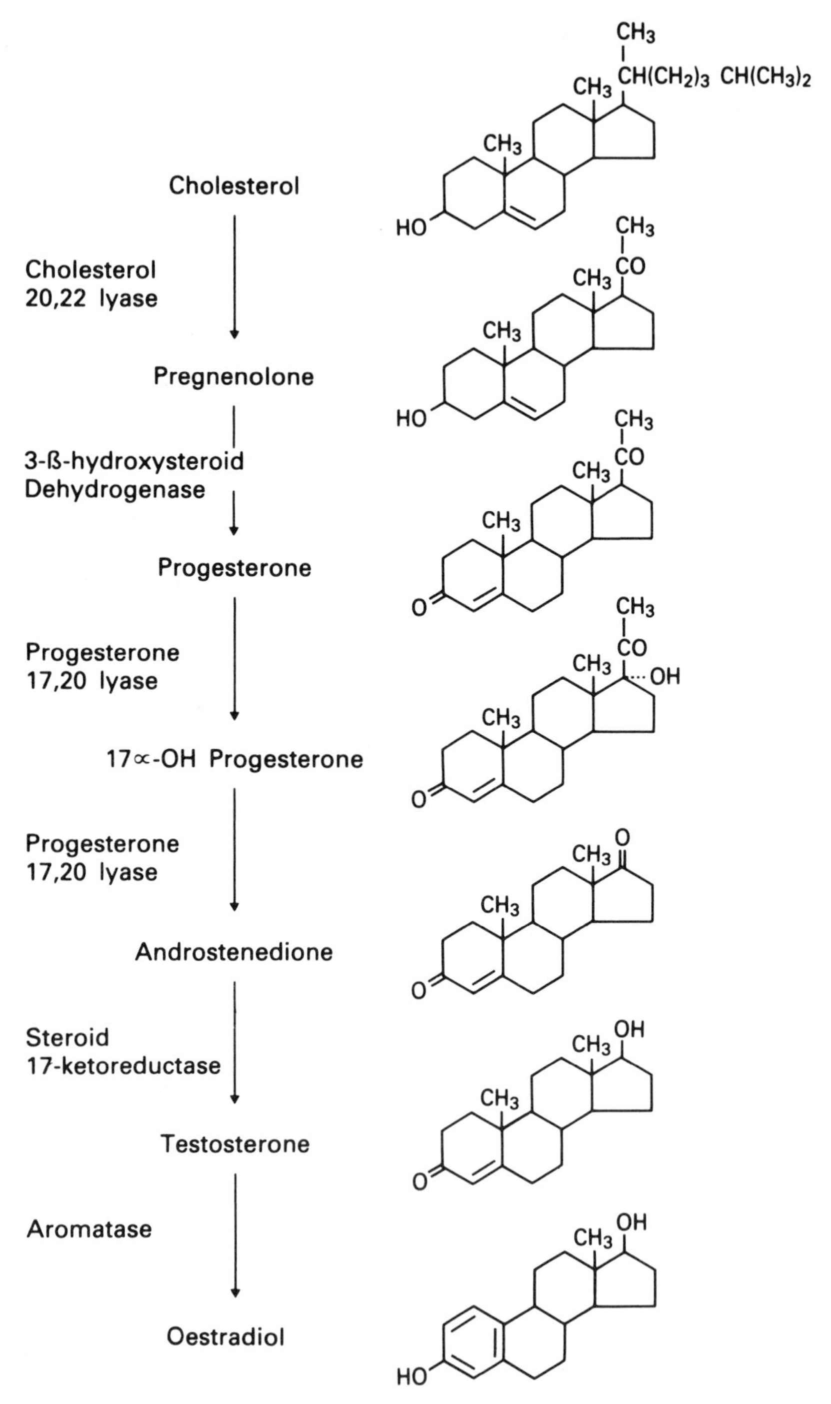

Figure 12.3 Sex hormone production from cholesterol.

calcium channel in the cell membrane. Calcium is also released from intracellular stores as normal for the phospholipase C pathway and acts as the second messenger via calmodulin (section 9.1). Protein kinase C is also activated but its role is not entirely clear, although it may be required for the expression of the LH β subunit gene. The gonadotrophins LH and FSH are stored in secretory granules which move to the cell surface, fuse with the cell membrane and discharge their contents (reviewed in Conn and Crawley, 1991).

The GnRH receptor has been cloned and expressed in *Xenopus* oocytes using as a detector the calcium-binding protein aequorin, which emits light when calcium is bound. The receptor showed little homology with other G-protein receptors, although there are structural similarities such as N-linked glycosylation sites near the extracellular N-terminus. There were two potential sites for phosphorylation by cyclic-AMP-dependent kinase in the first intracellular loop and a protein kinase C site in the third intracellular loop. There was no aspartate in the third transmembrane domain and, most unusual of all, there was no cytoplasmic C-terminal domain. Normally this area controls G-protein signalling and desensitization. In view of the importance of desensitization for this receptor (see below) it will be very interesting to discover which residues are involved (Reinhart *et al.*, 1992).

The part of the molecule that is responsible for binding to the receptor (the R site) is made up of residues from the C and N terminal ends, namely $PyroGlu^1$ and Gly^{10}, and modification of these residues destroys binding. On the other hand, His^2 and Trp^3 appear to be essential to stimulate secretion of LH (M site) which occurs as a consequence of receptor occupancy. Agonists of GnRH such as the peptide leuprolide contain both M and R sites. Gly^{10} is substituted by an ethylamide group and a D-amino acid is in position 6 which reduces proteolysis and increases receptor and plasma protein affinity. Recently, compounds have been developed with a bulky hydrophobic group at position 6 to increase the affinity still further. Half-lives *in vivo* are thus increased as well as potency. In addition, calcium is required for agonists to show full activity (Conn *et al.*, 1985), as one would expect for a secretory process.

Furthermore, agonist action induces another effect of importance to the therapeutic use of GnRH and its analogues, namely a surge of gonadotrophin release followed by increasing insensitivity to GnRH and its analogues. Desensitization can occur as a result of several mechanisms, including a reduction in the number of available receptors on the cell surface because of internalization, uncoupling of receptors from the signal pathways and an exhaustion of gonadotrophin stores inside the cell. Desensitization occurs in cultured pituitary cells (Catt *et al.*, 1985) and also *in vivo* and is calcium-independent - unlike LH release. Some of the internalized receptors may eventually be returned to the cell surface (Hazum and Conn, 1988).

Antagonists, however, merely block GnRH binding to the receptor and do not cause down-regulation. In contrast to agonist structure, GnRH antagonists contain only the R site. In addition to the substitutions for agonists, hydrophobic amino acids are substituted in positions 1 to 3. The antagonists

Gonadorelin (GnRH):

pyroGlu - His - Trp - Ser - Tyr - Gly - Leu - Arg - Pro - $GlyNH_2$

Leuprolide

pyroGlu - His - Trp - Ser - Tyr - D leu - Leu - Arg - Pro $CONHC_2H_5$

developed to date have shown unacceptable side-effects such as histamine release but other more specific compounds are under investigation (Conn and Crawley, 1991).

The conditions for which GnRH or its analogues are finding, or are likely to find, therapeutic usage are numerous (Conn and Crawley, 1991). The major use of the native hormone is to induce ovulation in women who suffer from infertility because of insufficient gonadotrophin. This has given rise, on occasion, to multiple pregnancies because the high level of hormones released from the pituitary causes a large number of follicles to mature at the same time. Other hormone abnormalities in men, such as delayed puberty (hypogonadotrophic hypogonadism) and delayed descent of the testicles (cryptorchidism) have also been shown to respond to treatment. These latter treatments will probably need to be given for life, however, as the therapeutic effects are reversible if drug administration is stopped.

Low doses of drug administered in pulses, mimicking the release of the natural hormone, produce an elevation in gonadotrophins. If, however, the drug is given systemically in higher doses, for example as a single daily subcutaneous injection, after a transient increase the pituitary is desensitized as a result of receptor down-regulation, and the levels of FSH and LH fall sharply. As a consequence, leuprolide and other GnRH agonists are, paradoxically, being used to treat conditions where elevated GnRH levels are the cause of the condition, notably the condition where GnRH levels rise very early in life and induce sexual maturation (centrally derived precocious puberty).

Furthermore, there are some tumours that are dependent on sex steroids: some breast cancers in women are dependent on oestrogen and 80 per cent of prostate cancers require testosterone (Vance and Smith, 1984), and these have responded to agonist therapy. A particular advantage that leuprolide therapy has for prostate cancer over the usual treatment with diethylstilboestrol, is the lack of feminizing and cardiovascular side-effects seen with the latter drug. Castration can also be used but is irreversible, and thereby unacceptable to many patients. Santen (1992) indicates that depot leuprolide (acetate form) given monthly is primary therapy for prostate cancer and may be combined with an androgen antagonist to combat the initial disease 'flare' (an agonist will initially produce a surge of gonadotrophins before desensitization occurs).

It has also been proposed that premenstrual syndrome results from a cyclic condition of elevated release of (or abnormal sensitivity to) LH and FSH (Coulson, 1986). The cyclic condition is characterized, amongst other symptoms, by weight gain, mood swings, irritability and inability to concentrate. Trials have shown that GnRH analogues were effective in relieving this condition (Muse, 1992), but further trials are required before the results can be confirmed, since the placebo effect is very marked in this condition.

Questions

1. What enzymes do insulin action (a) stimulate (b) inhibit?
2. How does the insulin receptor transmit the signal to the cell interior?
3. How do sulphonylurea drugs combat diabetes?
4. What is structurally unusual about the gonadorelin receptor compared with other 7-transmembrane receptors?
5. Why are gonadorelin agonists used to antagonize conditions which depend on high gonadorelin levels?

References

American Diabetic Association, 1979, *Diabetes,* **28**, 1027–30.

Asch, R.H., Balmaceda, R., Borghi, M.R., Niesvisky, R., Coy, D.H. and Schally, A.V., 1983, *J. Clin. Endocrinol. Metab.,* **57**, 367–72.

Bressler, R., 1978, *Drugs,* **17**, 461–70.

Catt, J.K., Loumaye, E., Wynn, P.C., Iwashita, M., Hirota, K., Morgan, R.O. and Chang, J.P., 1985, *J. Steroid Biochem.,* **25**(5B), 677–89.

Conn, P.M. and Crawley, W.F., 1991, *New Engl. J. Med.,* **324**, 93–103.

Conn, P.M., Staley, D., Jinnah, H. and Bates, M., 1985, *J. Steroid Biochem.,* **23**(5B), 703–10.

Coulson, C.J., 1986, *Med. Hypothes.,* **19**, 243–56.

Funkenstein, B., Waterman, M.R., Masters, B.S.S. and Simpson, E.R., 1983, *J. Biol. Chem.,* **258**, 10187–91.

Garrino, M.G., Meissner, H.P. and Henquin, J.C., 1986, *Eur. J. Pharmacol.,* **124**, 309–16.

Geisen, K., Hitzel, V., Okomonopoulos, R., Punter, J., Weyer, R. and Summ, H.-D., 1985, *Arzneim. Forsch./Drug Res.,* **35**, 707–12.

Goeddel, D.V., Kleid, D.G., Bolivar, F., Heyneker, H.L., Yansura, D.G., Crea, R., Hirose, T., Kraszewski, A., Itakura, K. and Riggs, A.D., 1979, *Proc. Nat. Acad. Sci. (U.S.A.),* **76**, 1106–10.

Hazum, E. and Conn, P.M., 1988, *Endocr. Rev.,* **9**, 379–87.

Jackson, J.E. and Bressler, R., 1981, *Drugs,* **22**, 211–45.

Johnson, I.S., 1983, *Science,* **219**, 632–7.

Kreisberg, R.A., 1985, *Ann. Intern. Med.,* **102**, 125–6.

Lebovitz, H.E., 1984, *Diabetes Care,* **7** (Suppl. 1), 67–71.

Lebovitz, H.E., 1992, *Drugs,* **w44**, Suppl. 3, 21–8.

Lincoln, D.W., Fraser, H.M., Lincoln, G.A., Martin, G.B. and McNeilly, H.S., 1985, *Rec. Progr. Hormone Res.,* **41**, 369–411.

Lockwood, D.H. and Amatruda, J.M., 1983, *Amer. J. Med.,* **75**, Suppl. 5B, 23-31.

Longman, S.D. and Hamilton, T.C., 1992, *Med. Res. Rev.,* **12**, 73-148.

Malchoff, C.D., Huang, L., Gillespie, N., Palasi, C.V., Schwartz, C.F.W., Cheng, K., Hewlett, E.L. and Larner, J., 1987, *Endocrinology,* **120**, 1327-37.

Maurer, A.C., 1979, *Amer. Sci.,* **67**, 422-31.

Montgomery, R., Dryer, R.L., Conway, T.W. and Spector, A.A., 1983, *Biochemistry: a case-orientated approach* (St. Louis: C.V. Mosby) Ch. 7.

Muse, K.N., 1992, *Clin Obstet. Gynecol.,* **35**, 658-66.

Nathan, D.M., 1992, *Drugs,* **44** (Suppl 3), 39-46.

Pierce, J.G. and Parsons, T.F., 1981, *Annu. Rev. Biochem.,* **50**, 465-95.

Pollet, R.J., 1983, *Amer. J. Med.,* **75**, Suppl. 5B, 15-22.

Raskin, P. and Unger, R.H., 1978, *New Engl. J. Med.,* **209**, 433-6.

Reinhart, J., Mertz, L.M. and Catt, K.J., 1992, *J. Biol. Chem.,* **267**, 21281-4.

Romero, G., Luttrell, L., Rogol, A., Keller, K., Hewlett, E. and Larner, J., 1988, *Science,* **240**, 509-11.

Roth, J., Kahn, C.R., Lesnick, M.A., Gorden, P., De Meyts, P., Megyesi, K., Neville, D.M., Gavin, J.R., Soll, A.H., Freycher, P., Goldfine, I.D., Bar, R.S. and Archer, J.A., 1975, *Rec. Progr. Hormone Res.,* **31**, 95-128.

Santen, R.J., 1992, *J. Clin. Endocrinol. Metab.,* **75**, 685-9.

Schally, A.V., 1978, *Science,* **202**, 18-28.

Sung, C.K., 1992, *J. Cell Biochem.,* **48**, 26-32.

Unger, R.H. and Orci, L., 1981a, *New Engl. J. Med.,* **304**, 1518-24.

Unger, R.H. and Orci, L., 1981b, *New Engl. J. Med.,* **304**, 1575-80.

Vance, M.A. and Smith, J.A., 1984, *Clin. Pharmacol. Ther.,* **36**, 350-4.

Yip, C.C., 1992, *J. Cell Biochem.,* **48**, 19-25.

Yoon, J.-W., Austin, M., Onodera T. and Notkins, A.L., 1979, *New Engl. J. Med.,* **300**, 1173-9.

Appendix

Quantification of ligand–macromolecule binding

The binding of a ligand to a macromolecule, whether it be receptor, enzyme or just a binding protein, is governed by the Law of Mass Action, which states that the rate of a reaction is proportional to the concentrations of the reactants. A variety of plotting techniques have been used to elucidate the binding of ligands to macromolecules. It is useful to show how ligand binding to macromolecules, enzyme kinetics and drug-receptor binding interrelate and, in addition, to describe some other methods of plotting such as the Hill and Scatchard plots.

A.1 Enzyme kinetics: I_{50} or K_i?

Much of the study of enzyme inhibition depends on measurement of K_i (the dissociation constant of the enzyme-inhibitor complex). Another measure of inhibition is IC_{50}; the concentration of inhibitor required to reduce the reaction velocity to 50 per cent of its value in the absence of inhibitor. K_i and IC_{50} are not the same, except in very unusual circumstances, and furthermore, the IC_{50} values will vary if different substrate conditions are used (see Cheng and Prusoff, 1973, for a detailed discussion).

Considering the simplest case of an enzyme reaction with one substrate, S; if the concentration of substrate is very much higher than that of enzyme, the initial velocity V_0 is given by:

$$V_0 = \frac{V_{max} \times [S]}{K_m + [S]} \qquad (1)$$

where V_{max} = maximum velocity; K_m = Michaelis constant of the substrate which equals the concentration of substrate that produces the half-maximal rate.

(a) In the most common form of competitive inhibition, described by the following equation, the substrate and inhibitor compete for the same active site, but with different affinities for it:

$$\begin{array}{l} E + S \rightleftharpoons ES \rightleftharpoons E + P \\ \quad\ \ \upharpoonleft\downharpoonright \\ \quad\ EI \end{array} \qquad (2)$$

Here EI = enzyme–inhibitor complex and P is product. The equation governing the reaction rate is:

$$V_i = \frac{V_{max} \times [S]}{K_m\left(1 + \frac{[I]}{K_i}\right) + [S]} \qquad (3)$$

where V_i is the initial velocity in the presence of inhibitor at concentration *i* and K_i is the dissociation constant of the enzyme–inhibitor complex. If the initial velocity is one-half V_0, the inhibitor concentration *i* by definition equals the IC_{50} with $V_i = \frac{1}{2}V_0$, so after rearrangement:

$$IC_{50} = K_i\left(1 + \frac{[S]}{K_m}\right) \qquad (4)$$

Clearly, IC_{50} is dependent on substrate concentration. Since different laboratories are unlikely to use the same substrate concentrations, their IC_{50} values are not a sound basis for comparison of different inhibitors. Only if the substrate concentration is low compared with the K_m does IC_{50} approximate to K_i. A low concentration (less than one-tenth of the K_m will also render the assay more sensitive because the effect of an inhibitor will thereby be magnified, but this might lead to problems in detecting product formation (Bush, 1986).

(b) In the case of non-competitive inhibition

$$\begin{array}{lll} E + S \rightleftharpoons & ES \rightleftharpoons E + P & \\ I \downarrow\uparrow K_{iS} & I \downarrow\uparrow K_{iI} & \\ EI & ESI & \end{array} \qquad (5)$$

where K_{iS} is the affinity of the inhibitor for full enzyme and K_{iI} for the enzyme–substrate complex.

It can be shown that:

$$IC_{50} = \frac{(K_m + [S])}{\left(\frac{K_m}{K_{iS}} + \frac{[S]}{K_{iI}}\right)} \qquad (6)$$

If the inhibitor binds equally well to the enzyme as to the enzyme–substrate complex, $K_{iS} = K_{iI} = K_i$ and the equation simplifies to:

$$IC_{50} = K_i \qquad (7)$$

Alternatively, if as is often the case $[S] \gg K_m$ and $K_m/[S] \ll K_{iS}/K_{iI}$, the equation reduces to:

$$IC_{50} = \frac{K_{iS}}{K_m/[S] + K_{iS}/K_{iI}} \tag{8}$$

again, if $K_m/[S] \ll K_{iS}/K_{iI}$, then:

$$IC_{50} = K_i \tag{9}$$

In the case of uncompetitive inhibition, provided that $[S] \gg K_m$, IC_{50} is independent of substrate concentration and equal to K_i. It is not always convenient to maintain a substrate concentration that is very much higher than the K_m because this may mask the effect of weaker inhibitors. Further derivations for two-substrate reactions may be found in Cheng and Prusoff (1973).

In practical terms there is not always enough time to measure sufficient data points in order to define a K_i. If a large number of compounds have to be tested, as is normally the case, it is often expedient to establish IC_{50}, and, provided that the compounds all inhibit the enzyme in a similar fashion, the IC_{50} values can be used for ranking purposes. Subsequently, when a few compounds are chosen for development more work can be carried out in order to estimate their K_i values and define the type of inhibition.

A further consideration in carrying out enzyme reactions with inhibitors *in vitro* is that the enzyme concentration is set at a rate-limiting level (very much less than the substrate concentration) in order that Michaelis–Menten kinetics should be applicable. This usually requires enzyme concentrations below 10^{-7} M and sometimes as low as 10^{-10} M. This can be quite unlike the normal physiological condition where reaction rates may be regulated by substrate availability rather than by enzyme concentration. Indeed, in the glycolytic pathway of the yeast *Saccharomyces carlsbergensis*, the molarities of only two enzymes, phosphofructokinase and adenylate kinase, lie below 10^{-5} M (Hess *et al.*, 1969). If we consider the 'concentration' of enzyme active sites, the range for all of the enzymes measured is between $2{\cdot}5 \times 10^{-5}$ and $2{\cdot}0 \times 10^{-4}$ M. The substrate concentrations vary in the same range.

Although enzymes in the glycolytic pathway may be exceptionally highly concentrated, there is no reason to suppose that other pathways may not be similar and, in addition, may show molecular organization to facilitate the passage of substrate along the pathway. It is therefore a matter of some surprise that inhibitors often do show effects *in vivo* that can be related to data obtained *in vitro*. That they do so makes life easier for the medicinal scientist.

One of the important features of enzyme inhibition is whether the inhibition is reversible or not. Irreversible inhibition is often characterized by inhibition that increases over time because it is a consequence of a chemical reaction that requires covalent bonds to be broken. Consequently, the inhibited enzyme will not recover during dialysis, unlike a reversibly-inhibited enzyme. Irreversible inhibition is frequently missed if considerable pre-incubation is not a normal part of the assay protocol, and often appears to be non-competitive inhibition because a portion of the enzyme is effectively put out of action by covalent modification (Bush, 1986). The order of addition of substrate, inhibitor and enzyme may be crucial but is frequently unreported. In fact, a plot of log(remaining activity) against time should be a straight line for this type of inhibition. Kinetically, the reactions are normally exemplified as follows (e.g. Coulson and Smith, 1979):

$$E + I \rightleftharpoons EI \rightarrow EI^* \qquad (10)$$

where EI* represents the inactivated enzyme. Irreversible inhibitors include amino-acid-modifying agents such as iodoacetamide, active-site-directed inhibitors and suicide inactivators such as the monoamine oxidase inhibitors.

A.2 Drug binding to receptors

Just as enzymologists use a plot of substrate concentration against reaction rate to quantify an enzyme reaction, pharmacologists also need to quantify the response of an organ preparation to a drug. For drug–receptor binding:

$$D + R \rightleftharpoons DR \rightarrow \text{Effect} \qquad (11)$$

where D and R represent drug and receptor, respectively. The equation governing the reaction is then:

$$E = \frac{[D]E_{max}}{([D] + K_d)} \qquad (12)$$

where E represents the effect produced by a given drug concentration [D], E_{max} is the maximum effect possible under the conditions and K_d is the dissociation constant of the drug–receptor complex. This clearly bears a formal similarity to the Michaelis–Menten equation of enzyme kinetics (equation 1).

The method of plotting is different from enzyme kinetics in that log(drug concentration) is plotted against response which normally gives a sigmoid type of curve (Fig. A.1). If two drugs are compared by this approach and produce parallel lines it is likely, but not certain, that the compounds are acting in a similar way at the receptor. If they are not parallel the interpretation is more complex.

One way of analysing such data is by a Schild plot. This depends on the fact that when an antagonist inhibits the response of a given tissue to an agonist competitively, the log(dose)–response curve is shifted to the right so that one obtains a plot similar to Fig. A.1, with the tissue showing the same response at a higher agonist concentration (x'_A) as it did previously at the original concentration (x'_A). It has been shown (Arunlakshana and Schild, 1959) that:

$$\frac{x'_A}{x_A} = 1 + \frac{x_B}{K_B} \qquad (13)$$

where x_B is the concentration of antagonist and K_B its binding constant. If the ratio between these agonist concentrations giving the same response is defined as R, then:

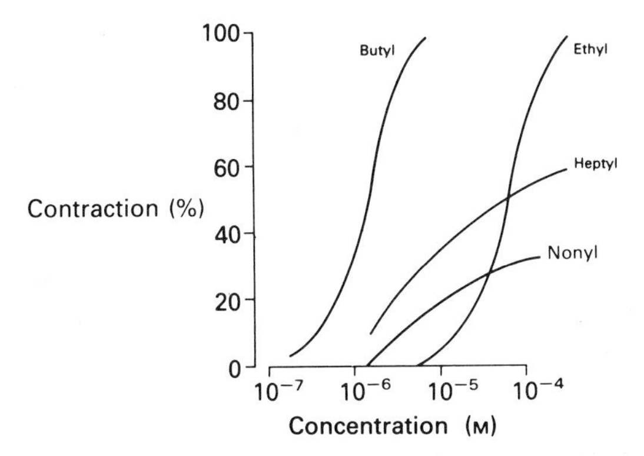

The ethyl- and butyl-trimethylammonium salts give parallel dose-response curves acting as agonists. The heptyl and nonyl analogues, however, produce curves with totally different shapes. This is interpreted as being the result of a gradual shift towards antagonism. After Stephenson, 1956.

Figure A.1 Log dose-response curve for contraction of guinea-pig ileum by alkyltrimethylammonium ions.

$$R = 1 + \frac{x_B}{K_B} \tag{14}$$

and

$$\log (R - 1) = \log x_B - \log K_B \tag{15}$$

The Schild plot is log(R–1) against –log x_B which will give a straight line with slope close to unity and a negative intercept on the x axis equal to –log K_B (Fig. A.2).

The binding of a drug to a receptor normally depends on the dose in a linear fashion. The magnitude of the tissue response is, however, often a non-linear function of the drug concentration and reaches a maximum value, in fact a maximum tissue response can occur with only a fraction of the receptors occu-

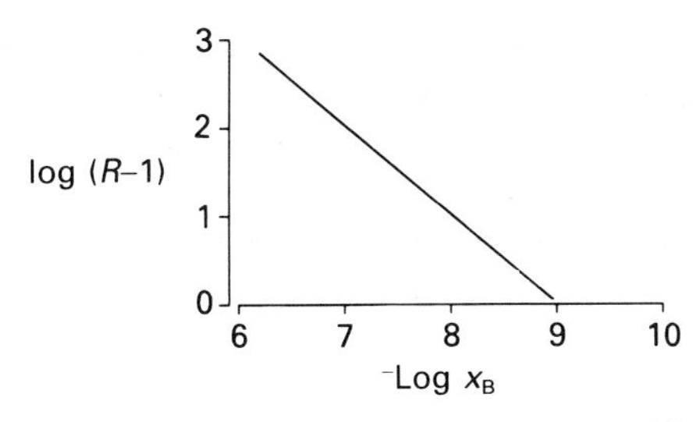

Antagonism by ligand B of a hypothetical agonist (A) at a specific receptor. The negative log of the binding constant for the antagonist is given by the intercept on the x axis (i.e. 10^{-9}M).

Figure A.2 Schild plot.

pied. Accordingly, it is often assumed that the tissue contains a finite number of receptors with which the drug can interact; the tissue response then depends on how many receptors are occupied by the drug which in turn depends on the affinity of the receptor for the drug.

The concept of efficacy is also useful here. Ligands are ranked on a scale from 0 to 1: an antagonist will bind to a receptor and elicit no tissue response at all, efficacy = 0, whereas a full agonist will be ranked at 1. Partial agonists will be ranked in between 0 and 1. These assumptions underlie much of what is discussed in this book, and have been found to hold up reasonably well in practice.

A.3 Ligand-protein binding

In order to complete this analysis, a third type of interaction must be considered - one where the binding of a ligand does not necessarily lead to any further effect. The binding of drugs to serum albumin comes into this class (Jusko and Gretch, 1976). In order to derive a mathematical formulation for the interaction a number of criteria have to be obeyed:

1. The binding sites must not interact with each other.
2. The binding must be specific and reversible.
3. A steady state must be reached, with equilibrium between free and bound ligand.

If we consider the simplest case of a macromolecule (M) with one binding site for one molecule of ligand L, we have the following equilibrium:

$$\mathrm{M} + \mathrm{L} \rightleftharpoons \mathrm{ML} \tag{16}$$

The dissociation constant for this equilibrium is given by:

$$K_{\mathrm{d}} = \frac{[\mathrm{M}]\,[\mathrm{L}]}{[\mathrm{ML}]} \tag{17}$$

where square brackets denote concentrations and [L] is free ligand. If we consider the number of moles of ligand bound (r) to one mole of macromolecule we have:

$$r = \frac{[\mathrm{L}]_{\mathrm{bound}}}{[\mathrm{M}]_{\mathrm{total}}} = \frac{[\mathrm{ML}]}{([\mathrm{M}] + [\mathrm{ML}])} \tag{18}$$

whence

$$r = \frac{[\mathrm{L}]}{(K_{\mathrm{d}} + [\mathrm{L}])} \tag{19}$$

If we consider the more usual case of n identical but independent sites we have the following formula:

$$r = \frac{n\,[\mathrm{L}]}{(K_d + [\mathrm{L}])} \tag{20}$$

This can be rearranged in three ways to give derivations that should yield straight lines when the following graphs are drawn; (a) $1/r$ against $1/[\mathrm{L}]$; (b) $r/[\mathrm{L}]$ against r, and (c) $[\mathrm{L}]/r$ against $[\mathrm{L}]$. The number of binding sites and the binding constant can be measured from the plot.

(a) $1/r = 1/n + K_d/n\,[\mathrm{L}]$ (21)

(b) $r/[\mathrm{L}] = n/K_d - r/K_d$ (22)

(c) $n\,[\mathrm{L}]/r = [\mathrm{L}] + K_d$ (23)

Each of these three deviations has been used to study ligand binding to macromolecules. The second is probably the best known and is attributed to Scatchard (1949). It is the best form mathematically, in that the plot does not rely heavily on measurements taken at low ligand concentrations which are subject to the greatest error in determination. The other two plots divide by r which will magnify the effect of inaccuracies. Furthermore, as Scatchard himself noted, 'double reciprocal plots tend to tempt straight lines where none exist'. It is an interesting exercise to plot one set of data both by double-reciprocal plot (equation 21) and by the Scatchard plot (equation 22) and compare the figures for n and K obtained.

One major use of the Scatchard plot is in studying hormone binding to receptor. The plot will also show up situations where more than one set of binding sites exist with different binding constants. The plot then gives two straight lines with different slopes connected by a boundary region where one curve blends into the other (Fig. A.3).

If we look at these equations in the light of enzyme kinetics and drug receptor binding, it is clear that there are considerable formal similarities - despite the fact that conversion to product occurs with an enzyme and pharmacological effects result from the binding of a ligand to a receptor, whereas no further change may result as a consequence of ligand binding to protein - for example the binding of drugs to serum albumin (Jusko and Gretch, 1976). The key similarity is that all of these interactions obey the Law of Mass Action.

Another method of plotting is due to Hill (1910) who derived an equation to explain the binding of oxygen to haemoglobin. This plot is of value when interaction between sites is suspected with the binding of one ligand to a macromolecule either facilitating (positive cooperativity) or hindering (negative cooperativity) the binding of a second. We take equation (18) and transform it into the Hill equation:

$$r/(n - r) = [\mathrm{L}]/K_d \tag{26}$$

We can call the fraction of the binding sites occupied (Z), where $Z = r/n$. Then the left-hand side of equation (24) becomes:

$$Z/(1 - Z) = [\mathrm{L}]/K_d \tag{27}$$

A. One set of binding sites.

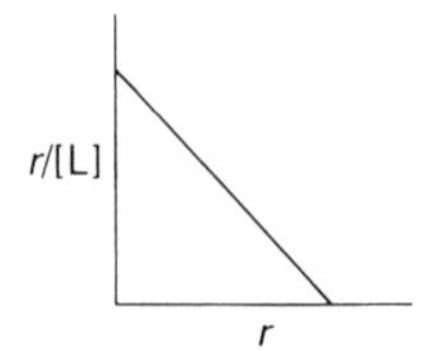

r is the number of moles of ligand bound to one mole of macromolecule. [L] is the free ligand concentration. The intercept on the x axis is n (the number of binding sites). The slope is $-1/K_d$.

B. Two sets of binding sites.

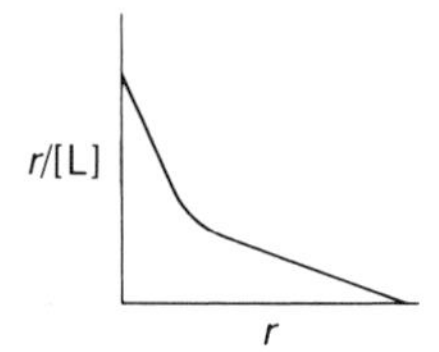

The intercept on the x axis is now the sum of the number of binding sites at each of the two sets of site, n_1+n_2 but the two slopes are not simply the negative reciprocals of the respective binding constants. They are rather more complex (see Feldman, 1972 for a mathematical analysis).

Figure A.3 Scatchard plot.

A plot of log $\{Z/(1 - Z)\}$ against log [L] will give a straight line with a slope of 1 if there is no cooperativity. If there is cooperativity, however, the slope in the central portion of the curve will be greater than 1 for positive, and less than 1 for negative, cooperativity. This is because at low concentrations of drug the first binding site is being filled and at high concentrations the last; it is only in the middle that sites already filled can influence those unfilled (Fig. A.4). In fact, a Scatchard plot will also indicate cooperativity by being curved.

It is interesting to note the number of double-reciprocal plots used for enzyme kinetics whereas for ligand binding the Scatchard plot is usually favoured. This is not consistent but it may allow data to be made presentable. Even the Scatchard plot may produce straight lines that could be the consequence of obedience to a more complex equation as discussed by Klotz (1983).

References

Arunlakshana, O. and Schild, H.O., 1959, *Brit. J. Pharmacol.,* **14**, 48–58.
Bush, K., 1986, *Drugs Exptl. Clin. Res.,* **12**, 565–76.
Cheng, Y.-C. and Prusoff, W.H., 1973, *Biochem. Pharmacol.,* **22**, 3099–108.
Coulson, C.J. and Smith, V.J., 1979, *Enzyme Microbial Technol.,* **1**, 193–6.
Hess, B., Boiteux, A. and Kruger, J., 1969, *Adv. Enzyme Regul.,* **7**, 149–67.

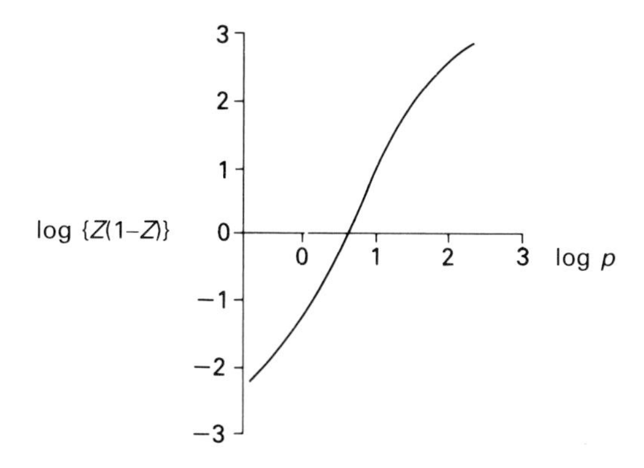

This is a Hill plot for haemoglobin in equilibrium with oxygen (p is the partial pressure of oxygen). The slope in the middle portion of the curve (the Hill coefficient) approximates to 2·8 whereas at both ends it is lower. The total number of binding sites for oxygen in haemoglobin is 4, but this figure is not reached because to derive the Hill equation infinite cooperativity is assumed. Since this is an ideal situation the Hill coefficient is always less than the actual number of sites and in practice indicates a minimum number, in this case 3. After Koshland (1970).

Figure 0.4 Hill plot.

Hill, A.V., 1910, *J. Physiol.*, **40**, iv–vii.

Jusko, W.J. and Gretch, M., 1976, *Drug. Metab. Rev.*, **5**, 43–140.

Klotz, I., 1983, *Trends Pharmacol. Sci.*, **4**, 253–5.

Koshland, D.E., 1970, In: *The Enzymes*, Vol. 1, 3rd Ed, Ed. P.D. Boyer (New York: Academic Press), p. 341.

Scatchard, G.S., 1949, *Ann. N.Y. Acad. Sci.*, **51**, 660–72.

Index